Tribology in Sustainable Manufacturing

Tribology in Sustainable Manufacturing compiles the fundamentals of friction in manufacturing processes and the application of tribology in advanced manufacturing. Covering topics such as 3D printing, green lubrication, laser sintering and Industry 4.0, the book promotes cost-effective and environmentally friendly manufacturing processes.

In an effort to reduce energy consumption, production time and costs, while simultaneously improving plant productivity, sustainable tribology plays a key role in modern manufacturing processes. With a focus on broadening the application of tribology in sustainable manufacturing, the book integrates cutting edge research from international contributors. It further covers machine learning, micro-machining, friction stir welding and metal forming. It also discusses the tribological properties of advanced materials and coatings, and how to model tribology in manufacturing processes.

This book will be of interest to engineers and students in the fields of machining, tribology, additive manufacturing, surface engineering and coating.

Manufacturing Design and Technology
Series Editor
J. Paulo Davim

This series will publish high quality references and advanced textbooks in the broad area of manufacturing design and technology, with a special focus on sustainability in manufacturing. Books in the series should find a balance between academic research and industrial application. This series targets academics and practicing engineers working on topics in materials science, mechanical engineering, industrial engineering, systems engineering, and environmental engineering as related to manufacturing systems, as well as professions in manufacturing design.

Advanced Manufacturing and Processing Technology
Edited by Chander Prakash, Sunpreet Singh, J. Paulo Davim

Non-Conventional Hybrid Machining Processes: Theory and Practice
Edited by Rupinder Singh, J. Paulo Davim

Functional Materials and Advanced Manufacturing: 3-Volume Set
Edited by Chander Prakash, Sunpreet Singh, J. Paulo Davim

Industry 4.0: Challenges, Trends, and Solutions in Management and Engineering
Edited by Carolina Machado, J. Paulo Davim

Sustainable Manufacturing for Industry 4.0: An Augmented Approach
Edited by K. Jayakrishna, Vimal K.E.K., S. Aravind Raj, Asela K. Kulatunga, M.T.H. Sultan, J. Paulo Davim

Industrial Tribology: Sustainable Machinery and Industry 4.0
Edited by Jitendra Kumar Katiyar, Alessandro Ruggiero, T. V. V. L. N. Rao J. Paulo Davim

Innovative Development in Micromanufacturing Processes
Edited by Pawan Kumar Rakesh, J. Paulo Davim

Tribology in Sustainable Manufacturing
Edited by Jitendra Kumar Katiyar, T. V. V. L. N. Rao Ahmad Majdi Abdul Rani, Mohd Hafis Sulaiman, J. Paulo Davim

For more information about this series, please visit: https://www.routledge.com/Manufacturing-Design-and-Technology/book-series/CRCMANDESTEC

Tribology in Sustainable Manufacturing

Edited by
Jitendra Kumar Katiyar
T. V. V. L. N. Rao
Ahmad Majdi Abdul Rani
Mohd Hafis Sulaiman
J. Paulo Davim

CRC Press
Taylor & Francis Group
Boca Raton London New York

CRC Press is an imprint of the
Taylor & Francis Group, an **informa** business

First edition published 2024
by CRC Press
2385 NW Executive Center Drive, Suite 320, Boca Raton FL 33431

and by CRC Press
4 Park Square, Milton Park, Abingdon, Oxon, OX14 4RN

CRC Press is an imprint of Taylor & Francis Group, LLC

ISBN: 978-1-032-42631-0 (hbk)
ISBN: 978-1-032-42632-7 (pbk)
ISBN: 978-1-003-36357-6 (ebk)

DOI: 10.1201/9781003363576

Typeset in Times
by SPi Technologies India Pvt Ltd (Straive)

Contents

Preface

The word *sustainability* connotes surprising challenges. Sustainability has gained importance due to its breakthrough research in every field. Similarly, manufacturing plays a significant role in numerous areas related to human living. Hence, sustainable manufacturing is of growing interest to researchers. It is defined as developing products that use non-polluting fabrication processes, save energy and resources, and are cost effective and safe for humans. In any fabrication process, tribology has played an essential function in energy saving and improving the life of a system. It studies friction, wear and lubrication between two interacting bodies in a relative motion. Noticeable is that the tribological performance of any system varies with regard to external parameters. These parameters are contact pressure, contact behavior, the temperature of the system and environment, and sliding speed at the interface. Therefore, over the past few decades, it has been recognized as essential in all manufacturing processes/operations. This creates the need for a book that compiles the complete knowledge of tribology in advanced manufacturing systems.

Therefore, the prime objective of this book is to broaden the concept of tribology in sustainable manufacturing. The book compiles the fundamentals of slidingolling friction in manufacturing processes obtained due to relative movements. It has also been noticed that the productivity and efficiency of any manufacturing process widely depend upon friction and wear of components. Hence, controlled manufacturing processes improve tribological performance, decreasing manufacturing costs and saving energy. In addition, the book includes and discusses the application of tribology in advanced manufacturing processes such as additive manufacturing, 3D printing, the laser sintering process, and more.

Furthermore, the book explores tribology in green lubrication and Industry 4.0, that is, the new era of manufacturing industries. All technological enhancements improve the life of components, and provide minor energy consumption, reduce production time and cost, thus improving the productivity of the overall manufacturing system: the present book will be of benefit to engineering students, research scientists and persons in the industry.

Editors

Dr. Jitendra Kumar Katiyar is presently working as a research assistant professor in the Department of Mechanical Engineering at the SRM Institute of Science and Technology Kattankulathur Chennai, India. His research interests include: modern manufacturing techniques, tribology of carbon materials, polymer composites, self-lubricating polymers, lubrication tribology, and coatings for advanced technologies. He obtained his bachelor's from UPTU Lucknow with honours in 2007. He received his master's from the Indian Institute of Technology Kanpur in 2010 and his PhD from the same institution in 2017. He is a life member of the Tribology Society of India, the Malaysian Society of Tribology, the Institute of Engineers, India, and The Indian Society for Technical Education (ISTE). He has authored/co-authored/published more than 65+ articles in reputed journals, 40+ articles in international/national conferences, 20+ book chapters, and 15+ books published by various publishers. He has served as guest associate editor in many mechanical engineering and tribology journals.

Further, he is on the editorial board of *Tribology Materials, Surfaces and Interfaces* and is a review editor for *Frontier in Mechanical Engineering*. He is an active reviewer for 20+ reputed materials, manufacturing, and tribology journals, and has delivered over 50+ invited talks on various research fields related to tribology, composite materials, surface engineering, and machining. He has organized 7+ FDP/short term courses and the International Tribology Research Symposium in tribology.

Professor T. V. V. L. N. Rao earned a PhD in tribology of fluid film bearings from the Indian Institute of Technology Delhi in 2000, and an MTech in mechanical manufacturing technology from the National Institute of technology (formerly known as Regional Engineering College) Calicut in 1994. Prof. Rao's current research interests are in tribology, lubrication, bearings, machining, and manufacturing. He has authored (and co-authored) over 140 publications to date. He has secured (as PI and Co-PI) several research grants from the Ministry of Higher Education, Malaysia and a research grant from The Sumitomo Foundation, Japan. Dr. Rao is an associate editor for the *Journal of Engineering Tribology*, has sat on the editorial board of *Tribology & Lubrication Technology* (2021, 2024), is a review editor for *Frontiers of Mechanical Engineering: Tribology* and *Frontiers in Chemistry: Nanoscience*

and Materials Research Communications, and a member of the editorial board of *Pusat Publikasi Ilmiah ITS*. He has been a guest associate editor for various mechanical engineering and tribology journals.

Prof. Rao is an executive member (2015–2023) of the Malaysian Tribology Society. He is a member of the Society of Tribologists and Lubrication Engineers, the Malaysian Tribology Society and the Tribology Society of India. He is also a reviewer of reputed journals associated with tribology and manufacturing. Rao is currently a professor and Dean of the Faculty of Engineering, Assam down town University. Prior to this, he served as professor and dean (2022–2023) at The Assam Kaziranga University; professor (2022) at The Assam Kaziranga University; professor (2021) then professor and head (2021) at MITS Madanapalle; research associate professor at SRMIST (2017–2020); visiting faculty at LNMIIT (2016–2017); associate professor at Universiti Teknologi PETRONAS (2010–2016); and assistant professor in BITS Pilani at the Pilani (2000–2004, 2007–2010) and Dubai (2004–2007) campuses.

Dr. Ahmad Majdi Abdul Rani has a PhD in mechanical and manufacturing engineering from Loughborough University, UK; an MSc in industrial engineering and a BSc in manufacturing from Northern Illinois University, USA. His research interests are in biomedical engineering, mechanical and manufacturing engineering and tribology. He is currently supervising more than 20 postgraduate (PhD and MSc) students. He recently received the Leader in Innovation Fellowship – Newton Award, UK. He has secured three patents and won more than 15 Gold awards in exhibitions such as ITEX, MaGRIs (MOSTI), MARS, PENCIPTA, IME, SPDEC, CoRIC, and more as well as several research grants from the Ministry of Higher Education, Malaysia (FRGS, PRGS, ERGS, MyBrain) and Universiti Teknologi PETRONAS (YUTP, I-GEN, STIRF) as PI and co-PI. He has also authored over 150 publications to date and has been associated with the Department of Mechanical Engineering at Universiti Teknologi PETRONAS since 1998 and was Head of the Department of Mechanical Engineering at Universiti Teknologi PETRONAS from Please remove the space between the year 2009-2011.

Dr. Mohd Hafis Sulaiman earned his PhD in mechanical engineering from Technical University of Denmark (DTU), after getting an MEng and BEng in mechanical engineering from Universiti Teknologi Malaysia (UTM). His research interests lie in the mechanical and manufacturing engineering field with an emphasis on metal forming, tribology, coating, and machine design. He has authored and co-authored over 70 publications to date, and has secured research grants from international, industry, and national funds such as Taiho Kogyo Tribology

Research Foundation (TTRF), Ministry of Higher Education, Malaysia, Malaysia Technical University Network (MTUN), International Islamic University Malaysia (IIUM), and Universiti Malaysia Perlis (UniMAP). He worked on a research project funded by the Danish Council for Independent Research, entitled *Environmentally Benign Sheet Metal Forming Tribology Systems* (SHETRIB). He has received numerous awards, such as the Young Tribologist Research and the Young Tribologist Best Paper awards; and travel grants from the Japanese Society of Technology on Plasticity (JSTP) and Nichidai Corporation, from Otto Mønsteds Fond, and from the International Federation for the Promotion of Mechanism and Machine Science (IFToMM). He won a crowdsourced Subject Matter Expert (SMX) Solutions Programme from the Steinbeis Malaysia Foundation and Agensi Inovasi Malaysia for solving industrial precision stamping and his forging of a motorcycle sprocket. He received his professional engineer qualification (Ir.) from the Board of Engineers Malaysia (BEM) and is a registered chartered engineer (C. Eng) with the Engineering Council, UK. Dr. Sulaiman is a qualified evaluating panelist for the Engineering Technology Accreditation Council (ETAC), which provides quality assurance in engineering higher education in Malaysia. He is also an executive member of the Malaysia Tribology Society (MYTRIBOS) (2020–2021). Currently, he is an assistant professor in the Department of Manufacturing and Materials Engineering, Kulliyyah of Engineering, International Islamic University of Malaysia (IIUM).

Professor J. Paulo Davim is a full professor at the University of Aveiro, Portugal. He holds several distinguished emeritus professorships at a number of universities, colleges, and institutes in China, India, and Spain. He received his PhD in mechanical engineering in 1997, MSc in mechanical engineering (materials and manufacturing processes) in 1991, BEng in mechanical engineering in 1986 from the University of Porto (FEUP), title of aggregate (Full Habilitation) from the University of Coimbra in 2005, and DSc (higher doctorate) from London Metropolitan University in 2013. He is a senior chartered engineer at the Portuguese Institution of Engineers with an MBA and has specialist titles in engineering and industrial management as well as metrology. Prof. Davim is also a Eur Ing at FEANI (Brussels) and a fellow (FIET) at IET (London). He has more than 30 years of teaching and research experience in manufacturing, materials, mechanical, and industrial engineering, with a special emphasis on machining and tribology. In addition, he is interested in management, engineering education, and higher education for sustainability. He has guided large numbers of postdoc, PhD, and master's students as well as coordinated and participated in several financed research projects. Dr. Davim is the recipient of several scientific awards and honors and has previously worked as a projects evaluator for the European Research Council (ERC) and other international research agencies as well as an examiner of PhD theses for many universities in different countries. He is the editor-in-chief of several international journals, a guest editor of journals, and books, and a book series editor, and a scientific advisor for many international journals and conferences. Presently, he is an editorial board member of 30 international journals and acts as reviewer for more than 100 prestigious Web of Science journals. In addition, Prof. Davim has also published as editor (and co-editor) more than 200 books and as author (and co-author) more than 15 books, 100 book chapters, and 500 articles in journals and conferences. He was listed in the World's Top 2% of Scientists in a Stanford University study.

Contributors

S. Arokya Agustin
Department of Mechanical
  Engineering
SRM Institute of Science and
  Technology
Kattankulathur, India

Y. Alex
School for Advanced Research in
  Petrochemicals: Laboratory for
  Advanced Research in Polymeric
  Materials (LARPM)
Central Institute of Petrochemicals
  Engineering and Technology
  (CIPET)
Bhubaneswar, India

Jencen Mathai Arivannoor
Gulf Oil International Global R&D
  Centre
Chennai, India

Utpal Barman
Faculty of Computer Technology
Assam down town University
Guwahati, India

Subho Chakraborty
Research and Development
Tata Steel Limited
Jamshedpur, India

C. T. Chidambaram
Gulf Oil International Global R&D
  Centre
Chennai, India

Nidhin C. Divakaran
School for Advanced Research in
  Petrochemicals: Laboratory for
  Advanced Research in Polymeric
  Materials (LARPM)
Central Institute of Petrochemicals
  Engineering and Technology (CIPET)
Bhubaneswar, India

Jaharah A. Ghani
Department of Mechanical and
  Material Engineering
Faculty of Engineering and Built
  Environment
Universiti Kebangsaan Malaysia
Selangor, Malaysia

Nurul Nadia Nor Hamran
Faculty of Manufacturing and
  Mechatronics
Universiti Malaysia Pahang
Pahang, Malaysia

A. Arul Jeyakumar
Department of Mechanical Engineering
SRM Institute of Science and
  Technology
Kattankulathur, India

P. V. Joseph
Gulf Oil International Global R&D
  Centre
Chennai, India

Anup A. Junankar
Poornima University Jaipur
Jaipur, India

Jitendra Kumar Katiyar
Department of Mechanical
    Engineering
SRM Institute of Science and
    Technology
Kattankulathur, India

Yashpal Kaushik
Poornima University Jaipur
Rajasthan, India

C. Shravan Kumar
Department of Mechanical
    Engineering
SRM Institute of Science and
    Technology
Kattankulathur, India

S. Vinith Kumar
Gulf Oil International Global R&D
    Centre
Chennai, India

Sanjay Kumar
Gulf Oil International Global R&D
    Centre
Chennai, India

Sanjeev Kumar
Department of Mechanical
    Engineering
Subharti Institute of Technology and
    Engineering
Meerut, India

Vincent Martin
Gulf Oil International Global R&D
    Centre
Chennai, India

Patrick J. Masset
Faculty of Materials Science and
    Engineering
Warsaw University of Technology
Warsaw, Poland

Saurabh Mishra
Centre for Advanced Studies
Dr APJ Abdul Kalam Technical
    University
Lucknow, India

Smita Mohanty
School for Advanced Research in
    Petrochemicals: Laboratory for
    Advanced Research in Polymeric
    Materials (LARPM)
Central Institute of Petrochemicals
    Engineering and Technology (CIPET)
Bhubaneswar, India

Ponnekanti Nagendramma
CSIR-Indian Institute of Petroleum
Dehradun, India

Ankit Pandey
CSIR-Indian Institute of Petroleum
Uttarakhand, India

Ashwin Pandit
Quality Assurance Department
Tata Steel Limited
Jamshedpur, India

P. Pranav
Advanced Tribology Research Centre
Department of Mechanical
    Engineering
College of Engineering Trivandrum
Trivandrum, India

Jayant K. Purohit
Banasthali Vidyapith Tonk
Tonk, India

Ahmad Majdi Abdul Rani
Department of Mechanical
    Engineering
Universiti Teknologi PETRONAS,
    Bandar Seri Iskandar
Perak, Malaysia

S. Rani
Advanced Tribology Research Centre
Department of Mechanical
    Engineering
College of Engineering Trivandrum
Trivandrum, India

T. V. V. L. N. Rao
Faculty of Engineering
Assam down town University
Guwahati, India

Anjan Ray
CSIR-Indian Institute of Petroleum
Dehradun, India

Krishnkant Sahu
Mechatronics Engineering Department
Sharad Institute of Technology
College of Engineering
Ichalkaranji, India

Atul Pratap Singh
CSIR-Indian Institute of Petroleum
Dehradun, India

Shailesh Kumar Singh
Advanced Tribology Research Centre
CSIR – Indian Institute of Petroleum
Dehradun, India

Ujjawal Singh
Centre for Advanced Studies
Dr APJ Abdul Kalam Technical
    University
Lucknow, India

Vivek Kumar Singh
Department of Mechanical
    Engineering
Indian Institute of Technology Bombay
Mumbai, India

Edla Sneha
Advanced Tribology Research Centre
Department of Mechanical
    Engineering
College of Engineering Trivandrum
Trivandrum, India

M. H. Sulaiman
Department of Mechanical and
    Manufacturing Engineering
Universiti Putra Malaysia (UPM)
Selangor, Malaysia

Ananthan D. Thampi
Advanced Tribology Research Centre
Department of Mechanical
    Engineering
College of Engineering Trivandrum
Trivandrum, India

Adesh Kumar Tomar
Department of Mechanical
    Engineering
Graphic Era (Deemed a University)
Dehradun, India

Saurabh Vashistha
Academy of Scientific and Innovative
    Research (AcSIR)
CSIR-HRDC Campus
Ghaziabad, India

S. Vignesh
Gulf Oil International Global R&D
    Centre
Chennai, India

List of Abbreviations

AcSIR	Academy of Scientific and Innovative Research
AIP	American Institute of Physics
AM	Additive manufacturing
AMC	Aluminum matrix composites
AOCS	American Oil Chemists Society
APISH	American Petroleum Institute (SH stated for the category of oil)
APJ	Avul Pakir Jainulabdeen
ASM	American Society for Metals
ASTM	American Society for Testing and Materials
AW	Anti-wear
BA	Boric acid
BEM	Board of Engineers Malaysia
BG	Butylene glycol
BGDC	Butylene glycol dicaprylate
BGDO	Butylene glycol dioleate
BGE	Butylene glycol based esters
BITS	Birla Institute of Technology and Science
BL	Biodegradable lubricants
BUE	Built-up edges
CAGR	Compound annual growth rate
CBN	Cubic boron nitride
CCC	Copyright Clearance Center
CCTO	Calcium–copper–titanate nanoparticles
CEng	Chartered engineer
CF	Carbon fiber/cutting fluids
CFD	Computational fluid dynamics
CIPET	Central Institute of Petrochemicals Engineering and Technology
CNT	Carbon nanotube
COC	Cleveland Open Cup
COD	Chemical Oxygen Demand
COF	Coefficient of friction
COOH	Carboxylic acid
COT	Clean oil tank
CRC	Chemical Rubber Company
CRCA	Cold Rolled Close Annealed steel
CS	Cold-sprayed
CuO	Copper Oxide
CZ	Contaminated zone
DED	Direct energy deposition
DIN	Deutsches Institut für Normung
DLC	Diamond-like carbon

DLIP	Direct laser interference patterning
DLP	Digital Light Processing
DM	Demineralized
DMLS	Direct melting laser sintering
DOC	Dissolved organic carbon
DOE	Design of experiment
DOT	Dirty oil tank
DTU	Technical University of Denmark
EAFSW	Electrically Assisted Friction Stir Welding
EBM	Electron beam melting
EDA	Electric discharge alloying
EDAX	Energy Dispersive X-ray Analysis
EDS	Energy Dispersive Spectroscopy
EHL	Elastohydrodynamic Lubrication
EP	Extreme-pressure
EPA	Environmental Protection Agency
ERC	European Research Council
ETAC	Engineering Technology Accreditation Council
EU	European Union
EWT	Entropy Weighted Technique
FEANI	European Federation of National Engineering Associations
FEM	Finite Element Method
FEUP	Faculty of Engineering of the University of Porto
FG	Functionally graded
FGMs	Functionally graded materials
FIET	Fellow of the Institution of Engineering and Technology
FM	Friction modifier
FSP	Friction stir processing
FSPD	Friction stir powder deposition
FSW	Friction Stir Welding
FTIR	Fourier-transform infrared spectroscopy
GDQ	Generalized differential quadrature
GHG	Green House gas
GIL	Green Ionic Liquids
GnP	Graphene Platelets
GNP	Graphene nao-platelets
GO	Graphene oxide
HEA	High entropy alloy
HOOT	Hot oil oxidation test
HOSO	High oleic sunflower oil
HPDC	High pressure die casting
HPP	High-performance polymers
HSE	Health, safety and environment
HSS	High-speed steel

IFToMM	International Federation for the Promotion of Mechanism and Machine
IIUM	International Islamic University Malaysia
ILSAC	International Lubricant Standardization Advisory Committee
IOP	Institute of Physics
IR	Infrared
ISTE	Indian Society for Technical Education
ITEX	International Tundra EXperiment
JSTP	Japanese Society of Technology on Plasticity
KOH	Potassium hydroxide
LaF	Lanthanum Fluoride
LCA	Life Cycle Assessment
LCT	Life Cycle Tribology
LoC	Library of Congress
LPBF	Laser powder bed fusion
LSEM	Large strain extrusion machining
LSS	Lap Shear Strength
LST	Laser surface texturing
MATEC	Materials science, Engineering and Chemistry
MATLAB	Matrix Laboratory
MCDM	Multi-criteria decision-making
MCEM	Monte Carlo Expectation-Maximization
MEA	Monoethanolamine
MEK	Methyl ethyl ketone
MITS	Madanapalle Institute of Technology & Science
MJO	Modified jatropha oil
MMC	Metal Matrix Composite
MML	Mechanical mixture layer
MnS	Manganese Sulfide
MoM	Metal-on-metal
MQL	Minimum Quantity Lubricant
MQLPO	MQL palm oil
MQLSE	MQL synthetic ester
MTM	Mini Traction Machine
MTUN	Malaysia Technical University Network
MWF	Metal Working Fluid
MWSD	Mean wear scar diameter
NCZ	Non-contaminated zone
NFMQL	Nanofiller Minimum Quality Lubrication
NMR	Nuclear Magnetic Resonance
OECD	Organization for Economic Cooperation and Development
OFHC	Oxygen-free high conductivity
OLS	Ontology Search
PAG	Polyalkylene Glycols

PAOS	Polyalphaolefins
PBF	Powder bed fusion
PbS	Lead sulphide
PEEK	Poly-ether ether-ketone
PEI	Polyetherimide
PI	Project Investigator
PLA	Polylactic acid
PM	Powder metallurgy
PMCC	Pensky-Martens Closed Cup
PPD	Pour point depressant
PTFE	Polytetrafluoroethylene
RA	Reserve Alkalinity
RHA	Rice husk ash
SCP	Sustainable Consumption and Production
SEM	Scanning electron microscope
SEM	Scanning Electron Microscopy
SHETRIB	Environmentally Benign Sheet Metal Forming Tribology Systems
SiC	Silicon carbide
SL	Sheet lamination/solid lubricant
SLIPS	Slippery liquid-infused porous surface
SLM	Selective laser melting
SLS	Selective Laser Sintering
SMX	Subject Matter Expert
SPIF	Single point incremental forming
SQ	Square
SR	Surface Roughness
SRM	Sri Ramaswamy Memorial
SRR	Slide-to-roll ratio
SRV	Schwingung (Oscillating), Reibung (Friction), Verschleiß (Wear)
STL	Stereolithography
SZ	Stir zone
TAN	Total Acid Number
TBN	Total Base Number
TCT	Twist Compression Test
TGA	Thermogravimetric Analysis
TiB	Titanium Boride
TIF	Three intermittent flat faces
TMP	Trimethylolpropane
TRS	Tool rotational speed
TTA	Tool tilt angle
TTRF	Taiho Kogyo Tribology Research Foundation
TTS	Tool Traverse Speed
TTT	Tapping torque test
UHMWPE	Ultra-High Molecular Weight Polyethylene
UN	United Nations

UniMAP	Universiti Malaysia Perlis
UPTU	Uttarpradesh Technical University
USDA	United States Department of Agriculture
UTM	Universiti Teknologi Malaysia
UV	Ultra violet
VO	Vegetable oil
WASPAS	Weighted Aggregated Sum Product Assessment
WC	Tungsten carbide
WPT	Weighted product technique
WSD	Wear scar diameter
WST	Weighted sum technique
XRD	X-ray Diffraction
XRF	X-ray fluorescence
ZDDP	Zinc dialkyldithiophosphates
ZnO	Zinc oxide

1 Recent Advances of Tribology in Sustainable Manufacturing

Jitendra Kumar Katiyar
SRM Institute of Science and Technology, Kattankulathur, India

Ahmad Majdi Abdul Rani
Universiti Teknologi PETRONAS, Bandar Seri Iskandar, Perak, Malaysia

M. H. Sulaiman
Universiti Putra Malaysia (UPM), Selangor, Malaysia

Utpal Barman
Assam down town University, Guwahati, India

Patrick J. Masset
Warsaw University of Technology, Warsaw, Poland

T. V. V. L. N. Rao
Assam down town University, Guwahati, India

1.1 INTRODUCTION

Sustainable manufacturing can be defined as a method for manufacturing that minimizes waste and reduces the environmental impact (Rao, 2013). The major principles to be considered in sustainable manufacturing are reducing the resource utilization in the process, using more environment-friendly materials, reducing the energy requirement for manufacturing processes, reducing all forms of waste, reusing and recycling material. Sustainable manufacturing integrates the design of products and processes with the aim of ultimately reducing environmental impact while minimizing waste and maximizing the resource efficiency (Rao, 2013).

Bell (1992) emphasized the potential technical and economic benefits of applying multiple surface engineering technologies to enhance the performance of engineering components. Abdollah (2017) highlighted the recent methods,

DOI: 10.1201/9781003363576-1

1

Tribology in Sustainable Manufacturing
-Friction and wear in sustainable manufacturing
-Lubrication in sustainable manufacturing

Friction and wear in Sustainable Manufacturing	Lubrication in Sustainable Manufacturing
-Additive manufacturing	-Biolubricants
-Laser assisted manufacturing	-Nanobiolubricants
-Metal working	-Surface texturing
-Powder metallurgy	-Surface coating
-Casting	-Minimum quantity lubrication
-Friction stir processing	-Cryogenic machining

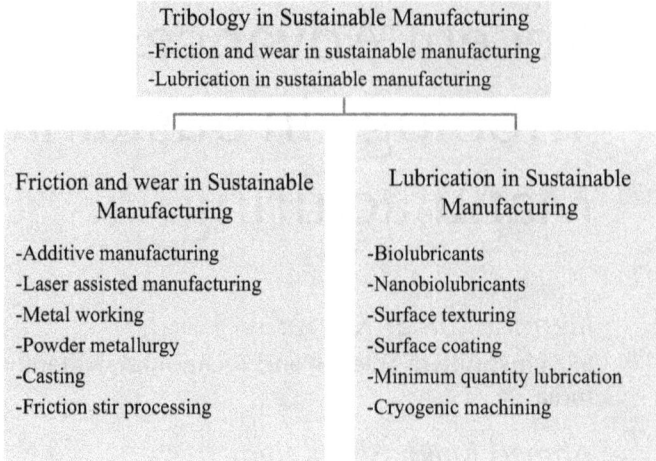

FIGURE 1.1 Recent advances of tribology in sustainable manufacturing: An overview.

analysis, designs, materials, processes and economics in tribology for sustainable manufacturing. Tribological considerations related to sustainable manufacturing can reduce product development time and cost, extend product life and reduce energy consumption. Suh (2021) underlined the significant relationship between tribology and design that determines the performance of all tribological systems. Potential tribological system failures may be avoided by rational and creative design.

Energy savings, environmental concerns, economic considerations and safety aspects of our society all emphasize the importance of controlling energy dissipative phenomena such as friction, wear and lubrication.

An overview of the recent advances of tribology in sustainable manufacturing is shown in Figure 1.1.

1.2 TRIBOLOGY IN SUSTAINABLE MANUFACTURING

Nosonovsky and Bhushan (2010) formulated the principles of green tribology. Nosonovsky and Bhushan (2010) further defined areas of green tribology as (i) biomimetics, (ii) environment-friendly lubrication, and (iii) renewable energy. Rao et al. (2021) highlighted the significance of green tribology in tribological systems for sustainable development and for significant friction and wear reduction in the areas of green lubricants and composites, textured surfaces and green machining. Woydt (2021) presented a broad range of solutions to reduce CO_2 and increase sustainability. CO_2 and friction reduction and sustainability and wear protection represent separate technology routes for reducing CO_2 emissions and increasing sustainability. A third of the world's energy consumption is attributed to friction. Thus, reducing friction improves fuel economy and reduces greenhouse

TABLE 1.1

Overview of Recent Advances in Tribology in Sustainable Manufacturing

Tribology in Sustainable Manufacturing	Authors
Formulation of the principles and areas of green tribology	Nosonovsky and Bhushan, (2010)
Areas of recent developments in green tribology are in green lubricants, green composites, texture surfaces and green machining	Rao et al. (2021)
The importance of tribology for reducing friction, wear and CO_2 emissions for sustainability	Woydt (2021)

gas emissions and reducing wear in tribological mechanisms lowers consumption of resources and increases global sustainability.

An overview of recent advances of tribology in sustainable manufacturing is presented in Table 1.1.

1.3 FRICTION AND WEAR IN SUSTAINABLE MANUFACTURING

Zhu et al. (2021) highlighted the state-of-the-art recent advances of friction and wear of additive manufacturing (AM). The challenges and opportunities for future tribology research in this exciting field of AM are presented. Norani et al. (2021) provided a systematic review on the friction and wear properties of 3D printed polymers. The specialized and functional 3D printed parts are designed and manufactured using polymer based materials for use in engineering and medical industries. Renner et al. (2021) presented a review of corrosion and wear as a surface failure of alloys fabricated through AM. The processing conditions have a profound influence on microstructure and anisotropic behavior of corrosion and wear of alloys. The major types of wear in the alloys are abrasive and adhesive wear. The abrasive wear is related to contact type and hardness of surfaces in contact, while adhesive wear depends on material microstructure and hardness. Additive manufacturing fabricated alloys have superior anti-wear properties compared to cast counterparts due to the inherent microstructure that hinders the wear of the alloys. The nanoparticles incorporated in the metal matrix are an effective way to improve AM fabricated alloys. The inclusion of nanoparticles enhances the hardness and wear resistance of the alloys.

Recent advances in minimizing friction and wear in sustainable manufacturing are presented in Table 1.2.

1.3.1 ADDITIVE MANUFACTURING

Additive manufacturing technologies, such as electron beam melting (EBM), direct melting laser sintering (DMLS) or selective laser melting (SLM), are increasingly being adopted for the production of near-net-shape products (Bordin et al., 2015). The complex and intricate shapes necessitate the near-net-shape

TABLE 1.2

Recent Advances of Friction and Wear in Sustainable Manufacturing

Friction and Wear Minimization in Sustainable Manufacturing	Authors
Recent advances of tribology research related to AM	Zhu et al. (2021)
A systematic review of the mechanical and tribological properties of 3D-printed polymers	Norani et al. (2021)
Review on corrosion and wear of additive manufactured alloys	Renner et al. (2021)

product realization through AM; these technologies are explored for producing customized components for aerospace, biomedical, automotive and food industries. Despite the reduction of manufacturing steps, finishing operations are still required for geometrical tolerances and surface characteristics that are difficult to achieve solely with AM technology. Zhu et al. (2016) presented a study of friction and wear of 316L stainless steel processed by SLM. The SLM-processed samples have pores and fine grains in the microstructure and showed improved friction and wear performance. Gu et al. (2018) investigated the friction and wear properties of in situ Al-matrix composites processed by SLM. The wear rate and coefficient of friction (COF) increased with increasing load, due to the higher contact stress and larger extent of particle fracturing. The wear rate and COF significantly reduced with increasing sliding speed, due to the formation of Al-oxide layer and the transfer of Fe-oxide layer from the counterface to the worn surface. The wear mechanisms of SLM-processed Al-matrix composites under various loads are abrasive and oxidative wear, whereas the wear mechanisms at different sliding speeds are oxidative and delamination wear. Sanjeev et al. (2019) investigated the wear of AM stainless steel (SS) fabricated by the laser beam powder bed fusion (LB-PBF) process. The wear rate for LB-PBF is found to be proportional relative to the applied load. The trend reversed for the lubricated condition with low wear rate due to the change in the dominant wear mechanism from adhesive to surface fatigue and abrasion.

Alvi et al. (2020) studied the high-temperature wear behavior of SLM 316L stainless steel (SS). The lower wear rate in SLM 316L SS at higher temperatures (300 °C and 400 °C) is due to its stable microstructure and the formation of stable oxides. The high temperature, asymmetric thermal gradients and fast cooling rates of SLM tailor the microstructure and improve the properties of processed materials. Anandakrishnan et al. (2020) investigated the wear of AM Inconel 718 samples through direct metal laser sintering process at different build orientations. The load is the most influencing parameter on the wear rate. Hanon et al. (2020) examined the tribological properties considering the impact of print orientation on 3D-printed polymers (polylactide acid and polyethylene terephthalate-glycol). The polymer specimens were fabricated horizontally and vertically by fused deposition modeling. The horizontal print orientation assists in reducing friction and wear. Kang and EL Mansori (2020) investigated the tribological behavior of

SLM-processed hypereutectic Al-Si alloy with silicon content >13 wt% Si. Al-Si alloys have low density, high strength, and are of low cost. The SLM-processed samples of high relative density with an ultra-fine microstructure showed higher wear resistance (with a lower wear rate). Mandal et al. (2020) presented a study of graphene metal matrix nano-composites, the Gr/SS 316L composite, prepared via SLM. The grapheme (Gr) reinforced SS 316L metal matrix composite exhibited a self-lubricating property with high hardness, high wear resistance and low COF. The high wear resistance of Gr-reinforced SS 316L metal matrix composite is due to the addition of Gr into matrix. Marquer et al. (2020) presented the tribological behavior of 3D-printed SLM parts with different orientations under severe loading conditions. The SLM parts showed higher a friction coefficient and lower wear. Yang et al. (2020) presented a study of SLM SS 316L samples fabricated using different energy densities by varying hatch spacing and scanning speed. The parameters of the SLM process impact the microstructure of parts, which further affects their wear characteristics. The tailoring of SLM process parameters improves surface properties of the fabricated parts, which results in enhancement of wear resistance. The increasing energy density and reducing hatch spacing, results in decreasing friction of cellular structures. The alignment of columnar structures in a constant tilt angle from the build direction results in a reduction of wear rates. Zhang et al. (2020) presented the friction and wear characteristics of polylactic acid (PLA) with different infill, printing direction against sliding direction. Superior friction and wear properties of PLA specimens were obtained for an optimum infill density of printed parts. Iakovakis et al. (2020) presented the development of carbide-rich tool steels using electron beam melting. Advances in AM present the opportunity to design tool steels with higher carbon contents to improve wear resistance with low friction. The increased carbide content and further matrix strengthening shifts the mechanism from micro-cutting to micro-plowing. Oxidative wear with adhered iron oxide is the wear mechanism for carbide-rich tool steels. An optimum percentage of carbides pushed into the matrix provides superior friction and wear performance.

Iakovakis et al. (2021) studied the dry sliding wear and friction performance of AM-produced hybrid-cemented carbide, consisting of Cr-rich carbides with WC carbides in the cobalt (Co) matrix. The hybrid-cemented carbides with a relatively high Co content provide high resilience to wear. The dominant wear mechanism is abrasion and adhesion through the removal of the Co matrix, surface oxidation, strain-hardening of the Co matrix and tribolayer formation. Tungsten carbides (WC) improve wear resistance and hardness, while the Co matrix improves the overall toughness of the material. Cemented carbides which improve hardness and toughness are wear-resistant materials due to their microstructure. Murashima et al. (2021) demonstrated a novel morphing surface by using a diaphragm structure with AM. The shape of the morphing surface changes reversibly depending on the air pressure. The surface also exhibits interesting tribological characteristics with varied friction coefficients for concave and convex surfaces. The properties of varied friction coefficients occur due to change in the real contact area depending on surface morphology. Thasleem et al. (2021) investigated the influence of various post-processing methods such as heat treatment and electric discharge alloying (EDA) wear behavior of SLM AlSi$_{10}$Mg alloy. The EDA treated SLM AlSi$_{10}$Mg showed a higher hardness, lower wear rate and COF due to the excessive formation of wear-resistant oxides and glaze layers. The prominent wear mechanisms observed are abrasive wear, adhesive wear, oxidation wear, and surface delamination. Turalıoğlu et al. (2021) studied the tribological behavior of lubricating surfaces designed with micro-channels by SLM with different geometries and impregnated with oil. The impregnated oil reached the surface of the samples which exhibited the highest wear resistance. Yu et al. (2021) investigated SLM-fabricated CNTs reinforced with aluminum matrix nano-composites of customized, tailored hybrid materials with unique hierarchical microstructures. The SLM-produced CNTs reinforced with aluminum matrix nano-composites have superior hardness, strength, frictional and wear behavior.

Kang et al. (2022)investigated the microstructure, wear behavior and friction-induced microstructure evolution of Ti6Al4V samples prepared using laser powder bed fusion (LPBF) and laser directed energy deposition (L-DED). Su et al. (2022) investigated the influence of direct aging and solution-aging treatments on a laser powder bed fusion-processed Ti-6Al-4V alloy. A higher aging-hardening response of the alloy after the solution treatment followed by water quenching resulted in higher wear resistance. Yang et al. (2022) evaluated the tribomechanical characteristics of different CoCrWAlNixAly alloys obtained by direct laser deposition. The effects of varying aluminum and nickel concentrations on the formation of wear-resistant mechanical mixture layer (MML) were investigated. The MML at the surfaces consisting of nano-oxides formed by the aluminum addition was beneficial for lubrication and wear resistance.

Recent advances in friction and wear in additive manufacturing are presented in Table 1.3.

1.3.2 LASER-ASSISTED MANUFACTURING

The laser-assisted manufacturing technique can be applied to small components with complex geometries. The laser-cladding process can be used to improve the high-temperature behavior of the components with dense coatings free of pores and cracks. Pereira et al. (2015) evaluated dry friction and wear behavior at low and high temperatures of NiCoCrAlY and CoNiCrAlY laser-clad coatings. The friction coefficient reduces with high temperature for the NiCoCrAlY coating, while the friction coefficient of CoNiCrAlY coating significantly increases at high temperature. The main wear mechanisms are abrasion and adhesion, caused by oxidized particles on the contact surface. A reduction in wear rate at high temperature was obtained for coatings tested, resulting in improved wear characteristics. Yan et al. (2021) investigated multiple NiCrBSi-WC coatings manufactured on steel substrate using laser cladding with good morphology and microstructure. The coatings, as well as the interfaces at the coating/substrate, showed superior tribological properties of both coefficients of friction and wear rate. Olofsson et al. (2021) presented a study on the formation of a surface layer on gray cast iron

TABLE 1.3
Recent Advances in Friction and Wear in Additive Manufacturing

Friction and Wear Minimization in Additive Manufacturing	Authors
Friction and wear of 316L stainless steel prepared by selective laser melting (SLM)	Zhu et al. (2016)
Friction and wear properties of in situ Al-matrix composites processed by selective laser melting (SLM)	Gu et al. (2018)
Wear of AM stainless steel (SS) fabricated by a laser beam powder bed fusion (LB-PBF) process	Sanjeev et al. (2019)
High-temperature wear behavior of SLM 316L stainless steel (SS)	Alvi et al. (2020)
Wear of AM Inconel 718 through direct metal laser sintering process at different build orientations	Anandakrishnan et al. (2020)
Impact of 3D-printed polymers print orientation on tribological properties	Hanon et al. (2020)
SLM-processed hypereutectic Al-Si alloy with silicon content	Kang and EL Mansori (2020)
Graphene metal matrix nano-composites, Gr/SS 316L composites, via SLM	Mandal et al. (2020)
3D-printed SLM parts with different build orientations	Marquer et al. (2020)
SLM SS 316L samples fabricated using different energy densities by varying hatch spacing and scanning speed	Yang et al. (2020)
Polylactic acid (PLA) with different infill, printing direction against sliding direction	Zhang et al. (2020)
Carbide-rich tool steels developed using electron beam melting	Iakovakis et al. (2020)
Hybrid-cemented carbide (CrC-rich WC–Co cemented carbides) developed using EBM	Iakovakis et al. (2021)
Novel morphing surface structure by using a diaphragm structure with AM	Murashima et al. (2021)
Influence of post-processing methods on the wear behavior of SLM AlSi10Mg alloy	Thasleem et al. (2021)
Lubricating surfaces designed with micro-channels by SLM with different geometries and impregnated with oil	Turalıoğlu et al. (2021)
SLM CNTs reinforced aluminum matrix nano-composites	Yu et al. (2021)
Dry sliding wear behavior of L-PBF and L-DED processed Ti6Al4V samples	Kang et al. (2022)
Direct aging and solution-aging treatments on an L-PBF-processed Ti-6Al-4V alloy	Su et al. (2022)
Tribomechanical evaluations of different CoCrWAlNixAly alloys obtained by direct laser deposition	Yang et al. (2022)

disc brake rotors through laser cladding using a stainless steel powder. The refurbished brake rotor yielded higher friction and demonstrating an effective way to reducing particulate matter emissions. The potential of reusing brake disks is evaluated considering energy and environmental impacts from a sustainability perspective. The remelting of the disc brake rotors for replacement results in a huge waste of energy and an increase in the CO_2 footprint. Liang et al. (2021) designed and produced a high entropy alloy coating of $AlCrFeNiW_{0.2}Ti_{0.5}$ on Q235 steel by laser cladding for marine applications. The $AlCrFeNiW_{0.2}Ti_{0.5}$ coating showed an anomalous 'sunflower-like' morphology. The coating showed excellent tribological performance due to the generation of metal oxides, hydroxides and protective

TABLE 1.4

Recent Advances of Friction and Wear in Laser-assisted Manufacturing

Friction and Wear in Laser-assisted Manufacturing	Authors
Dry friction and wear behavior at low and high temperatures of NiCoCrAlY and CoNiCrAlY laser-clad coatings	Pereira et al. (2015)
Multiple NiCrBSi-WC coatings manufactured on steel substrate using laser cladding	Yan et al. (2021)
Gray cast iron disc brake rotors refurbished by a surface layer with laser cladding using a stainless steel powder	Olofsson et al. (2021)
High entropy alloy coating of $AlCrFeNiW_{0.2}Ti_{0.5}$ designed and produced on Q235 steel via laser cladding	Liang et al. (2021)
High entropy alloy coating with interesting 'island-like' microstructure designed and produced by laser cladding	Liang et al. (2023)

tribofilm on the formation on the $AlCrFeNiW_{0.2}Ti_{0.5}$ coating. Liang et al. (2023) designed and produced a high entropy alloy coating with an 'island-like' microstructure on Q235 steel by laser cladding. The W, Al and Ti elements are added in the high entropy alloy coating based on their outstanding characteristics of improving hardness and wear resistance. Wear resistance is usually closely related to the hardness of specimens. The tribological properties of Q235 steel substrate were greatly improved by high entropy alloy (HEA) coating, with the lowest friction coefficient and wear rate in de-ionized water. Thus, HEA coating is a promising choice of wear-resistant material under water lubrication.

Recent advances of friction and wear in laser-assisted manufacturing are presented in Table 1.4.

1.3.3 Metal Working

The metalworking becomes more attractive when metallic parts must be produced economically in high numbers. Lenard (2000) presented an overview of tribology in the rolling process. The tribological mechanisms, productivity and quality improvements, friction, lubrication and heat transfer at the surface of contact in the rolling are discussed. Fülöp (2001) presented a general overview of tribology of sheet metal forming. The challenges in the sheet metal forming and system development are as follows: lubricant selection, interaction between lubricants and surface parameters, and lubricant performance evaluation. The growing sustainability considerations are environmental considerations, health concerns, safety awareness, post-operation cleanliness of parts, lubricant quantity minimization and relative cost reduction. Engel (2006) incorporated the size effect in friction in the microforming process based on the general friction law developed by Wanheim/Bay. Friction is considerably affected when the forming process is scaled down, yielding distinctly increased friction in the microforming process. Tribology plays an important role in the feasibility and quality of microforming. Iglesias et al. (2010) studied the tribological behavior of nanostructured

TABLE 1.5

Recent Advances in Friction and Wear in Metal Working

Friction and Wear in Metal Working	Authors
An overview of tribology in the rolling process considering friction, lubrication and heat transfer	Lenard (2000)
A general overview of tribology of sheet metal forming	Fülöp (2001)
Size effect in friction taken into account based on the general friction law by Wanheim/Bay	Engel (2006)
Nanostructured oxygen-free high conductivity (OFHC) copper by large strain extrusion machining	Iglesias et al. (2010)
An overview of the temperature, coating and lubrication effects on the tribological characteristics in hot forming	Dohda et al. (2015)

oxygen-free high conductivity (OFHC) copper manufactured by large strain extrusion machining (LSEM). The deformation process in LSEM combines microstructure refinement and dimensional control. The wear resistance of this anisotropic extruded material depends on the sliding direction. The highest wear resistance for an elongated grain structure in the extrusion direction is observed for nanostructured copper material. Dohda et al. (2015) presented an overview of the temperature, coating and lubrication effects on the tribological characteristics in hot forming (rolling and stamping). The review also showed the great potential in further innovations in tribology, such as tribometers for different metal forming processes at elevated temperatures.

Recent advances in friction and wear in metal working are presented in Table 1.5.

1.3.4 Powder Metallurgy

Powder metallurgy (PM) manufacturing is a low cost large-scale manufacturing process that produces a broad range of materials through sintering. Furthermore, sintering temperature is lower than melting temperature, promoting ceramics, hard metals and composites. The manufacturing of complex shaped components with surface pores using sintering techniques avoids additional machining operations, thus reducing costs and material waste. Ravindran et al. (2013a, 2013b) investigated the tribological properties of Al (aluminum) based Gr (graphite) and SiC (silicon carbide) reinforced composites manufactured by PM. The SiC-reinforced hybrid composites exhibited a lower wear loss compared to the Al–Gr composites under dry sliding conditions. With increasing SiC content, the wear resistance increased monotonically with hardness. The addition of both SiC and graphite significantly improved the wear resistance of aluminum composites which can be considered for high strength and wear-resistant applications. Shaikh et al. (2019) evaluated the aluminum (Al) matrix composites (AMCs) reinforced with rice husk ash (RHA) and SiC using the PM process. The hardness and

TABLE 1.6

Recent Advances of Friction and Wear in Powder Metallurgy

Friction and Wear in Powder Metallurgy	Authors
Tribological studies of aluminum (Al)-based graphite (Gr) and silicon carbide (SiC) by powder metallurgy (PM)	Ravindran et al. (2013a, 2013b)
Aluminum matrix composites (AMCs) reinforced with rice husk ash (RHA) and silicon carbide (SiC) using PM process	Shaikh et al. (2019)
Effect of surface pores in sintered materials for superior performances in lubricated point contacts	Boidi et al. (2021)
In situ carbide/CoCrFeNiMn composites with excellent anti-wear properties fabricated by PM process	Ye et al. (2023)

wear properties of the AMCs reached optimum with an increase in RHA. Boidi et al. (2021) evaluated the effect of surface pores in different sintered materials for obtaining superior performances in lubricated point contacts under different sliding-rolling conditions. Steel powder is sintered to promote both porosity and pore dimensions reduction. The random micro-irregularities in sintered materials change lubricant conditions and potentially improve the efficiency of tribological systems. The decrease of porosity in sintered materials generally improves tribological performance and promotes friction reduction. Ye et al. (2023) investigated in situ a carbide/CoCrFeNiMn composite, a high entropy alloy, with excellent anti-wear properties fabricated by flake PM process. The hard carbide phase, Cr_7C_3, is formed in the soft matrix via in situ reaction at the GO/HEA interface using graphene oxide (GO) and CoCrFeNiMn as powders.

Recent advances of friction and wear in powder metallurgy are presented in Table 1.6.

1.3.5 Casting

Energy consumption as well as the capital involved in casting processes has a major influence upon the costs of cast products. The efficiency of the casting processes and surface quality of the cast products need continuous tribological improvements for sustainability. Salas et al. (2003) investigated the design of a multilayer coating system that encompasses an increased die surface performance and die life. The coatings on steel substrates of die casting are aimed at improving the tribological performance of dies for the aluminum foundry. The wear and adhesion properties are related by the brittleness or toughness of the coating and by the nature of the substrate/coating interface. Joshi et al. (2004) presented a model for soldering growth and dissolution in liquid aluminum and die steel molds. The physicochemical interactions between cast metal and die steel interface lead to aluminum–iron–silicon intermetallic formation. The mass loss of the die steel was due to dissolution of the intermetallic layer. The diffusion barrier treatment, nitriding, applied at the interface prevents soldering formation and provides reduced friction and adhesion at the interface. Gecu and Karaaslan (2019) investigated the effect of volume friction of vacuum-assisted melt infiltration casting to produce A356 aluminum matrix composites with commercially pure titanium (CP-Ti) contents. Strongly bonded $TiAl_3$ phase at the interface layer formed at the interface between A356 and CP-Ti phases provided better wear resistance. Abbasipour et al. (2019) investigated the tribological characteristics of A356 aluminum alloy castings and A356–CNT nano-composite castings. The wear loss, wear rate and friction coefficient characteristics of nano-composites with the addition of CNTs were remarkably improved. The abrasion was the dominant wear mechanism of the nano-composite samples. Sardar et al. (2019) investigated the tribological properties of Al-Si-Mg matrix composites reinforced with Al_2O_3 particles by enhanced stir casting with increasing particle content. The hardness of composite rose in spite of a marginal increase in porosity. The composites exhibited significantly reduced wear rate and COF with reinforcement content followed by grit size. The mechanisms of abrasion in composites were observed to be mainly delamination with limited micro-plowing. Mohanavel et al. (2020) presented the manufacturing of aluminum composites (AMCs) with AA6351 (Al-Si-Mg) as matrix material, and silicon nitride (Si_3N_4) particles as reinforcement material via a stir casting technique. Aatthisugan et al. (2022) investigated the tribological characteristics of AZ91 magnesium alloy-based metal matrix composites with boron carbide (B4C) particles reinforcements made of bottom pouring stir casting. The dry sliding wear tests of hard reinforcements on the casted AZ91D-B4C magnesium composite samples showed improved wear resistance.

Recent advances in friction and wear in casting are presented in Table 1.7.

TABLE 1.7
Recent Advances in Friction and Wear in Casting

Friction and Wear in Casting	Authors
Design of a multilayer coating system that encompasses increased die surface performance and die life	Salas et al. (2003)
A model for soldering growth and dissolution in liquid aluminum and die steel molds	Joshi et al. (2004)
Vacuum-assisted melt infiltration casting for aluminum matrix composites of commercially pure titanium (CP-Ti)	Gecu and Karaaslan (2019)
A356 aluminum alloy castings and A356–CNT nano-composite castings	Abbasipour et al. (2019)
Al–Zn–Mg–Cu matrix composites reinforced with Al_2O_3 particles by enhanced stir casting	Sardar et al. (2019)
Al-Si-Mg as matrix material and silicon nitride (Si_3N_4) particles as reinforcement material via stir casting technique	Mohanavel et al. (2020)
AZ91 magnesium alloy-based metal matrix composites with boron carbide (B4C) particles reinforcements made of bottom pouring stir casting	Aatthisugan et al. (2022)

1.3.6 Friction Stir Processing

Friction stir processing (FSP) helps reduce the spacing between particles (i.e., porosity and voids) while improving the microstructure, and even inducing a phase transformation. The surface composites of solid lubricant reinforcement particles along with abrasive ceramics have gained attention for their enhancement of tribological properties. Aldajah et al. (2009) presented an application of FSP to carbon steel for enhancement of near-surface material properties. FSP transformed the original pearlite microstructure to martensite, resulting in a significant increase in surface hardness. The reduced plasticity of the near-surface material resulted in an improvement of friction and wear behavior. Pezeshkian et al. (2018) investigated the fabrication of a surface composite of Ni particles on the surface of copper using FSP. The Cu plate specimens' grooves of different dimensions are filled with optimized percentage of Ni reinforcement particles. The micro hardness and wear properties were improved compared to the substrate. Sharma et al. (2019) presented a study on AA6061-T651 aluminum alloy surface composites with varying amounts of MoS2 using friction stir processing. The wear resistance of surface composites is improved due to MoS2 solid lubricant. Ralls et al. (2021) highlighted the key factors that enable FSP technology for steel metallurgical improvement. An analysis of the FSP processing variables is presented in relation to steel metallurgical improvement for enhancement of tribological, corrosion and erosion properties. FSP has been increasingly utilized owing to its cost effective and negative environmental impact. It induces large amounts of strain and plastic deformation causing dynamic recrystallization which enhances the properties (tribological, corrosion and erosion). Wu et al. (2022) summarized state-of-the-art research on FSP technology and various types of reinforcement particles to produce surface material matrix composites (MMC). The effect of FSP parameters and types of reinforcement particles on the microstructure, mechanical and tribological properties are discussed. Friction stir processing promotes the development of surface MMCs by embedding various reinforcement particles for enhanced hardness and tribological characteristics. Ralls et al. (2023) investigated the tribological and corrosion properties of cold-sprayed (CS) 316L SS deposits using FSP. The enhancement in tribological properties such as the reduction in friction and wear rate of the CS deposits after FSP is due to the closure of pores and localized grain refinement.

Recent advances in friction and wear in friction stir processing are presented in Table 1.8.

1.4 LUBRICATION IN SUSTAINABLE MANUFACTURING

There is growing importance in the environmental aspects relating to manufacturing with more stringent enforcement of environmental legislation. Stanford and Lister (2002) reported feasible methods of reducing the consumption of cutting fluid and efficient methods of cutting fluid utilization (minimum quantity lubricant delivery). The consideration of dry cutting technologies arises from the

TABLE 1.8
Recent Advances in Friction and Wear in Friction Stir Processing

Friction and Wear in Friction Stir Processing	Authors
FSP applied to carbon steel for enhancement of near-surface material properties	Aldajah et al. (2009)
Surface composite of Ni particles on the surface of copper	Pezeshkian et al. (2018)
Surface composites of aluminum alloy with MoS2	Sharma et al. (2019)
Key factors which enable FSP technology for steel metallurgical improvement	Ralls et al. (2021)
State-of-the-art research on FSP technology and various types of reinforcement particles to produce surface MMC	Wu et al. (2022)
Cold-sprayed (CS) 316L SS deposits	Ralls et al. (2023)

TABLE 1.9
Recent Advances in Lubrication in Sustainable Manufacturing

Lubrication in Sustainable Manufacturing	Authors
Feasible methods of cutting fluid consumption reduction, minimum quantity lubricant delivery and dry cutting technologies	Stanford and Lister (2002)
A comprehensive review of sustainable machining techniques (dry, minimum quantity lubrication (MQL) and cryogenic)	Ahmad et al. (2021)

difficulties experienced in removing cutting fluids from the metal cutting process. Ahmad et al. (2021) presented a comprehensive review using sustainable machining techniques: dry, minimum quantity lubrication and cryogenic. The sustainable machining techniques are studied with respect to tool wear, cutting force, surface roughness, chips formation and economics of material removal rate.

Recent advances in lubrication in sustainable manufacturing are presented in Table 1.9.

1.4.1 BIOLUBRICANTS

The use of sustainable biolubricants will help conserve the environment. Rajewski et al. (2000) focused on the types of lubricants found in the food industry, their benefits and potential drawbacks. The regulations and requirements for lubricants in food manufacturing and processing equipment have evolved to include a wide variety of synthetic-based lubricants. Maleque et al. (2003) presented a critical review on the advantages and manufacturing processes of vegetable oil-based additives with applications. A study on palm oil methyl ester as an additive is presented. The utilization of vegetable oil additives reduces wear and friction and improves lubrication. Han et al. (2005) analyzed the application of water vapor as a coolant and lubricant in green cutting. The cooling and lubricating properties of water vapor are due to excellent penetration and low shearing strength. The cutting forces and friction coefficient are reduced with the application of water vapor as coolant and lubricant. Salih et al. (2011) outlined the products with improved physicochemical and tribological properties as good biolubricant base stocks. The modifications in the epoxidation, esterification and acylation reactions are used to produce oleic acid-based triester derivatives. An increase in mid-chain substituent length improves the pour point, and gives excellent lubricity and anti-wear properties. Razak et al. (2015) elucidated the performance of palm lubrication between the femoral head and the acetabular cup metal-on-metal (MoM) hip replacements. The use of palm olein, palm kernel oil, and palm fatty acid distillate as lubricants for hip implants has shown tremendous potential. Reeves et al. (2015) conducted experimental studies on the friction and wear properties of bio-based lubricants such as avocado, canola (rapeseed), corn, olive, peanut, safflower, sesame and soybean. Based on the mechanisms governing the improved tribological performance, the effects of fatty acid composition and thermal response of the natural oils, the avocado oil showed superior friction and wear performance. Rao et al. (2018) presented an overview of tribological applications of biolubricants in Malaysia and Japan under the categories of base stock, mixtures and additives. The biolubricants with additives provide excellent lubricant performance, support the dependence on renewable resources and increase applications in industrial markets. Chan et al. (2018) assessed the tribological performance of biolubricant base stocks and their additives. The basic parameters and molecular structure of biolubricants for high tribological performance based on the friction coefficient and wear measurements are discussed. These additives, such as plant-derived eco-friendly compounds, polymers, particulate/layered materials and ionic liquids, are also discussed. Noor El-Din (2018) presented the formulation of a castor oil-based metalworking fluid as the base oil optimized with nonionic surfactants and additives. Castor oil-based MWF with monoethanolamine (MEA) as a corrosion inhibitor is highly stable. Pranav et al. (2021) provided a review of various bio-based cutting fluids (CFs) and eco-friendly additives for sustainability in machining. Durango-Giraldo et al. (2022) presented an overview on the processing and tribological performance of palm oil as a biolubricant in industrial applications. The concerns related to palm oil as an alternative biolubricant are due to deforestation and loss of biodiversity. The significant value of palm oil as a commodity to replace petroleum-based products is derived from a more controlled extraction, a better use of the extracted oil and a higher performance of the extracted products. Lee et al. (2022) presented a bibliometric analysis to elucidate the potential of vegetable oil-based biolubricants. The correlation between fatty acid composition and its tribological performance is recommended for biolubricants to be used in a wide range of applications.

Recent advances in lubrication with biolubricants are presented in Table 1.10.

1.4.2 Nano-biolubricants

Nano-biolubricants are gaining more significance because of their excellent tribological performance, biodegradability, renewability and minimal environmental

TABLE 1.10

Recent Advances in Lubrication with Biolubricants

Lubrication with Biolubricants	Authors
Types of lubricants in the food industry, benefits and potential drawbacks	Rajewski et al. (2000)
A critical review of vegetable-based lubricant additives	Maleque et al. (2003)
Water vapor as a coolant and lubricant in green cutting	Han et al. (2005)
Products with improved physicochemical and tribological properties as good biolubricant base stocks	Salih et al. (2011)
Palm lubrication between the femoral head and the acetabular cup MoM hip replacements	Razak et al. (2015)
Friction and wear properties of bio-based lubricants	Reeves et al. (2015)
Tribological applications review of biolubricants in Malaysia and Japan	Rao et al. (2018)
Tribological performances of biolubricant base stocks and their additives	Chan et al. (2018)
Castor oil-based metalworking fluid optimized with nonionic surfactants and additives	Noor El-Din (2018)
A review of bio-based cutting fluids (CFs) with eco-friendly additives	Pranav et al. (2021)
A review on the processing and tribological performance of palm oil as a biolubricant in industrial applications	Durango-Giraldo et al. (2022)
A bibliometric analysis to elucidate the potential of vegetable oil-based biolubricants	Lee et al. (2022)

impact. Nano-biolubricant selection extends benefits in terms of energy and environment conservation in addition to the positive impact on cost, performance and life. Reeves et al. (2013) evaluated the boron nitride particle size effects on the tribological performance of canola oil-based lubricant mixtures. The nanometer-sized particle mixtures outperformed micron/submicron-sized particle combinations in terms of friction and wear performance. The nanometer-sized particles were able to coalesce better in the asperity valleys to enhance tribological properties. The biolubricants were subsequently explored for energy conservation, sustainability and lubricant market potential. Shaari et al. (2015) investigated the tribological properties of palm oil biolubricant modified with titanium dioxide (TiO_2) nanoparticles as additives. The palm oil biolubricant samples with TiO_2 nanoparticle additives produced the lowest COF and wear scar diameter. Gupta and Harsha (2017) investigated tribological properties of castor oil with calcium–copper–titanate nanoparticles (CCTO) and zinc dialkyldithiophosphate (ZDDP) additives. Tribological tests revealed an improvement in the anti-wear and extreme-pressure (EP) properties of castor oil with additives. An optimum concentration of CCTO nanoparticles and ZDDP additives showed improvement in COF and wear properties. Chowdary et al. (2021) reviewed tribological and thermophysical performance of biolubricants for eco-friendly automobile engines. Pawar et al. (2022) presented a review on the advancements in the synthesis, characterization and performance of nano-biolubricants. This review elaborates on

TABLE 1.11

Recent Advances in Lubrication with Nano-biolubricants

Lubrication with Nano-biolubricants	Authors
Boron nitride particle size effect on the tribological performance of canola oil-based lubricant mixtures	Reeves et al. (2013)
Tribological properties of palm oil biolubricant modified with titanium oxide (TiO$_2$) nanoparticles as additives	Shaari et al. (2015)
Castor oil with calcium–copper–titanate nanoparticles (CCTO) and zinc dialkyldithiophosphate (ZDDP) additives	Gupta and Harsha (2017)
Reviewed tribological and thermophysical performance of biolubricants for eco-friendly automobile engines	Chowdary et al. (2021)
Reviewed the advancements in the synthesis, characterization and performance of nano-biolubricants	Pawar et al.(2022)
Castor oil biolubricant using Fe$_3$O$_4$ nanoparticles and ethylene glycol in a transesterification process	Ahmad et al. (2022)

the dispersion stability, size, shape, concentration, morphology and mechanism of nanoparticles and their effects on tribological properties. Vegetable oils have significant prospects to be engineered as sustainable lubricants with nanoparticle additives. Ahmad et al. (2022) synthesized and evaluated the potential of castor oil biolubricant using iron oxide (Fe$_3$O$_4$) nanoparticles as an additive and ethylene glycol in a transesterification process. The synthesized castor oil nano-biolubricant showed a decreased COF and wear.

Recent advances in lubrication with nano-biolubricants are presented in Table 1.11.

1.4.3 SURFACE TEXTURING

Surface texturing has emerged as a viable option resulting in significant improvement in tribological characteristics of components. Textures functionality is ensured by appropriate optimization of its geometrical parameters. Etsion (2005) presented a review of state-of-the-art laser surface texturing (LST) potential in various lubricated contacts. Laser surface texturing produces a number of microdimples, each of which can serve either as a micro-hydrodynamic bearing, a micro-reservoir or a micro-trap for wear debris. Coblas et al. (2015) proposed a general classification of texture fabrication methods, with a pertinent description of advantages and drawbacks of their application in mechanical systems. Arslan et al. (2016) presented an overview of the techniques to manufacture micro-/nano-textures for tribological and machining applications. Texture quality is significantly influenced by the manufacturing processes. The tribological characteristics are remarkably improved by surface texturing: friction and wear reduction and lubrication enhancement. Machining performance is improved by texturing: lubrication enhancement, tool-chip contact area, wear, friction and cutting force reduction, and wear debris trapping. Niketh and Samuel (2017) investigated the effectiveness of micro textures on drill tools for the sustainable machining of Ti-6Al-4V. The micro textures on the drill tools minimize the cutting forces and minimize the energy loss by reducing frictional forces at the contact surfaces. Grützmacher et al. (2019) explored the potential of multi-scale surface texturing for optimizing friction and wear in tribological systems. State-of-the-art knowledge about the multi-scale surface texture mechanisms is summarized and future research directions outlined. Mao et al. (2020) reviewed the recent advancements of LST for enhanced tribological properties. The effects of laser processing parameters on the texture features for the enhancement of tribological properties are highlighted. Laser surface texturing has gained significance due to its superior flexibility and high accuracy. Vishnoi et al. (2021) reviewed surface-texturing methods on metals and non-metals to enhance tribological performance. The fabrication, characterization, advantages, disadvantages and applications of textured surfaces are presented. Surface texturing has considerable effects on the improvement of friction, wear and lubrication properties. Allen and Raeymaekers (2021) presented a critical review of the literature on textured orthopedic prosthetic hip implant surfaces. The challenges in manufacturing must be overcome to create texture features on orthopedic biomaterial implant surfaces. The patterned micro textural features on the prosthetic implants reduce wear. Gupta et al. (2021) focused on an investigation of sustainable textured-turning tool performance of Inconel-718 under the influence of dry, cryogenic and MQL+ hBN solid lubricants. The textured-turning tool with MQL+ hBN solid lubricants provides superior machining performance. The application of a textured tool reduces abrasion, rubbing and microchipping with low wear marks. The surface-textured tools have less area of contact between the interfaces, entrapped debris within the recesses and fluid layer formation on the rake surface of tool.

Recent advances in lubrication with surface texturing are presented in Table 1.12.

1.4.4   SURFACE COATING

The successful application of coatings on tools and components is an efficient way of improving their tribological properties. A thin layer of coating on a surface changes the properties and controls tribological properties. Holmberg et al. (2000) presented the fundamentals of coating tribology with a holistic approach. The tribological parameters analyzed are coating-to-substrate hardness, film thickness, surface roughness and debris. The important influence of thin tribo- and transfer-layers is highlighted. The mechanisms of the lubricated boundary and reaction films are discussed. Hogmark et al. (2000) presented tribological design considerations of coating composites and methods of their evaluation. The purpose of applying tribological coatings is increased lifetime and productivity, prolonged tool life, improved wear resistance, lowered friction, reduced energy consumption, improved lubrication and/or cooling. The properties of the coating, the substrate and the interface play significant roles in avoiding premature failure of the tribological coatings. Cutting tools continue to be the leading application

TABLE 1.12

Recent Advances in Lubrication with Surface Texturing

Lubrication with Surface Texturing	Authors
A review of state-of-the-art laser surface texturing (LST) potential in various lubricated contacts	Etsion (2005)
A general classification of texture manufacturing techniques	Coblas et al. (2015)
An overview of the techniques to manufacture micro-/nano-textures for tribological and machining applications	Arslan et al. (2016)
Application of micro textures on drill tools for the sustainable machining of Ti-6Al-4V	Niketh and Samuel (2017)
Potential of multi-scale surface texturing for optimizing friction and wear in tribological systems	Grützmacher et al. (2019)
Review of the recent advancements of LST for enhanced tribological properties	Mao et al. (2020)
Review of surface-texturing methods on metals and non-metals to enhance tribological performance	Vishnoi et al. (2021)
A review of literature on textured orthopedic prosthetic hip implant surfaces	Allen and Raeymaekers (2021)
Textured-turning tool performance of Inconel-718 under the influence of dry, cryogenic and MQL+ hBN solid lubricants	Gupta et al. (2021)

of tribological coatings. The coating of a relatively small size of cutting tools in batches will keep the cost within reasonable limits. Holmberg et al. (2007) discussed the tribological mechanisms, scale effects and parameters of coated surfaces. The dominant mechanisms considering scale effects of rough and smooth surfaces are illustrated for diamond- and DLC-coated surfaces in a tribological contact. The systematic surface design for tribological applications is modeled using 3D FEM to analyze deformations, stresses and strains. Wood (2010) studied the tribology of thermal-sprayed WC–Co based coatings. The tribological and corrosion performance of these coatings is related to their mechanical and corrosion properties, deposition parameters, microstructure and composition.

Jähnig and Lasagni (2020) introduced a strategy for improvement of friction and wear properties using a selective laser treatment of a hydrogen-free diamond-like carbon layer (DLC) in the forming tools. Hydrogen-free DLCs are deposited on steel surfaces and micro-structured subsequently using direct laser interference patterning (DLIP). The characteristics are relevant for improving tribological performance without using lubrication (lubricant substitution) in forming processes. Frutos et al. (2023) proposed a novel design of laser-textured patterns with superior tribological performance. The laser-textured patterns with high anchor densities promote the mechanical bonding of highly abrasive sintered diamond/carbide reinforcement. The higher nucleation of TiC/Fe_3C on the diamond surfaces promotes increased bonding within the matrix, preventing its graphitization, and detachment. Reduction in the coefficient of friction was observed for coatings reinforced with diamonds under dry conditions. Thicker coatings show superior

TABLE 1.13

Recent Advances of Lubrication with Surface Coating

Lubrication with Surface Coating	Authors
Fundamentals of coating tribology in a holistic approach	Holmberg et al. (2000)
Tribological design considerations of coating composites and methods of evaluation	Hogmark et al. (2000)
Tribological mechanisms, scale effects and parameters of coated surfaces	Holmberg et al. (2007)
Tribology of thermal-sprayed WC–Co based coatings	Wood (2010)
Selective laser treatment of a hydrogen-free diamond-like carbon layer (DLC) in forming tools	Jähnig and Lasagni (2020)
Highly abrasive sintered diamond/carbide reinforcement with superior tribological performance	Frutos et al. (2023)

tribological performance (scratch/wear resistance) than thinner ones because they have a higher load-carrying capacity.

Recent advances in lubrication with surface coating are presented in Table 1.13.

1.4.5 Minimum Quantity Lubrication

Minimum quantity lubrication (MQL) possesses better economic benefits considering the minimum amount of used lubricants. Additionally, the MQL approach has a noticeable benefit in terms of minimizing the cutting forces, improving the operating life of tools and the surface quality of the machined surfaces. Carou (2015) presented a review of the MQL system in turning operations considering surface quality or tool life. The comprehensive review is useful for performance improvement in metalworking firms in terms of cost, environmental impact and safety. Khan et al. (2019) investigated the superiority of nanofluids-based MQL (NFMQL) in comparison to MQL of the manufacture of AISI D2 steel during face milling. Sustainable machining processes such as NFMQL reduce the resource utilization with savings in energy consumption and cost. Ibrahim et al. (2020) evaluated palm oil-based nanofluids containing different percentages of graphene nano-platelets (GNPs) utilizing the MQL model. The palm oil+GNPs could lessen the specific cutting energy due to reduction in the friction coefficient. Palm oil is a promising cooling and lubrication medium owing to the high content of triglycerides and good polarity. It forms a good physical adsorption film and exhibits lubrication effects because of its higher viscosity. The sustainable nanofluid+MQL significantly increases the tool life and reduces machining costs. Khaliq et al. (2020) analyzed the lubrication and cooling effects under MQL in the micro-milling of additively manufactured Ti–6Al–4V using coated tungsten carbide micro endmills. The MQL in micro-milling is applied to reduce the tool wear rate for cost savings and to improve the surface finish of the machined part.

Lizzul et al. (2020) evaluated milling of laser powder bed-fused (LPBF) Ti_6Al_4V parts with different build-up orientations (anisotropy) under MQL conditions. An extended tool life is necessary to ensure high quality machined surfaces and to save costs. The tool life of manufactured samples with horizontal build-up orientations is higher than vertically manufactured parts.

Danish et al. (2022) investigated the tool wear and surface roughness in micromilling of additively fabricated Inconel 718 under the effect of dry, MQL and flood conditions. The governing wear mechanisms of the cutting tools under the lubricating/cooling MQL and flood medium are abrasive and adhesive wear. The MQL and flood medium provide significant advantages such as improving tool life and surface finish with the cooling and lubrication effects. Wang et al. (2022) analyzed the cooling/lubrication mechanism, tool wear and surface quality in turning under MQL, nanofluids, cryogenic medium, ultrasonic vibration and textured tools for difficult-to-machine materials. The MQL technology accounts for sustainable manufacturing, and is an effective way of meeting carbon neutrality requirements in manufacturing. Şen (2022) investigated the effectiveness of single point incremental forming (SPIF) by an MQL-assisted process. Lubrication is a prerequisite for SPIF processes with acceptable surface qualities. The surface quality can be improved by the MQL-assisted SPIF process by high pressure spraying of a lubricant into the contact zone. The environment and health considerations are improved and production costs minimized with the MQL technique in SPIF processes.

Recent advances in lubrication under minimum quantity lubrication are presented in Table 1.14.

TABLE 1.14

Recent Advances in Lubrication under Minimum Quantity Lubrication

Lubrication under Minimum Quantity Lubrication	Authors
Review of the MQL system in turning operations	Carou (2015)
MQL and nanofluids-based MQL (NFMQL) sustainable lubrication methods	Khan et al. (2019)
Eco-friendly nano cutting fluids to boost the sustainability of the machining processes	Ibrahim et al. (2020)
Dry and MQL conditions under the micro-milling process	Khaliq et al. (2020)
Laser powder bed-fused (LPBF) Ti_6Al_4V anisotropy of a part during milling operations under MQL conditions	Lizzul et al. (2020)
Micro-milling of additively fabricated Inconel 718 using dry, MQL and flood cooling medium	Danish et al. (2022)
Turning under MQL, nanofluids, cryogenic medium, ultrasonic vibration and textured tools	Wang et al. (2022)
Formation of sheet metals into truncated forms by an MQL-assisted single point incremental forming (SPIF) process	Şen (2022)

1.4.6 CRYOGENIC MACHINING

Lubrication systems such as MQL, cryogenic cooling, etc. are introduced to lower the friction between surfaces and to enhance tribological performance for sustainable manufacturing. The innovative strategies in cooling/lubrication of machining processes provide solutions toward sustainability with economic viability. Bordin et al. (2015) evaluated tool wear mechanisms in turning the Ti_6Al_4V alloy under dry and cryogenic conditions using PVD-coated tungsten carbide inserts. The cryogenic cooling strategy reduced the tool adhesive wear. Cryogenic machining is an efficient solution for machining biomedical implants by reducing cleaning and sterilizing costs as liquid nitrogen gasifies during machining. Bruschi et al. (2016) presented the effect of the machining and cooling strategies on the wear behavior of additive manufactured EBM Ti_6Al_4V titanium alloys used for biomedical applications. The adoption of cryogenic cooling during machining significantly affected the Ti_6Al_4V surface properties in terms of lower friction coefficient, improving wear performance. Muhamad et al. (2020) investigated the wear mechanisms of a multilayered TiAlN/AlCrN-coated carbide tool during the milling of AISI 4340 steel under cryogenic machining. The causes of tool wear were abrasion and adhesion wear on the flank face. Cryogenic machining through the application of liquid nitrogen prolongs the coated carbide tool life at higher cutting speed. Korkmaz et al. (2022) investigated the influence on lubrication regime and tribological characteristics of additively manufactured 316 steel under dry, MQL, cryogenic and cryo+MQL conditions. The cryo+MQL combination is helpful for providing a good lubrication film, consequently improving tribological properties. The hybrid cryo+MQL system increases viscosity and improves tribological properties due to a change in the boundary lubrication regime to mixed and elasto-hydrodynamic regimes. Demirsöz et al. (2022) investigated the tribological characteristics of additively manufactured 316L stainless steel under dry, MQL, cryogenic and cryo+MQL conditions. The cooling and lubrication combination of cryo+MQL conditions is helpful for improving the tribological performance, minimizing wear depth and wear rates.

Recent advances of lubrication under cryogenic machining are presented in Table 1.15.

TABLE 1.15
Recent Advances of Lubrication under Cryogenic Machining

Lubrication under Cryogenic Machining	Authors
Tool wear mechanisms in turning under dry and cryogenic conditions	Bordin et al. (2015)
Machining and cooling strategies on the additive manufactured Ti_6Al_4V for biomedical applications	Bruschi et al. (2016)
Multilayered TiAlN/AlCrN-coated carbide tool for milling of AISI 4340 steel under cryogenic machining	Muhamad et al. (2020)
Lubrication regime relation with tribological properties under dry, cryogenic, MQL, cryo+MQL conditions	Korkmaz et al. (2022)
Tribological properties under dry, MQL, cryogenic, cryo+MQL conditions	Demirsöz et al. (2022)

1.5 CONCLUSIONS AND FUTURE SCOPE

Traditional manufacturing can no longer meet the requirements of environmental protection policies, laws and standards. In the context of carbon neutrality requirements, traditional manufacturing must be transformed into sustainable manufacturing. Sustainable manufacturing involves the manufacturing of a product without any detriment to the environment. The tribology of sustainable manufacturing is a decisive factor in the production cycle, cost and quality of parts. Sustainable manufacturing principles impact cost and energy saving, and environmental and health friendly considerations. Energy conservation, economics and environment-friendly considerations have taken a global direction toward sustainable manufacturing. This review discusses the state-of-the-art tribological methods that may co-contribute to sustainable manufacturing processes.

REFERENCES

Aatthisugan, I., Murugesan, R. and Rao, T. V. V. L. N. (2022). Influence of boron carbide content on dry sliding wear performances of AZ91D magnesium alloy, *Proceedings of the Institution of Mechanical Engineers, Part J: Journal of Engineering Tribology*, https://doi.org/10.1177/13506501221109033

Abbasipour, B., Niroumand, B., Vaghefi, S. M. M. and Abedi, M. (2019). Tribological behavior of A356–CNT nano-composites fabricated by various casting techniques, *Transactions of Nonferrous Metals Society of China*, 29 (10), 1993–2004.

Abdollah, M. F. B. (2017). Special issue on 50 years of tribology – Sustainable manufacturing, *Industrial Lubrication and Tribology*, 69 (3), 333–333.

Ahmad, A. A., Ghani, J. A. and Haron, C. H. C. (2021). Green lubrication technique for sustainable machining of AISI 4340 alloy steel, *Jurnal Tribologi*, 28, 1–19.

Ahmad, U., Naqvi, S. R., Ali, I., et al. (2022). Biolubricant production from castor oil using iron oxide nanoparticles as an additive: Experimental, modelling and tribological assessment, *Fuel*, 324 (Part A), 124565.

Aldajah, S. H., Ajayi, O. O., Fenske, G. R. and David, S. (2009). Effect of friction stir processing on the tribological performance of high carbon steel, *Wear*, 267 (1–4), 350–355.

Allen, Q. and Raeymaekers, B. (2021). Surface texturing of prosthetic hip implant bearing surfaces: A review, *Journal of Tribology*, 143 (4), 040801.

Alvi, S., Saeidi, K. and Akhtar, F.(2020). High temperature tribology and wear of selective laser melted (SLM) 316L stainless steel, *Wear*, 448–449, 203228.

Anandakrishnan V., Sathish S., Muthukannan, D., Dillibabu V. and Balamuralikrishnan N. (2020). Dry sliding wear behavior of Inconel 718 additively manufactured by DMLS technique, *Industrial Lubrication and Tribology*, 72 (4), 491–496.

Arslan, A., Masjuki, H. H., Kalam, M. A. et al. (2016). Surface texture manufacturing techniques and tribological effect of surface texturing on cutting tool performance: A review, *Critical Reviews in Solid State and Materials Sciences*, 41 (6), 447–481.

Bell, T. (1992).Towards designer surfaces, *Industrial Lubrication and Tribology*, 44 (1), 3–11.

Boidi, G., Profito, F. J., Kadiric, A., Machado, I. F. and Dini, D. (2021). The use of powder metallurgy for promoting friction reduction under sliding-rolling lubricated conditions, *Tribology International*, 157, 106892.

Bordin, A., Bruschi, S., Ghiotti, A. and Bariani, P. F. (2015). Analysis of tool wear in cryogenic machining of additive manufactured Ti6Al4V alloy, *Wear*, 328–329, 89–99.

Bruschi, S., Bertolini, R., Bordin, A., Medea, F. and Ghiotti, A. (2016). Influence of the machining parameters and cooling strategies on the wear behavior of wrought and additive manufactured Ti_6Al_4V for biomedical applications, *Tribology International*, 102, 133–142.

Carou, D., Rubio, E. M. and Davim, J. P. (2015). A note on the use of the minimum quantity lubrication (MQL) system in turning, *Industrial Lubrication and Tribology*, 67 (3), 256–261.

Chan, C.-H., Tang, S. W., Mohd, N. K. et al. (2018). Tribological behavior of biolubricant base stocks and additives,*Renewable and Sustainable Energy Reviews*, 93, 145–157.

Chowdary, K., Kotia, A., Lakshmanan, V., Elsheikh, A. H. and Ali, M. K. A. (2021). A review of the tribological and thermophysical mechanisms of bio-lubricants based nanomaterials in automotive applications, *Journal of Molecular Liquids*, 339, 116717.

Coblas, D. G., Fatu, A., Maoui, A. and Hajjam, M. (2015). Manufacturing textured surfaces: State of art and recent developments. *Proceedings of the Institution of Mechanical Engineers, Part J: Journal of Engineering Tribology*, 229 (1), 3–29.

Danish, M., Aslantas, K., Hascelik, A., Rubaiee, S. et al. (2022). An experimental investigations on effects of cooling/lubrication conditions in micro milling of additively manufactured Inconel 718, *Tribology International*, 173, 107620.

Demirsöz, R., Korkmaz, M. E. and Gupta, M. K. (2022). A novel use of hybrid Cryo-MQL system in improving the tribological characteristics of additively manufactured 316 stainless steel against 100 Cr_6 alloy, *Tribology International*, 173, 107613.

Dohda, K., Boher, C., Rezai-Aria, F. and Mahayotsanun, N. (2015). Tribology in metal forming at elevated temperatures, *Friction*, 3, 1–27.

Durango-Giraldo, G., Zapata-Hernandez, C., Santa, J. F. and Buitrago-Sierra, R. (2022). Palm oil as a biolubricant: Literature review of processing parameters and tribological performance, *Journal of Industrial and Engineering Chemistry*, 107, 31–44.

Engel, U. (2006). Tribology in microforming, *Wear*, 260 (3), 265–273.

Etsion, I. (2005). State of the art in laser surface texturing, *Journal of Tribology*, 127 (1), 248–253.

Frutos, E., Richhariya, V., Silva, F. S. and Trindade, B. (2023). Manufacture and mechanical-tribological assessment of diamond-reinforced Cu-based coatings for cutting/grinding tools, *Tribology International*, 177, 107947.

Fülöp, M. T-T. (2001). A general overview of tribology of sheet metal forming, *Journal for Technology of Plasticity*, 6 (2), 11–25.

Gecu, R. and Karaaslan, A. (2019). Volume fraction dependent wear behavior of titanium-reinforced aluminum matrix composites manufactured by melt infiltration casting, *Journal of Tribology*, 141 (2), 021603.

Grützmacher, P. G., Profito, F. J. and Rosenkranz, A.(2019). Multi-scale surface texturing in tribology—Current knowledge and future perspectives, *Lubricants*, 7 (11), 95.

Gu, D., Jue, J., Dai, D., Lin, K. and Chen, W. (2018). Effects of dry sliding conditions on wear properties of al-matrix composites produced by selective laser melting additive manufacturing, *Journal of Tribology*, 140 (2), 021605.

Gupta, M. K., Song, Q., Liu, Z., Singh, R., Sarikaya, M. and Khanna, N. (2021). Tribological behavior of textured tools in sustainable turning of nickel based super alloy, *Tribology International*, 155, 106775.

Gupta, R. N. and Harsha, A. P. (2017). Synthesis, characterization, and tribological studies of Calcium–Copper–Titanate nanoparticles as a biolubricant additive, *Journal of Tribology*, 139 (2), 021801.

Han, R., Liu, J. and Sun, Y. (2005). Research on experimentation of green cutting with water vapor as coolant and lubricant, *Industrial Lubrication and Tribology*, 57 (5), 187–192.

Hanon, M. M., Marczis, R. and Zsidai, L. (2020). Impact of 3D-printing structure on the tribological properties of polymers, *Industrial Lubrication and Tribology*, 72 (6), 811–818.

Hogmark, S., Jacobson, S. and Larsson, M. (2000). Design and evaluation of tribological coatings, *Wear*, 246 (1–2), 20–33.

Holmberg, K., Ronkainen, H., Laukkanen, A. and Wallin, K.(2007). Friction and wear of coated surfaces — Scales, modelling and simulation of tribomechanisms, *Surface and Coatings Technology*, 202 (4–7), 1034–1049.

Holmberg, K., Ronkainen, H. and Matthews, A. (2000). Tribology of thin coatings, *Ceramics International*, 26 (7), 787–795.

Iakovakis, E., Avcu, E., Roy, M. J., Gee, M., and Matthews, A. (2021). Dry sliding wear behaviour of additive manufactured CrC-rich WC-Co cemented carbides. *Wear*, 486–487, 204127.

Iakovakis, E., Roy, M. J., Gee, M., and Matthews, A. (2020). Evaluation of wear mechanisms in additive manufactured carbide-rich tool steels, *Wear*, 462–463, 203449.

Ibrahim, A. M. M., Li, W., Xiao, H., Zeng, Z., Ren, Y. and Alsoufi, M. S. (2020). Energy conservation and environmental sustainability during grinding operation of Ti–6Al–4V alloys via eco-friendly oil/graphene nano additive and minimum quantity lubrication, *Tribology International*, 150, 106387.

Iglesias, P., Bermúdez, M. D., Moscoso, W. and Chandrasekar, S. (2010). Influence of processing parameters on wear resistance of nanostructured OFHC copper manufactured by large strain extrusion machining, *Wear*, 268 (1–2), 178–184.

Jähnig, T. and Lasagni, A. F. (2020). Laser interference patterned ta-C-coated dry forming tools, *Industrial Lubrication and Tribology*, 72 (8), 1001–1005.

Joshi, V., Srivastava, A. and Shivpuri, R. (2004). Intermetallic formation and its relation to interface mass loss and tribology in die casting dies, *Wear*, 256 (11–12), 1232–1235.

Kang, N. and El Mansori, M. (2020). A new insight on induced-tribological behaviour of hypereutectic Al-Si alloys manufactured by selective laser melting, *Tribology International*, 149, 105751.

Kang, N., El Mansori, M., Feng, E., Zhao, C., Zhao, Y. and Lin, X. (2022). Sliding wear and induced-microstructure of Ti-6Al-4V alloys: Effect of additive laser technology, *Tribology International*, 173, 107633.

Khaliq, W., Zhang, C., Jamil, M. and Khan, A. M. (2020). Tool wear, surface quality, and residual stresses analysis of micro-machined additive manufactured Ti–6Al–4V under dry and MQL conditions, *Tribology International*, 151, 106408.

Khan, A. M., Jamil, M., Ul Haq, A., Hussain, S., Meng, L. and He, N. (2019). Sustainable machining. Modeling and optimization of temperature and surface roughness in the milling of AISI D2 steel, *Industrial Lubrication and Tribology*, 71 (2), 267–277.

Korkmaz, M. E., Gupta, M. K. and Demirsöz, R. (2022). Understanding the lubrication regime phenomenon and its influence on tribological characteristics of additively manufactured 316 Steel under novel lubrication environment, *Tribology International*, 173, 107686.

Lee, C. T., Lee, M. B., Mong, G. R. et al. (2022). A bibliometric analysis on the tribological and physicochemical properties of vegetable oil–based bio-lubricants (2010–2021), *Environmental Science and Pollution Research*, 29, 56215–56248.

Lenard, J. G. (2000). Tribology in metal rolling keynote presentation forming group F, *CIRP Annals*, 49 (2), 567–590.

Liang, H., Hou, J., Cao, Z. and Jiang, L. (2023). Interesting 'island-like' microstructure and tribological evaluation of $Al_{1.5}CrFeNiWTi_{0.5}$ high entropy alloy coating manufactured by laser cladding, *Tribology International*, 179, 108171.

Liang, H., Qiao, D., Miao, J., Cao, Z., Jiang, H. and Wang, T. (2021). Anomalous micro-structure and tribological evaluation of $AlCrFeNiW_{0.2}Ti_{0.5}$ high-entropy alloy coating manufactured by laser cladding in seawater, *Journal of Materials Science and Technology*, 85, 224–234.

Lizzul, L., Sorgato, M., Bertolini, R., Ghiotti, A. and Bruschi, S. (2020). Influence of additive manufacturing-induced anisotropy on tool wear in end milling of Ti_6Al_4V, *Tribology International*, 146, 106200.

Maleque, M. A., Masjuki, H. H. and Sapuan, S. M. (2003). Vegetable-based biodegradable lubricating oil additives, *Industrial Lubrication and Tribology*, 55 (3), 137–143.

Mandal, A., Tiwari, J. K., AlMangour, B. et al. (2020). Tribological behavior of graphene-reinforced 316L stainless-steel composite prepared via selective laser melting, *Tribology International*, 151, 106525.

Mao, B., Siddaiah, A., Liao, Y. and Menezes, P. L. (2020). Laser surface texturing and related techniques for enhancing tribological performance of engineering materials: A review, *Journal of Manufacturing Processes*, 53, 153–173.

Marquer, M., Laheurte, P., Faure, L. and Philippon, S. (2020). Influence of 3D-printing on the behaviour of Ti_6Al_4V in high-speed friction, *Tribology International*, 152, 106557.

Mohanavel, V., Ali, K. S. A., Prasath, S., Sathish, T. and Ravichandran, M. (2020). Microstructural and tribological characteristics of $AA6351/Si_3N_4$ composites manufactured by stir casting, *Journal of Materials Research and Technology*, 9 (6), 14662–14672.

Muhamad, S. S., Ghani, J. A., Haron, C. H. C. and Yazid, H. (2020). Wear mechanism of a coated carbide tool in cryogenic machining of AISI 4340, *Industrial Lubrication and Tribology*, 72 (4), 509–514.

Murashima, M., Imaizumi, Y., Kawaguchi, M. et al. (2021). Realization of a novel morphing surface using additive manufacturing and its active control in friction, *Journal of Tribology*, 143 (5), 051104.

Niketh, S. and Samuel, G. L. (2017). Surface texturing for tribology enhancement and its application on drill tool for the sustainable machining of titanium alloy, *Journal of Cleaner Production*, 167, 253–270.

Noor El-Din, M. R., Mishrif, M. R., Kailas, S. V., Suvin, P. S. and Mannekote, J. K. (2018). Studying the lubricity of new eco-friendly cutting oil formulation in metal working fluid, *Industrial Lubrication and Tribology*, 70 (9), 1569–1579.

Norani, M. N. M., Abdullah, M. I. H. C., Abdollah, M. F. B., Amiruddin, H., Ramli, F. R. and Tamaldin, N. (2021). Mechanical and tribological properties of FFF 3D-printed polymers: A brief review, *Jurnal Tribologi*, 29, 11–30.

Nosonovsky, M. and Bhushan, B. (2010). Green tribology: Principles, research areas and challenges, Article number 57. https://doi.org/10.1098/rsta.2010.0200

Olofsson, U., Lyu, Y., Åström, A. H. et al. (2021). Laser cladding treatment for refurbishing disc brake rotors: Environmental and tribological analysis, *Tribology Letters*, 69, 57.

Pawar, R. V., Hulwan, D. B. and Mandale, M. B. (2022). Recent advancements in synthesis, rheological characterization, and tribological performance of vegetable oil-based lubricants enhanced with nanoparticles for sustainable lubrication, *Journal of Cleaner Production*, 378, 134454.

Pereira, J., Zambrano, J., Licausi, M., Tobar, M. and Amigó, V. (2015). Tribology and high temperature friction wear behavior of MCrAlY laser cladding coatings on stainless steel, *Wear*, 330–331, 280–287.

Pezeshkian, M., Ebrahimzadeh, I. and Gharavi, F. (2018). Fabrication of Cu surface composite reinforced by Ni particles via friction stir processing: Microstructure and tribology behaviors, *Journal of Tribology*, 140 (1), 011607.

Pranav, P., Sneha, E. and Rani, S. (2021). Vegetable oil-based cutting fluids and its behavioral characteristics in machining processes: A review, *Industrial Lubrication and Tribology*, 73 (9), 1159–1175.

Rajewski, T. E., Fokens, J. S. and Watson, M. C. (2000). The development and application of synthetic food grade lubricants, *Industrial Lubrication and Tribology*, 52 (3), 110–116.

Ralls, A. M., Daroonparvar, M., Kasar, A. K., Misra, M. and Menezes, P. L. (2023). Influence of friction stir processing on the friction, wear and corrosion mechanisms of solid-state additively manufactured 316L duplex stainless steel, *Tribology International*, 178, 108033.

Ralls, A. M., Kasar, A. K. and Menezes, P. L. (2021). Friction stir processing on the tribological, corrosion, and erosion properties of steel: A review, *Journal of Manufacturing and Materials Processing*, 5 (3), 97.

Rao, P. N. (2013). Sustainable manufacturing – Principles, applications and directions. *28 National Convention of Production Engineers & National Seminar on "Advancements in Production and Operations Management"*, Institution of Engineers India, MNIT, Jaipur, 28.

Rao, T. V. V. L. N., Kasolang, S., Xie, G., Katiyar, J. K. and Rani, A. M. A. (2021). *Green Tribology: Emerging Technologies and Applications*. Emerging Materials and Technologies Series (Series Editor Prof. Boris I. Kharissov), CRC Press (Taylor & Francis).

Rao, T. V. V. L. N., Rani, A. M. A., Awang, M., Baharom, M. and Uemura, Y. (2018). An overview of research on biolubricants in Malaysia and Japan fortribological applications, *Jurnal Tribologi*, 18, 40–57.

Ravindran, P., Manisekar, K., Narayanasamy, R. and Narayanasamy, P. (2013b). Tribological behaviour of powder metallurgy-processed aluminum hybrid composites with the addition of graphite solid lubricant, *Ceramics International*, 39 (2), 1169–1182.

Ravindran, P., Manisekar, K., Rathika, P. and Narayanasamy, P. (2013a). Tribological properties of powder metallurgy – Processed aluminum self lubricating hybrid composites with SiC additions, *Materials & Design*, 45, 561–570.

Razak, D. M., Syahrullail, S., Sapawe, N., Azli, Y. and Nuraliza, N. (2015). A new approach using palm olein, palm kernel oil, and palm fatty acid distillate as alternative biolubricants: Improving tribology in metal-on-metal contact, *Tribology Transactions*, 58 (3), 511–517.

Reeves, C. J., Menezes, P. L., Jen, T.-C. and Lovell, M. R. (2015). The influence of fatty acids on tribological and thermal properties of natural oils as sustainable biolubricants, *Tribology International*, 90, 123–134.

Reeves, C. J., Menezes, P. L., Lovell, M. R. et al. (2013). The size effect of boron nitride particles on the tribological performance of biolubricants for energy conservation and sustainability, *Tribology Letters*, 51, 437–452.

Renner, P., Jha, S., Chen, Y., Raut, A., Mehta, S. G. and Liang, H. (2021). A review on corrosion and wear of additively manufactured alloys, *Journal of Tribology*, 143 (5), 050802.

Salas, O., Kearns, K., Carrera, S. and Moore, J. J. (2003). Tribological behavior of candidate coatings for Al die casting dies, *Surface and Coatings Technology*, 172 (2–3), 117–127.

Salih, N., Salimon, J. and Yousif, E. (2011). The physicochemical and tribological properties of oleic acid based triester biolubricants, *Industrial Crops and Products*, 34 (1), 1089–1096.

Sanjeev, K. C., Nezhadfar, P. D., Phillips, C., Kennedy, M. S., Shamsaei, N. and Jackson, R. L. (2019). Tribological behavior of 17–4 PH stainless steel fabricated by traditional manufacturing and laser-based additive manufacturing methods, *Wear*, 440–441, 203100.

Sardar, S., Karmakar, S. K. and Das, D. (2019). Microstructure and tribological performance of alumina–aluminum matrix composites manufactured by enhanced stir casting method, *Journal of Tribology*, 141 (4), 041602.

Şen, N., Şirin, S., Kıvak, T., Civek, T. and Seçgin, Ö. (2022). A new lubrication approach in the SPIF process: Evaluation of the applicability and tribological performance of MQL, *Tribology International*, 171, 107546.

Shaari, M. Z., Roselina, N. R. N., Kasolang, S., Hyie, K. M., Murad, M. C., Bakar, M. Z. A. (2015). Investigation of tribological properties of palm oil biolubricant modified nanoparticles, *Jurnal Teknologi (Sciences & Engineering)*, 76 (9), 69–73.

Shaikh, M. B. N., Arif, S., Aziz, T., Waseem, A., Shaikh, M. A. N. and Ali, M. (2019). Microstructural, mechanical and tribological behaviour of powder metallurgy processed SiC and RHA reinforced Al-based composites, *Surfaces and Interfaces*, 15, 166–179.

Sharma, D. K., Patel, V., Badheka, V., Mehta, K. and Upadhyay, G. (2019). Fabrication of hybrid surface composites AA6061/(B4C + MoS2) via friction stir processing, *Journal of Tribology*, 141 (5), 052201.

Stanford, M. and Lister, P. M. (2002). The future role of metalworking fluids in metal cutting operations, *Industrial Lubrication and Tribology*, 54 (1), 11–19.

Su, J., Xie, H., Tan, C., et al. (2022). Microstructural characteristics and tribological behavior of an additively manufactured Ti-6Al-4V alloy under direct aging and solution-aging treatments, *Tribology International*, 175, 107763.

Suh, N. P. (2021). A perspective on tribology and design, *Journal of Tribology*, 143 (5), 051701.

Thasleem, P., Kuriachen, B., Kumar, D., Ahmed, A. and Joy, M. L. (2021). Effect of heat treatment and electric discharge alloying on the tribological performance of selective laser melted $AlSi_{10}Mg$, *Journal of Tribology*, 143 (5), 051111.

Turalıoğlu, K., Taftalı, M. and Yetim, F. (2021). Determining the tribological behavior of 316L stainless steel with lubricating micro-channels produced by the selective laser melting (SLM) method. *Industrial Lubrication and Tribology*, 73 (5), 700–707.

Vishnoi, M., Kumar, P. and Murtaza, Q. (2021). Surface texturing techniques to enhance tribological performance: A review, *Surfaces and Interfaces*, 27, 101463.

Wang, X., Li, C., Zhang, Y., et al. (2022). Tribology of enhanced turning using biolubricants: A comparative assessment, *Tribology International*, 174, 107766.

Wood, R. J. K. (2010). Tribology of thermal sprayed WC–Co coatings, *International Journal of Refractory Metals and Hard Materials*, 28 (1), 82–94.

Woydt, M. (2021). The importance of tribology for reducing CO_2 emissions and for sustainability, *Wear*, 474–475, 203768.

Wu, B., Ibrahim, M. Z., Raja, S. et al. (2022). The influence of reinforcement particles friction stir processing on microstructure, mechanical properties, tribological and corrosion behaviors: A review, *Journal of Materials Research and Technology*, 20, 1940–1975.

Yan, X., Chang, C., Deng, Z. et al. (2021). Microstructure, interface characteristics and tribological properties of laser cladded NiCrBSi-WC coatings on PH 13–8 Mo steel, *Tribology International*, 157, 106873.

Yang, X., Li, C., Zhang, M. et al. (2022). Dry sliding wear behavior of additively manufactured CoCrWNixAly alloys, *Wear*, 496–497, 204285.

Yang, Y., Zhu, Y. and Yang, H. (2020). Enhancing wear resistance of selective laser melted parts: Influence of energy density, *Journal of Tribology*, 142 (11), 111701.

Ye, W., Xie, M. and Huang, Z. et al. (2023). Microstructure and tribological properties of in situ carbide/CoCrFeNiMn high entropy alloy composites synthesized by flake powder metallurgy, *Tribology International*, 181, 108295.

Yu, T., Liu, J., He, Y., Tian, J., Chen, M. and Wang, Y. (2021). Microstructure and wear characterization of carbon nanotubes (CNTs) reinforced aluminum matrix nano-composites manufactured using selective laser melting, *Wear*, 476, 203581.

Zhang, P., Hu, Z., Xie, H., Lee, G.-H. and Lee, C.-H. (2020). Friction and wear character-istics of polylactic acid (PLA) for 3D printing under reciprocating sliding condition, *Industrial Lubrication and Tribology*, 72 (4), 533–539.

Zhu, Y., Raeymaekers, B. and Michael Khonsari, M. (2021). Special issue on tribology of additive manufacturing, *Journal of Tribology*, 143 (5), 050201.

Zhu, Y., Zou, J., Chen, X. and Yang, H. (2016). Tribology of selective laser melting pro-cessed parts: Stainless steel 316L under lubricated conditions, *Wear*, 350–351, 46–55.

2 Significance of Additives on the Modern Casting Method

Shailesh Kumar Singh
Advanced Tribology Research Centre, CSIR – Indian
Institute of Petroleum, Dehradun, India

Saurabh Vashistha
Academy of Scientific and Innovative Research (AcSIR),
CSIR-HRDC Campus, Ghaziabad, India

Vivek Kumar Singh
Indian Institute of Technology Bombay, Mumbai, India

2.1 INTRODUCTION

Manufacturing is the process of converting raw material into a finished product, or it may be defined as a process of value in addition to raw material. The various manufacturing processes include casting, forming, fabrication, material removal process/machining, etc. Casting is the basic or primary manufacturing process, whereas forming, fabrication, and machining are the secondary manufacturing processes. The selection of a manufacturing process depends on various factors, such as the shape and size of the object as well as the number of components to be produced; properties, accuracy and surface finish required by the object; and cost of the object.

Casting may be defined as a process in which molten liquid metal solidifies in a predefined mould cavity. A basic flow diagram of sand casting is shown in Figure 2.1. In the case of sand moulding, a moulding flask keeps the sand mould intact. If it is made in two pieces, the top half is called 'Cope' and the bottom half 'Drag'. For complex and intricate castings, the flask can be made up of three pieces, in which case the intermediate one is called 'Cheek' [1]. The vertical channel created in a sand mould, which feeds the molten metal from the pouring basin to the runner channel, is called 'SPRUE.' A pouring basin helps avoid any sand erosion due to the turbulence of liquid metal entering the sprue channel. A strainer is placed in the sprue channel to filter large impurities [1]. A schematic diagram of sand casting, including the gating elements, is shown in Figure 2.2.

DOI: 10.1201/9781003363576-2

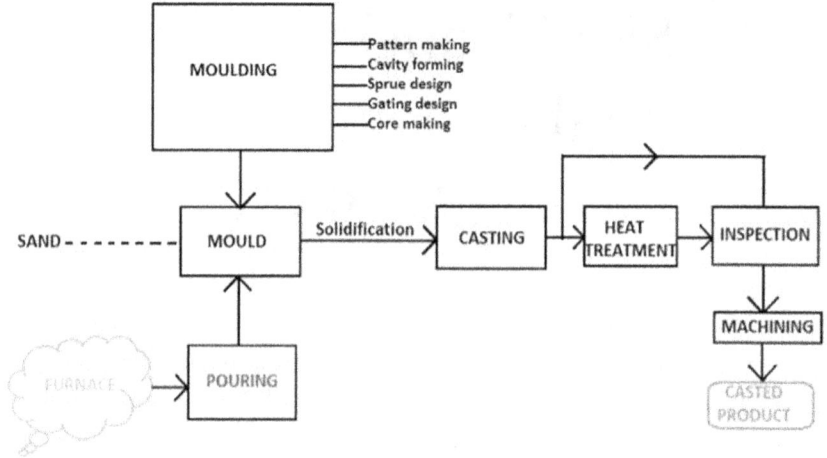

FIGURE 2.1   Flow diagram of sand casting.

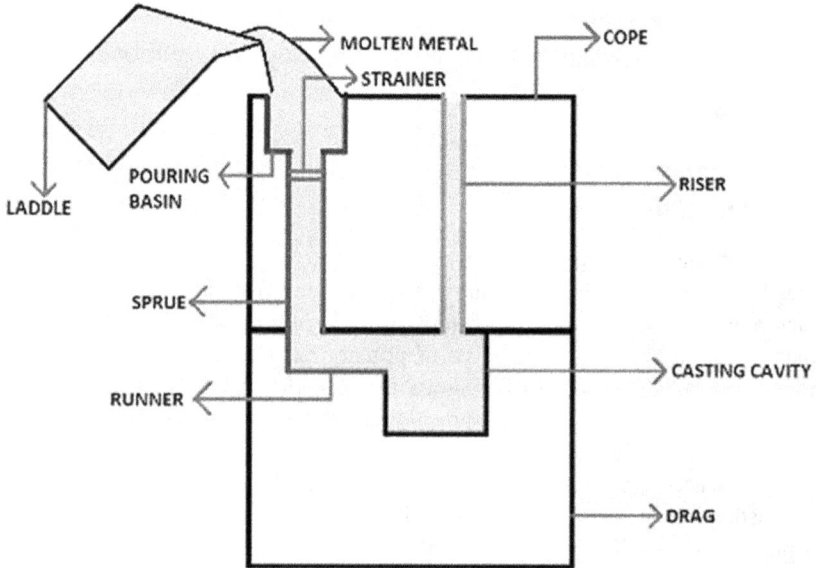

FIGURE 2.2   Gating elements of sand casting.

2.2   MODERN CASTING METHODS

Apart from sand casting, other techniques include:

- Investment casting
- Squeeze casting

- High pressure die casting
- Slush casting
- Rheocasting
- Thixocasting

Rheocasting, thixocasting, high pressure die casting (HPDC), investment casting, and squeeze casting fall under modern casting technologies. Compared to conventional castings, such as sand casting, the component obtained via this technique has a better surface finish and strength. However, the addition of an alloying element further enhances the strength and tribological properties.

Rheocasting is a semisolid die casting process, where the metal is kept in a state where it is neither fully solid nor liquid. This state is called slurry. In this method, molten metal is poured into the chamber to form the slurry, which is then pushed inside the die cavity to create the final component. The thixocasting technique is slightly different from rheocasting. Instead of slurry, the billet obtained from the mould is reheated to the semisolid temperature range and then injected into the die cavity, where pressure is applied to get the desired component.

High pressure die casting: In this process, molten metal is injected directly into the mould and then pressure is applied to fill the cavity. Here, there is no requirement to make a slurry or heat the billet, and hence it is a simpler process than rheocasting. The HPDC process is cheaper than thixocasting and is used for the mass production of aluminium die casting for various industries. It is a cost effective and efficient process for casting a large metal part.

In investment casting, the pattern is made of wax or suitable polymer and coated by dipping the material into the slurry of refractory material. When the coating hardens, the wax pattern is removed by heating and molten metal is poured into the cavity created by the wax pattern. After the metal solidifies within the mould, the metal casting is removed by breaking the refractory mould.

Investment casting is useful in the aerospace sector which requires intricately shaped components with high precision and tolerance and a good surface finish. However, the investment casting component has lower strength and poor tribological properties. So, the role of a suitable alloying element becomes essential to enhance or achieve these properties as per die casting.

2.3 ROLE OF ADDITIVES IN MANUFACTURING

Additives play an essential role in manufacturing processes. They are the metal constituents that are deliberately added to the significant or principal elements of materials to enhance the mechanical, tribological, electrical, or any other desired properties. Currently, alloy steels are the most widely used. Let us discuss the alloy steel to gain insight into the importance of additives in the manufacturing world.

When one or more alloying substances (e.g., Cr, Ni, Mo) other than carbon are added to steel to reach a desired property, the resulting steel is known as alloy steel.

1. Nickel: The addition of nickel results in increased toughness as well as strength of steel. The composition of this steel consists of 0.1–0.5% carbon and 2–5% nickel. With this composition, the steel's strength, hardness, ductility, and corrosion resistance improve drastically. The applications of this steel include boiler tubes, spark plugs, valves, etc. When a nickel steel alloy contains 36% nickel, it is known as invar which has an almost zero coefficient of thermal expansion and has excellent applications in the manufacturing of measuring instruments.

2. Chromium: When a combination of hardness and a high elastic limit is required, chromium is used as an alloying element to the steel. The corrosion resistance of the steel also gets improved. These steels generally contain 0.5–2.0% chromium and 0.1–1.5% carbon. The applications of these steels include balls and races of bearings, motor crankshafts, armour plates, gears, axles, etc.

3. Vanadium: Mixing a tiny amount of vanadium produces marked increases in tensile strength and elastic limit in low and medium carbon steels without loss of ductility. Chrome vanadium steel containing 0.5–1.5% chromium, 0.15–0.3% vanadium, and 0.13–1.1% carbon have excellent tensile strength, elastic limit, endurance limit, and ductility. These steels are frequently used for springs, shafts, gears, pins, and many drop-forged parts.

4. Manganese: Manganese contributes to improving steel strength in heat-treated conditions. Manganese alloy steel containing over 1.5% manganese with a carbon range of 0.40–0.55% is used extensively in gears, axles, shafts, and other parts where high strength combined with reasonable ductility is required. The principal uses of manganese steel are in machinery parts subjected to severe wear. These steels are all cast and ground to finish.

5. Silicon: This steel behaves like nickel steel. Its elastic limit is higher than that of ordinary steel. Silicon steel containing from 1–2% silicon and 0.1–0.4% carbon and other alloying elements is used for electrical machinery, valves in I.C. engines, springs, and corrosion resisting materials.

6. Molybdenum: A very small amount (0.15–0.3%) of molybdenum is generally used with chromium and manganese (0.5–0.8%) to make molybdenum steel. This steel possesses extra tensile strength and is used for plane fuselages and automobile parts. It can replace tungsten in high-speed steels.

7. Boron: Boron (0.003%) is a compelling hardenability carbon content, being from 250 to 750 times as effective as nickel, 75 to 125 times as effective as molybdenum, and about 100 times as powerful as chromium. Only a few thousandths of percent are sufficient to produce the desired effect in low carbon steel, but the results diminish rapidly with increasing carbon content. Since no carbide formation or ferrite strengthening is produced, improved machinability and cold-forming capability often result from the use of boron in place of other hardenability additions. It does not affect the tensile strength of steel.

8. Sulphur: Sulphur is an unwanted impurity in steel because it results in the formation of iron sulfides, resulting in cracking. However, when sulphur is

added along with manganese, it forms MnS, which improves the machinability of steels. Its content may vary from 0.06 to 0.3%.

9. Copper: Copper is used to increase the corrosion of the steel. The amount of copper that is added to steel generally varies between 0.1 and 0.5%. The hot working behaviour and the surface quality deteriorate in low-carbon steel.

2.4   ROLE OF ADDITIVES IN MODERN CASTING

Additives play an essential role in modern castings. First, let us take an example of slip casting. Slip casting is a colloidal and cost-effective method to fabricate the engineering of ceramics and their composites in aqueous suspension by pouring in porous mould and draining water from the slurry of a green body ready for the sintering process [2]. The properties of the composite materials manufactured by the method of slip casting directly relate to the rheological properties of the slurry. Several additives like binders and dispersants are mixed into the suspension. The binders used are kaolin and ball clay, whereas the dispersants used are Tiron, tripolyphsphate, and dolapix. The viscosity value decreases from the presence of binders and dispersants and, as a result, a higher value density of the fabricated products is achieved [2]. In a research study, the effect of titanium oxide (TiO_2) content on the tribological and mechanical behaviour of aluminium alloy matrix composites (AMCs) is investigated and the AMCs were made by using the stir casting method with different weight percentages (0, 5, 10 and 15) of titanium oxide particles. It is evident in the study that an increase in the TiO_2 content up to 10% also increases the various mechanical properties like compressive strength, hardness, flexural strength, tensile strength, and impact strength of the composite. The coefficient of friction and wear loss of the composite also reduces significantly with titanium oxide particles [3].

Aluminium-based composites can be developed by using various casting methods such as die casting, centrifugal casting, investment casting, mechanical stir casting, spark plasma sintering technique, and spray deposition. Research has found that the mechanical stir casting method is preferred. Furthermore, it has been found that carbide, nitride, agro-waste and industrial waste-based reinforcements help improve aluminium's tribological and mechanical behaviour [4].

2.5   EFFECT OF ADDITIVES IN CASTING FROM A TRIBOLOGICAL POINT OF VIEW

When cast parts are used in an application where they will be in relative motion to the mating parts, material loss due to wear from the mating surface is inevitable. Thus, it becomes significant to understand cast parts' tribological aspects to prolong service life. Some of the work relating to the tribological behaviour of nano-additives' reinforced cast parts are discussed here.

Wang et al. [5] prepared a Cr-Mo-Cu cast iron alloy with and without rare earth nano-additives to study the effect of graphite morphology on the hardness and wear characteristics. It was found that the two types of alloys showed a different microstructural response to the heat treatment in terms of graphite morphology, size, and amount. The bare cast alloy developed an irregular shape flake-like morphology, which weakens the metal matrix connection, while the nano-additive reinforced cast alloy depicted a vermicular-like morphology with an increase in the degree of curl and the amount of graphite, which resulted in an increase in hardness and grain refinement compared to the Cr-Mo-Cu cast alloy without nano-additives. This change in morphology and increase in hardness with the addition of nano-additives resulted in an improvement in wear performance. Kaleicheva et al. [6] studied the microstructural and tribological behaviour of austempered cast iron with TiCN + TiN nano-additives in it. The TiCN + TiN additives resulted in a decrease in the size of the graphite sphere and an increase in graphite that makes the alloy more resistant to wear. These TiCN + TiN nano-additives promote the conversion of metastable retained austenite to strain-induced martensite during wear testing, resulting in improved wear performance. Anasyida et al. [7] conducted a wear analysis for as-cast Al-12Si-4Mg alloy after adding a small percentage of cerium. It was found that micro-hardness and wear resistance are at the maximum when the alloy contains 2 wt% cerium, while beyond 2 wt%, micro-hardness and wear resistance start decreasing. Ali et al. [8] studied the wear behaviour of an Al6061 alloy prepared with a stir casting technique by adding different wt% of TiO_2 composites. That with 5 wt% TiO_2 showed the highest coefficient of friction, while the best wear performance was dictated by 2 wt% TiO_2. Thirugnanasambandam et al. [9] developed an Al6061 cast alloy by adding 2% SiC of 40 nm size to study the effect of ultrasonication, rheocasting and squeezing pressure on mechanical, tribological properties. The SiC particle resulted in a reduction of specific wear rate by about 80.3% as compared to the base metal.

2.6 CONCLUSIONS

Modern casting methods are gaining significant importance due to better surface finish and the ability to make intricately shaped components. However, strength and tribological properties need to be improved to match the forged parts. The choice and role of additives become essential to enhance mechanical and tribological properties. The choice of additive affects the material integrity, yielding improved corrosion, tribological, and mechanical properties. Hence, understanding the exact composition and selection of additives to improve properties is essential for casting metallic alloy using the modern casting method. This chapter contributes sufficient knowledge on the different types of additives and their role and impact on the modern casting method.

REFERENCES

[1] Rundman, K. B., 2000, *Metal Casting (Reference Book for MY4130)*.
[2] Shahrestani, S., Ismail, M. C., Kakooei, S., and Beheshti, M., 2020, "Effect of Additives on Slip Casting Rheology, Microstructure and Mechanical Properties of Si_3N_4/SiC Composites," *Ceram. Int.*, **46**(5), pp. 6182–6190.

[3] Alagarsamy, S. V., and Ravichandran, M., 2019, "Synthesis, Microstructure and Properties of TiO$_2$ Reinforced AA7075 Matrix Composites via Stir Casting Route," *Mater. Res. Express*, **6**(8), p. 86519.

[4] Bharti, C., Singh, A., Rahul, R., Sharma, D., and Dwivedi, S. P., 2021, "A Critical Review of Aluminium Based Composite Developed by Various Casting Technique with Different Reinforcement Particles to Enhance Tribo-Mechanical Behaviour," *Mater. Today Proc.*, **47**, pp. 4092–4097.

[5] Wang, Y., Pan, Z., Wang, Z., Sun, X., and Wang, L., 2011, "Sliding Wear Behavior of Cr-Mo-Cu Alloy Cast Irons with and without Nano-Additives," *Wear*, **271**(11–12), pp. 2953–2962.

[6] Kaleicheva, J., Mishev, V., Karaguiozova, Z., Nikolcheva, G., and Miteva, A., 2017, "Effect of Nanoadditives on the Wear Behavior of Spheroidal Graphite Cast Irons," *Tribol. Ind.*, **39**(3), pp. 294–301.

[7] Anasyida, A. S., Daud, A. R., and Ghazali, M. J., 2010, "Dry Sliding Wear Behaviour of Al-12Si-4Mg Alloy with Cerium Addition," *Mater. Des.*, **31**(1), pp. 365–374.

[8] Ali, A., 2021, "Wear Behavior of Al6061/TiO$_2$ Composites Synthesized by Stir Casting Process," *J. Adv. Eng. Trends*, **41**(2), pp. 113–125.

[9] Arunkumar, T., Selvakumaran, T., Subbiah, R., Ramachandran, K., and Manickam, S., 2021, "Development of High-Performance Aluminium 6061/SiC Nanocomposites by Ultrasonic Aided Rheo-Squeeze Casting Method," *Ultrason. Sonochem.*, **76**(June), p. 105631.

3 Impact of Additives in Tribology of Metal-Forming Processes

Ujjawal Singh and Saurabh Mishra
Centre for Advanced Studies, Dr APJ Abdul Kalam
Technical University, Lucknow, India

3.1 INTRODUCTION

Metal forming is one of the most effective procedures for manufacturing a wide range of items and is used in many industries, (Kumar et al. 2010). The material is subjected to plastic deformation to modify the shape of the component and convert it to the desired shape. The induced stress during plastic deformation fluctuates amid the material's yield strength and fracture strength. Metal forming aims to control the mechanical properties and product quality to meet market demand. Metal forming is categorized as (1) bulk forming; (2) sheet-metal forming, and (3) powder-metal forming, as shown in Figure 3.1. Bulk deformation, such as rolling, forging, extrusion, and drawing fields, refers to using raw materials for the forming, as they have a low surface area to volume ratio (Graff 2015). *Mechanical* components such as gears or bushed valves, engine parts, such as valves, connecting rods, and hydraulic valves, have been produced using forming presses with the help of tools and die. Sheet-metal forming employs hydraulic or pneumatic presses to bend, draw, shear, blank, punch sheets, plates, and strips. Bulk deformation forms raw materials with low surface area to volume ratios, such as rolling, forging, extrusion, and drawing; because of its features, a new class of forming methods known as powder forming is gaining popularity. One of the most significant advantages of powder forming is its ability to generate pieces very close to final dimensions with minimal material waste.

Extreme mechanical pressure is applied during the metal-forming process to cause plastic deformation. Furthermore, as previously stated, the process is influenced by various forces and stress systems. The formation of such a force and stress system during the metal-forming process means that heat accumulation between the tool and the workpiece must be released. The heat generated due to constant contact between the workpiece and the tool harms the quality of the

DOI: 10.1201/9781003363576-3

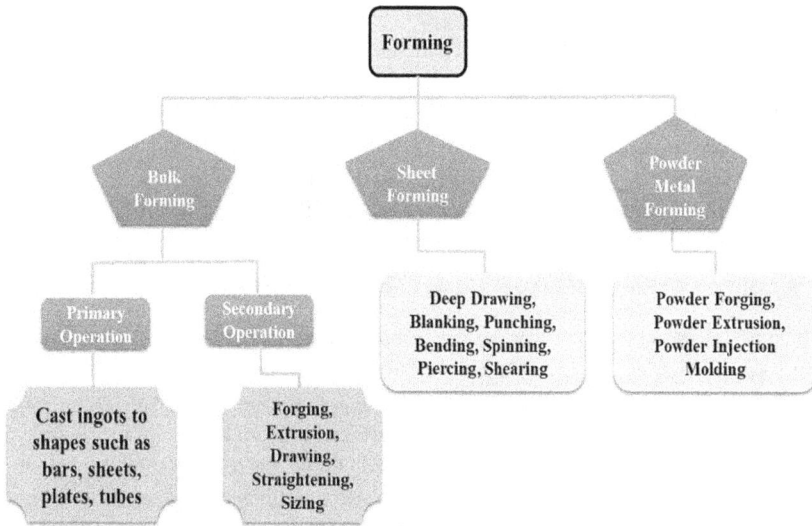

FIGURE 3.1   Different metal-forming processes.

finished product and the tool's life. Heat energy may affect the microstructure as heat plays a significant role at a lower dimension. Hence, this consequence of metal forming necessitates the use of lubricants throughout the process. It is therefore essential to use lubrication in the metal-forming process to ensure optimal tribological behaviour and heat dissipation in the forming area, resulting in a higher-quality finished product and extended tool life overall. Various lubricants have been employed in the industrial sector for many years, and scientists and researchers throughout the literature have documented numerous advancements. As lubrication directly affects material formability and defects throughout the forming procedure, the feasibility and productivity of a forming process are heavily reliant on the lubrication provided. Another critical component that can be influenced by lubricant performance is the quality of the formed part, tool life, and needed forming load (Weidel et al. 2010).

Water-, oil-, synthetic-, and solid-based lubricants are the four types of metal-forming fluids (Zareh-desari and Davoodi 2016). A suitable lubricant for a typical metal-forming process can be chosen by considering parameters such as contact pressure, temperature, sliding velocity, tool and workpiece material qualities, and surface roughness (Zareh-desari and Davoodi 2016, Lazzarotto et al. 1998). In recent years, the rising limits imposed by environmental authorities have compelled manufacturers to consider environmental impacts as a novel and critical element (Sanchez et al. 2010). Because conventional lubricants are frequently flammable, and the resulting contaminated waste from their disposal contains hazardous components, such as phosphorus, sulphur, chlorine, and zinc, researchers have concentrated on potential alternative practices (Rao and Xie 2006).

In this context, significant efforts are being made to develop novel coating processes and materials that would eliminate the need for lubricant in various metal-forming activities (Ghiotti and Bruschi 2011).

This chapter discusses nano-additives' role in the metal-forming process and how they can enhance the attributes of various metal-forming operations. The vast research developments in the additives used in the forming process and their impact on the quality of the end product and tool life are also covered.

3.2 ROLE OF ADDITIVES IN METAL-FORMING FLUIDS

Mechanical presses of one-ton capacity have been used to convert the material into precise shapes as part of the metal-forming process. Friction and heat are produced because of deformation. The fluid is essential to reduce friction and heat during the metal-forming process. Furthermore, additives in the fluid can change their properties, resulting in improved performance. Formulating the fluid for metal forming using certain additives is problematic because these fluids must function with, rather than against, the generated force and heat. The primary function of fluid with additives is demonstrated in Figure 3.2.

In typical metal-forming processes, the fluid lubricates, cools, washes, and protects the workpiece and tool used to create the finished product. To produce fluid with the desired characteristics for use in various metal-forming processes, all the formulation is done by precisely balancing base fluid and additives. Metal-forming fluids include both mineral and synthetic oils from seeds and vegetables. On the other hand, water-based synthetic lubricants have gained popularity in the metal-forming process. These developed fluids are eco-friendly and reduce the cost of the metal-forming process. Today's material combinations frequently necessitate more significant metal contact pressures, resulting in even more heat

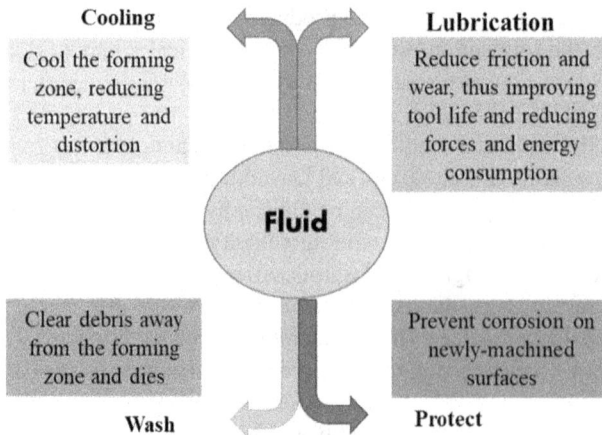

Cooling
Cool the forming zone, reducing temperature and distortion

Lubrication
Reduce friction and wear, thus improving tool life and reducing forces and energy consumption

Fluid

Clear debris away from the forming zone and dies

Prevent corrosion on newly-machined surfaces

Wash

Protect

FIGURE 3.2   Desired function required for fluid in the metal-forming process.

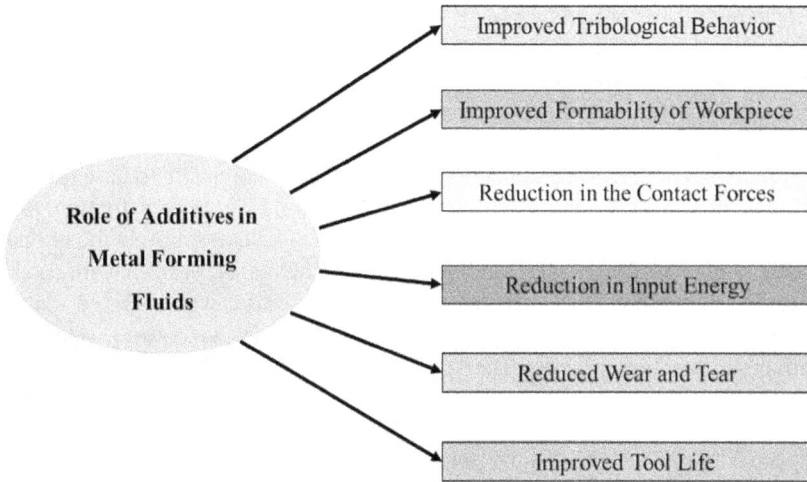

FIGURE 3.3 Role of additives in metal-forming fluids.

and friction. Advances in metal-forming fluids can mitigate this increase in metal contact force. Additives are essential for balancing the metal contact force and providing a tribological effect. Additives in forming fluids may impart specific improvements resulting in improved tribological behaviour, workpiece formability, reduced contact forces, reduced input energy for plastic deformation, reduced wear and tear, and improved tool life, as shown in Figure 3.3.

3.3 NANOTECHNOLOGICAL ASPECTS OF ADDITIVES IN METAL-FORMING PROCESSES

The use of nano-particles as lubricant additives has piqued the interest of tribologists and researchers in recent years. Extensive research has revealed that adding nano-particles to base oil/fluid can improve tribological performance (Gulzar et al. 2016; Shahnazar, Bagheri, and Abd Hamid 2016; Dai et al. 2016). These nano-lubricants are colloidal solutions that disperse nano-particles in base fluids such as mineral oils, liquid paraffin, and vegetable oils. A wide range of nano-particles has been investigated while exploring nano-lubricants' performance, including ZnO, CuO, MoS_2, WS_2, graphite, SiO_2, Al_2O_2, TiO_2, and PbS. The literature contains detailed assessments of nano-particles and their concentration levels for nano-lubricant formulation (Gulzar et al. 2016; Dai et al. 2016; Shahnazar, Bagheri, and Abd Hamid 2016). Due to their excellent physical and chemical properties, researchers and scientists are keen to use nano-materials in formulating metal-forming fluids/lubricants. The nano-scale particle size can enter quickly into the interface between the workpiece and the tool resulting in a reduced metal-to-metal contact area.

Nano-additives are employed in the lubricating oil to deal with the extreme pressure and anti-wear. As stated before, with the rapid advancements in nano-technology, nano-particles of various metals play a vital role in developing a new class of metal-forming fluids (Lee et al. 2009). Many studies have shown that lubricating oils containing a suitable proportion of nano-particles have significantly improved anti-wear and anti-friction properties compared to base oil fields (Peng et al. 2009). Huang et al. (2006) observed a 75% drop in friction while using graphite nano-sheets as additives in paraffin oil, demonstrating the ultimate tribological properties of graphite nano-sheets. CuO nano-particles were used to improve the tribological behaviour of coconut oil, and it was observed that the friction coefficient was the lowest and the surface was smooth (Thottackkad, Perikinalil, and Kumarapillai 2012). Prior to this, Battez et al. (2009) and Wu, Tsui, and Liu (2007) had investigated the effect on improving lubrication properties of CuO nano-particles in mineral and synthetic oil for metal-forming fluids. Xie et al. (2016) widely examined the influence of MoS_2 and SiO_2 nano-particles suspended in engine oil on the tribological behaviour of Mg alloy-steel contact using a tribometer. They reported excellent lubrication for both MoS_2 and SiO_2 with the optimal concentration of 1.0 and 0.7 wt%, respectively. The optimal concentration of nano-particles in base oil has also been investigated in research studies that vary with the characteristics of lubricants and test conditions (Luo et al. 2014; Jiao et al. 2011; Jatti and Singh 2015; Li et al. 2011). The nano-particles' structure, size, and shape can significantly affect nano-additives' performance in metal-forming fluids. It was observed that SiO_2 nano-particles (diameter ~ 58 nm) showed better tribological behaviour in carrying the load compared to giant sized nano-particles (Peng et al. 2010).

When mixed with a base oil, nano-particles impart the best tribological behaviour in metal-forming operations while simultaneously lubricating. As previously stated, these nano-particles enter through the workpiece and tool interface. The presence of nano-particles in the metal-to-metal contact area provides sliding action, which reduces force and hence the input energy for the metal-forming process. Furthermore, the nano-particles' tribological behaviour while lubricating provides a smooth surface and improves formability.

3.4 CONCLUSION

The various metal-forming activities necessitate using an appropriate metal-forming fluid for lubrication. Nano-additives are critical in overcoming friction, surface finish, and energy consumption difficulties in metal forming. Furthermore, nano-particles are essential for delivering excellent tribological behaviour in various metal-forming activities to fulfil efficient lubrication using vegetable and synthetic oils. To reduce or eliminate the problem of friction, researchers and industries utilise several types of lubricants which generate a thin film at the interface and aid in friction reduction. Nano fluid-based lubricants have grown in popularity recently due to their excellent tribological and physiochemical qualities.

REFERENCES

Battez, A. R. Hernández J. L. González D. Viesca E. Asedegbega Blanco, and A. Osorio. 2009. "Tribological Behaviour of Two Imidazolium Ionic Liquids as Lubricant Additives for Steel/Steel Contacts." *Wear* 266 (11–12): 1224–28. https://doi.org/10.1016/j.wear.2009.03.043

Dai, Wei, Bassem Kheireddin, Hong Gao, and Hong Liang. 2016. "Roles of Nano-particles in Oil Lubrication." *Tribology International* 102: 88–98. https://doi.org/10.1016/j.triboint.2016.05.020

Ghiotti, A., and S. Bruschi. 2011. "Tribological Behaviour of DLC Coatings for Sheet Metal Forming Tools." *Wear* 271 (9–10): 2454–58. https://doi.org/10.1016/j.wear.2010.12.043

Graff, K. F. 2015. Ultrasonic Metal Forming: Processing. In Juan A. Gallego-Juárez and Karl F. Graff (Eds) *Power Ultrasonics: Applications of High-Intensity Ultrasound*. Elsevier Ltd. https://doi.org/10.1016/B978-1-78242-028-6.00015-6

Gulzar, M., H. H. Masjuki, M. A. Kalam, M. Varman, N. W. M. Zulkifli, R. A. Mufti, and Rehan Zahid. 2016. "Tribological Performance of Nano-particles as Lubricating Oil Additives." *Journal of Nanoparticle Research* 18 (8): 1–25. https://doi.org/10.1007/s11051-016-3537-4

Huang, H. D., J. P. Tu, L. P. Gan, and C. Z. Li. 2006. "An Investigation on Tribological Properties of Graphite Nanosheets as Oil Additive." *Wear* 261 (2): 140–44. https://doi.org/10.1016/j.wear.2005.09.010

Jatti, Vijaykumar S., and T. P. Singh. 2015. "Copper Oxide Nano-Particles as Friction-Reduction and Anti-Wear Additives in Lubricating Oil." *Journal of Mechanical Science and Technology* 29 (2): 793–98. https://doi.org/10.1007/s12206-015-0141-y

Jiao, Da, Shaohua Zheng, Yingzi Wang, Ruifang Guan, and Bingqiang Cao. 2011. "The Tribology Properties of Alumina/Silica Composite Nano-particles as Lubricant Additives." *Applied Surface Science* 257 (13): 5720–25. https://doi.org/10.1016/j.apsusc.2011.01.084

Kumar, Apurv, P. Viswanath, K. Mahesh, M. Swati, P. M. Vinay Kumar, A. Abhijit, and Swadesh Kumar Singh. 2010. "Prediction of Springback in V-Bending and Design of Dies Using Finite Element Simulation." *International Journal of Materials and Product Technology* 39 (3–4): 291–301. https://doi.org/10.1504/IJMPT.2010.035804

Lazzarotto, L., L. Dubar, A. Dubois, P. Ravassard, J. P. Bricout, and J. Oudin. 1998. "A Selection Methodology for Lubricating Oils in Cold Metal Forming Processes." *Wear* 215 (1–2): 1–9. https://doi.org/10.1016/S0043-1648(97)00297-4

Lee, Chang Gun, Yu Jin Hwang, Young Min Choi, Jae Keun Lee, Cheol Choi, and Je Myung Oh. 2009. "A Study on the Tribological Characteristics of Graphite Nano Lubricants." *International Journal of Precision Engineering and Manufacturing* 10 (1): 85–90. https://doi.org/10.1007/s12541-009-0013-4

Li, Wei, Shaohua Zheng, Bingqiang Cao, and Shiyu Ma. 2011. "Friction and Wear Properties of ZrO_2/SiO_2 Composite Nano-particles." *Journal of Nanoparticle Research* 13 (5): 2129–37. https://doi.org/10.1007/s11051-010-9970-x

Luo, Ting, Xiaowei Wei, Xiong Huang, Ling Huang, and Fan Yang. 2014. "Tribological Properties of Al_2O_3 Nano-particles as Lubricating Oil Additives." *Ceramics International* 40 (5): 7143–49. https://doi.org/10.1016/j.ceramint.2013.12.050

Peng, D. X., Y. Kang, R. M. Hwang, S. S. Shyr, and Y. P. Chang. 2009. "Tribological Properties of Diamond and SiO_2 Nano-particles Added in Paraffin." *Tribology International* 42 (6): 911–17. https://doi.org/10.1016/j.triboint.2008.12.015

Peng, De Xing, Cheng Hsien Chen, Yuan Kang, Yeon Pun Chang, and Shi Yan Chang. 2010. "Size Effects of SiO_2 Nano-particles as Oil Additives on Tribology of Lubricant." *Industrial Lubrication and Tribology* 62 (2): 111–20. https://doi.org/10.1108/00368791011025656

Rao, K. P., and C. L. Xie. 2006. "A Comparative Study on the Performance of Boric Acid with Several Conventional Lubricants in Metal Forming Processes." *Tribology International* 39 (7): 663–68. https://doi.org/10.1016/j.triboint.2005.05.004

Sanchez, J. A., I. Pombo, R. Alberdi, B. Izquierdo, N. Ortega, S. Plaza, and J. Martinez-Toledano. 2010. "Machining Evaluation of a Hybrid $MQL-CO_2$ Grinding Technology." *Journal of Cleaner Production* 18 (18): 1840–49. https://doi.org/10.1016/j.jclepro.2010.07.002

Shahnazar, Sheida, Samira Bagheri, and Sharifah Bee Abd Hamid. 2016. "Enhancing Lubricant Properties by Nanoparticle Additives." *International Journal of Hydrogen Energy* 41 (4): 3153–70. https://doi.org/10.1016/j.ijhydene.2015.12.040

Thottackkad, Manu Varghese, Rajendrakumar Krishnan Perikinalil, and Prabhakaran Nair Kumarapillai. 2012. "Experimental Evaluation on the Tribological Properties of Coconut Oil by the Addition of CuO Nano-particles." *International Journal of Precision Engineering and Manufacturing* 13 (1): 111–16. https://doi.org/10.1007/s12541-012-0015-5

Weidel, S., U. Engel, M. Merklein, and M. Geiger. 2010. "Basic Investigations on Boundary Lubrication in Metal Forming Processes by in Situ Observation of the Real Contact Area." *Production Engineering* 4 (2): 107–14. https://doi.org/10.1007/s11740-009-0198-5

Wu, Y. Y., W. C. Tsui, and T. C. Liu. 2007. "Experimental Analysis of Tribological Properties of Lubricating Oils with Nanoparticle Additives." *Wear* 262 (7–8): 819–25. https://doi.org/10.1016/j.wear.2006.08.021

Xie, Hongmei, Bin Jiang, Junjie He, Xiangsheng Xia, and Fusheng Pan. 2016. "Lubrication Performance of MoS_2 and SiO_2 Nano-particles as Lubricant Additives in Magnesium Alloy-Steel Contacts." *Tribology International* 93: 63–70. https://doi.org/10.1016/j.triboint.2015.08.009

Zareh-Desari, Behrooz, and Behnam Davoodi. 2016. "Assessing the Lubrication Performance of Vegetable Oil-Based Nano-Lubricants for Environmentally Conscious Metal Forming Processes." *Journal of Cleaner Production.* https://doi.org/10.1016/j.jclepro.2016.07.040

4 Lubricant Performance Characteristics in Metalworking Manufacturing

*Sanjay Kumar, C. T. Chidambaram,
Vincent Martin, P. V. Joseph, and
Jencen Mathai Arivannoor*
Gulf Oil International Global R&D Centre, Chennai, India

4.1 INTRODUCTION

Manufacturing refers to activities that transform raw materials into new products, usually enhancing economic value. Manufacturing activities used to be performed by hand, but most manufacturing today is done by machines. Today, the manufacturing sector accounts for 17% of the global economy. Although the sector saw a decline in 2019 and 2020 owing to the COVID-19 pandemic, the manufacturing sector has recovered, growing by 7.2% in 2021 (UNIDO, 2022). Nearly 50% of the economy added in this sector is from industries which involve machining operations in their process. Machining operations are a critical cog in the manufacturing sector. Obtaining high productivity from these operations is necessary for overall economic development.

In the context of metal and alloy transformation, the manufacturing technology which is called metalworking is the transformation of raw metals into new products (Youssef and El-Hofy, 2008). This involves a large range of processes, including

- Machining – the shaping of metal into a desired product by removing unwanted material
- Forming – the mechanical deformation of the metal without loss of material or change in mass
- Casting – manufacturing of metal products by pouring liquid metal into molds and then cooling
- Welding – the joining of two metal products

DOI: 10.1201/9781003363576-4

Very often, the fabrication of a single metal product involves multiple metalworking operations in the production chain. Metalworking operations are broadly classified based on whether the process involves the loss of material or not. Further classification is based on the form of the raw material (such as bulk material, bars, etc.) and the nature of the product received (such as sheets, wires, rods, etc.). Machining operations are also classified based on whether the materials were mechanically removed using a harder tool, or alternate techniques – referred to as traditional and non-traditional machining processes, respectively. Figure 4.1 depicts the classification of metalworking operations, including the various types under each.

4.2 METALWORKING FLUIDS

Metalworking fluids (MWF) is a term applied to the fluids used in metalworking operations to improve the process. The primary purpose of metalworking fluid is to lubricate and reduce friction and cool the tool and workpiece. For this reason, metalworking fluids are referred to as lubricants and coolants. The secondary purpose is to enhance the surface integrity of the product or provide the necessary finish to it. In machining operations like turning, milling or drilling, MWF should also help remove the chips or swarf generated. The fluid is applied to the tool and workpiece by flooding or jetting, filtered and recirculated for use. In effect, metalworking fluids prevent damage to the tool and reduce wastage and energy consumption, eventually reducing operational and maintenance costs.

Many lubricants are used for different types of metalworking operations. However, the role of the lubricant – also called metalworking fluid (MWF) – in enhancing productivity of manufacturing operations is often overlooked. Selection of a suitable MWF ensures that the tool piece lasts longer and needs fewer changes, allowing for the machining operation to run for longer durations before maintenance, leading to an improvement in production. A suitable MWF also helps control the workpiece quality, increasing the yield of high-quality products. Although the cost of lubricant averages about 1.4% of the revenue from the operation, this can help significantly reduce the cost of tool replacement which averages around 6% of the revenue (Felix, 2019). In terms of costs to the company, the maintenance, usage and disposal of MWF accounts for about 17% of the manufacturing expenses. However, only 6% of this expense is from the purchase of the fluid (Mang and Dresel, 2007). The majority of lubrication expenses are from maintenance, disposal and plant investment – all of which can be reduced by the selection of the appropriate MWF, significantly boosting savings.

There is no universal lubricant that can work in all metalworking applications. Each application operates under a different range of load, speed, temperature, feed rate, materials and methodology. The tribology of these processes determines the type of fluid required and the properties the fluids must possess.

One way of classifying metalworking fluids is by its intended application. Cutting fluids and grinding fluids are used in the corresponding metal removal processes. These oils focus on both tool and workpiece protection and removing

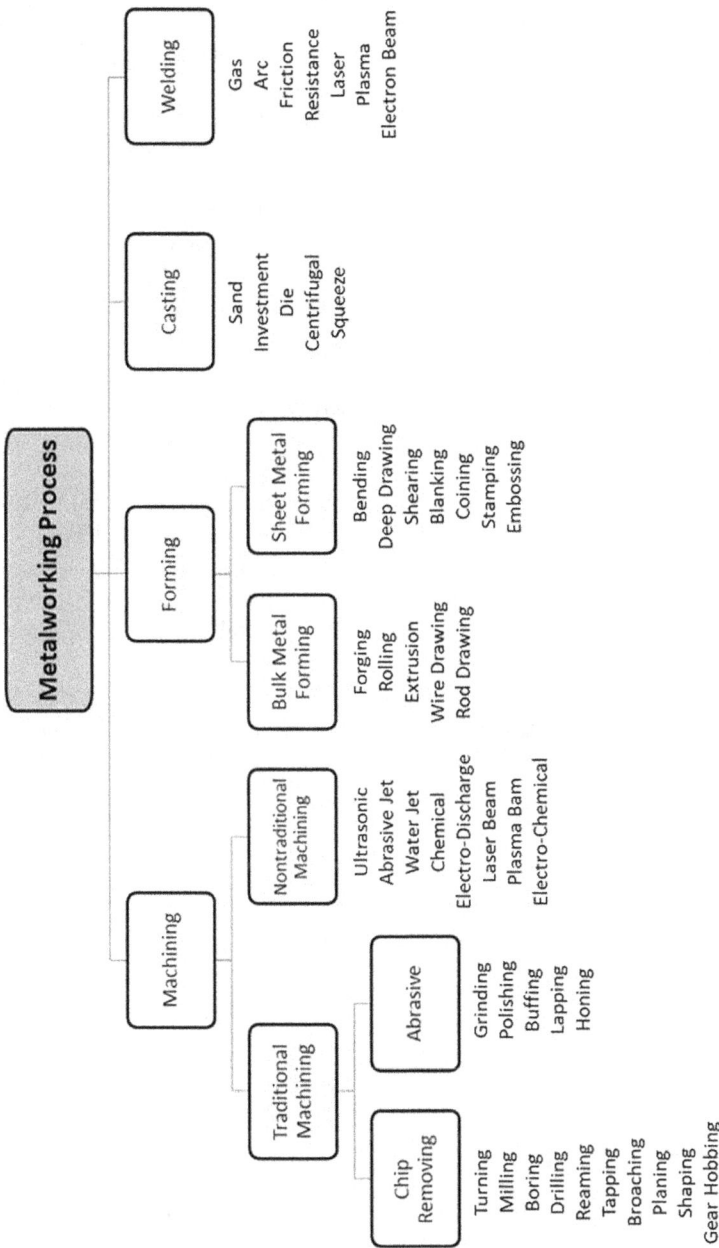

FIGURE 4.1 Types of metalworking processes.

the swarf. Forming applications require forming oils, which instead focuses a lot on the surface finish of the workpiece. Metal protective fluids such as rust preventives, heat-treatment fluids like quenching oils, and other oils such as mold release oils and cleaners, are also often classified as metalworking fluids. As a term and unless specified, metalworking fluids are used to refer to metal removal and metal forming fluids due to their popularity in the segment.

The most popular method of classifying metalworking fluids is by their composition – by the quantity of mineral oil and water in the MWF product. Standards such as ASTM D2881 provide a consolidated look at this type of classification and the related terminologies. ASTM D2881 also classifies solid lubricants for use in metalworking such as powders, vitreous materials, greases, pastes and solid films.

Neat oils refer to oils which are formulated from mineral oils and additives and used in metalworking as is. The product contains no water in it and cannot be emulsified. These fluids were historically the first type of product used in metalworking applications. They are preferred when lubricity and tool protection are the primary requirements of the MWF, and may not be very effective when increased cooling is required.

A subcategory of these products is straight synthetic fluids. These fluids are made from synthetic or Bio-based hydrocarbons, esters, polyalkylene glycols (PAGs) or even renewable vegetable oils. They contain no oil of petroleum origin or water but may contain other additives, and they are not diluted in water for use. Due to the higher cost in comparison to straight mineral oils, these are not very common.

Water-extendable metalworking fluids, also called water-based metalworking fluids or water-dilutable metalworking fluids, are used when both lubricity and cooling are required. These products are manufactured as undiluted concentrates. The end user emulsifies the product at the site and uses the emulsion as the coolant. The presence of water in the emulsion greatly increases the cooling ability. At the point of contact between the tool and the workpiece, the oil and additives in

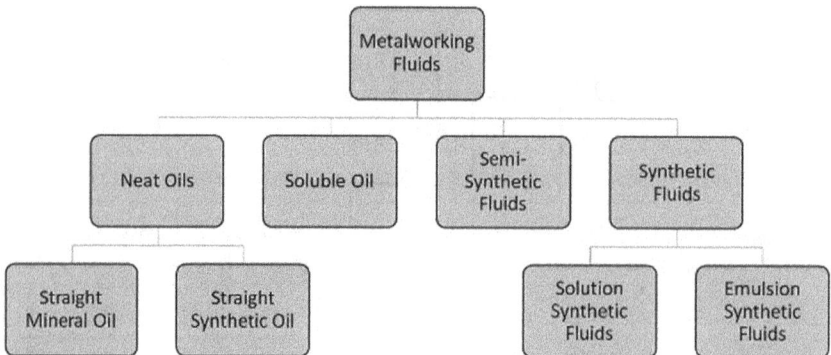

FIGURE 4.2   Types of metalworking fluids.

the dispersed phase provide lubricity. However, it also brings in a slew of concerns, including emulsion stability, corrosion, increased foam, and microbial growth.

Soluble oils, or emulsifiable oils, are the first type of these oils introduced. They contain over 30% of mineral oil in the concentrate and also contain emulsifiers and corrosion inhibitors. On blending with water, they create oil-in-water macro-emulsions. The emulsions are milky and opaque as the average particle size tends to be greater than 1 micron.

Synthetic fluids, sometimes called synthetic oils, were introduced to cater for applications with an even greater need for cooling. The concentrate does not contain a mineral oil of petroleum origin. Due to the reduced lubricity and higher costs, they are more commonly used in light- and medium-duty applications (Byers et al., 2018). There are two subcategories of synthetic water-extendable fluids

- Solution synthetic fluids are true solutions of water-soluble additives. On dilution with water, they further form a true solution with no micelles, which is used in the application. The diluted product is usually clear or very translucent.
- Emulsion synthetic fluid contains emulsifiers. On dilution, they form an emulsion with water. These products form milky emulsions similar to soluble oils.

Semi-synthetic metalworking fluids form the middle ground between soluble oils and synthetic water-extendable fluids. The concentrate contains less than 30% of oil. Water, emulsifiers, corrosion inhibitors, lubricity additives and other additives

TABLE 4.1
Type of Metalworking Fluids for Different Applications

Type of Operation	Metalworking Application	Neat Oils	Soluble Oils	Semi-Synthetic Fluids	Synthetic Fluids
Very Heavy Duty	Threading Tapping Reaming Broaching Gear hobbing	✓			
Heavy Duty	Drilling Sawing Tapping Grinding	✓	✓	✓	✓
Medium Duty	Turning Grinding	✓	✓	✓	✓
Light Duty	Milling Grinding	✓		✓	✓

TABLE 4.2
Summary of Components of a Metalworking Fluid, with the Common Chemistries and Their Application

Component	Primary Function	Chemistry	Application
Base oil	Carrier fluid for the additives Provide lubricity	Naphthenic oils Mineral oils (Gr. I, Gr. II) Vegetable oils PAGs	Neat oil Soluble oil Semi-synthetic oils
Lubricity additive (extreme pressure)	Provide lubricity at high loads and temperatures to prevent wear	Chlorinated paraffins Sulphurized olefins Sulphurized ester Phosphate esters Overbased sulphonates Ethylene oxide propylene oxide Copolymers	Neat oil Soluble oil Semi-synthetic oils Synthetic fluids
Lubricity additive (anti-wear and friction reducing)	Provide lubricity at medium and low loads to prevent wear and reduce friction	Oleic acid Vegetable oils Fatty acid esters Ethylene oxide propylene oxide Copolymers	Neat oil Soluble oil Semi-synthetic oils Synthetic fluids
Corrosion inhibitor	Protect tool and ferrous workpiece from rusting	Calcium sulphonates Sodium petroleum Sulphonates Monoethanolamine borate Triethanolamine Phosphate ester amines Phosphoric acid amines	Neat oil Soluble oil Semi-synthetic oils Synthetic fluids
Metal deactivator	Protect non-ferrous metals and alloys from staining and corrosion	Sodium tolyltriazole Phosphate ester amines Substituted alkyl thiadiazoles	Neat oil Soluble oil Semi-synthetic oils Synthetic fluids
Emulsifiers	Allow the product to emulsify in water by reducing the surface tension of the product	Sodium petroleum sulfonates Fatty alcohol ethoxylates Ethoxylated fatty amines Polyethylene glycol esters Amine soaps	Soluble oil Semi-synthetic oils Synthetic fluids
Couplers	Enhance the ability of emulsifiers and improve stability	Oleic acid Dicarboxylic acids Propylene glycol ethers	Semi-synthetic oils Synthetic fluids
Biocides	Prevent bacterial or fungal growth in the MWF	Triazine Oxazolidines Morpholines	Soluble oil Semi-synthetic oils Synthetic fluids
Anti-foam agents and defoamers	Destabilize foam or prevent foam formation	Silicones	Soluble oil Semi-synthetic oils Synthetic fluids
Reserve alkalinity Booster and pH booster	Neutralize corrosion inhibitors to enable them to work in alkaline mediums Increase alkalinity	Monoethanolamine Triethanolamine Diglycolamine	Soluble oil Semi-synthetic oils Synthetic fluids
Anti-mist additives	Minimize mist formation	Polyacrylates Polyisobutylene Ethylene oxide propylene oxide Copolymers	Neat oil Soluble oil Semi-synthetic oils Synthetic fluids
Anti-oxidant	Prevent oxidation of MWF in high-temperature or high-load applications	Aromatic amines Hindered phenols	Neat oils

constitute the rest. Semi-synthetic oils form translucent oil-in-water microemulsions on dilution. The particle size of the dispersed phase is less than 1 micron.

Bio-based metalworking fluids are a subset of MWFs, where a majority of the composition of the product is of biological origin or made from renewable sources. Neat oils and water-based fluids can both be bio-based. These products were devised owing to the environmental and health concerns of using conventional MWFs. The components, particularly the base oil, are biodegradable, and the additives are selected to have little toxicity. Vegetable oil-based formulations are shown to have improved performance owing to the lubricity of the fatty acids present in them, reducing machining force, temperatures and tool wear (Vasudevan et al., 2019). Although currently not mainstream, there are programs such as the EU Ecolabelling scheme and USDA Biopreferred to promote these lubricants for sustainable manufacturing. Bio-based metalworking fluids are described in further detail in the subsequent Chapter 5 on green lubrication.

4.3 PHYSICO-CHEMICAL CHARACTERISTICS OF METALWORKING FLUIDS

Gauging the performance of an MWF is not an easy task. Each metalworking process involves multiple parameters, each of which can significantly affect the operating conditions under which the MWF must perform. The metal being worked on, the metallurgy and shape of the tool, the speed of the tool or workpiece, the feed rate, etc. are some of the many variables that can affect this.

This makes it difficult to predict exactly how well an MWF will perform in an application. However, over the years, several parameters have been identified to help the formulator design the most appropriate fluid and to help the consumer identify the best fluid for their application. These parameters and characteristics are determined from laboratory evaluation and can help predict performance in the field.

Some of these are general physico-chemical properties of the lubricant in its neat or undiluted form and some are performance-related characteristics which may predict the actual performance to an extent. Some properties are specific to the type of fluid being evaluated. These properties can be used to identify its suitability to a particular metalworking operation, tool and metal, and also help to compare between different fluids.

4.3.1 Viscosity

This is one of the most fundamental properties of a metalworking fluid. Viscosity in essence is the resistance of a fluid to flow. Under constant stress, a fluid with a higher viscosity flows slower while a fluid with a lower viscosity flows faster. When two surfaces are in relative motion to each other, a high-viscosity fluid can offer better oil film thickness even at high speeds, which translates to better separation of the surfaces, better wear protection and better lubricity. However, more energy will be required to keep the surfaces in motion as fluid friction has to be overcome. Lower viscosity fluids can offer better cooling properties, as the heat transfer coefficient of a fluid is inversely proportional to the viscosity. They also spread and wet surfaces faster, requiring less energy to do so (Byers, 2018). However, a low-viscosity fluid may not be able to form the required film thickness, which reduces lubricity.

The Stribeck curve succinctly shows the effect of viscosity on lubrication. This is a plot of coefficient of friction (COF) against the speed, or dimensionless numbers such as Hersey number. For the same speed and load conditions, a change in viscosity can shift the lubrication regime as shown in Figure 4.3. At lower viscosity, film thickness will be low leading to surface-to-surface contact (boundary lubrication regime), which increases the COF. As the viscosity of the oil increases, the film thickness increases and reduces the COF. The COF reduces further as it passes through the mixed lubrication regime, enters the elasto-hydrodynamic lubrication regime and the hydrodynamic regime. As the viscosity of the fluid increases even further, fluid friction comes into play, increasing the COF.

In metalworking applications, high viscosity can help protect the tool piece, making it last longer, and also protect the surface of the workpiece, particularly in the case of very high-speed applications. However, a lower viscosity oil can improve the cooling of the tool and workpiece, requires less energy for circulation and offers better chip removal. The viscosity of a product must be determined by the equipment, the type of tool, the metallurgy of the workpiece and the working conditions, choosing an appropriate viscosity which can provide both adequate tool and workpiece protection, as well as adequate cooling.

Viscosity is determined not only for neat metalworking fluids such as cutting oils but also for the undiluted concentrates of soluble oils, semi-synthetic and

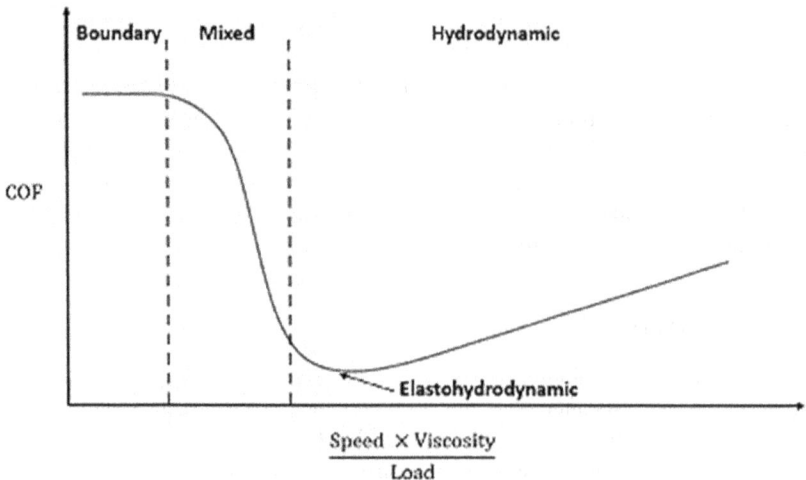

FIGURE 4.3 Pictorial representation of a typical Stribeck curve and the four lubrication regimes.

synthetic MWFs. For neat metalworking fluid, viscosity can help identify the most suitable fluid for an application. For water-extendable MWFs, viscosity is more of a benchmark or quality control parameter for the manufacturer.

4.3.2 STABILITY

Any MWF manufactured must be able to maintain its integrity during manufacture, transport, storage and during operation.

The different additives of the MWF should not separate and float or settle during storage and transport. In the case of semi-synthetics and synthetic products, even the concentrates are dispersions in themselves, making their stability as a concentrate even more critical. Considering the variety of components added into the formulations, several of which interact with each other antagonistically, it is necessary to evaluate the stability of the product at different temperatures and longer durations.

This is especially so for water-extendable MWFs which are used as dilute emulsions during the operation. Barring certain applications where meta-stability or transient stability of the dilute emulsions is required, emulsion stability is equally important to ensure proper lubricity. Equally relevant to concentrate stability is dilution stability. The stability of the emulsified product will determine the long-term performance of the MWF in manufacturing, and this stability is dependent on the components added to the concentrate and the hardness of the water used for dilution. A loss of stability of the emulsion in the field means frequent replacement of the emulsion, which leads to more downtime for maintenance and increased costs (Deluhery et al., 2005).

4.3.3 COMPOSITION

Additives are added to MWF to enhance, inhibit or add a variety of properties. Several anti-wear (AW), extreme pressure (EP) or friction modifier (FM) additives are added to enhance lubricity. These components may be sulphur, phosphorus, nitrogen, chlorine, or ester-based. It is often necessary to determine the amount of the additives added, both as part of formulating the MWF and as quality control during production.

4.3.4 CONCENTRATION

Water-extendable MWFs are used at varied dilutions. This dilution is often optimized for operation for maximum cooling and lubricity. However, in operation, the concentration of the MWF can change significantly. If the concentration of the product reduces, it can reduce lubricity and increase corrosion and microbial growth. On the other hand, increased concentration leads to foaming and increased safety hazard. The concentration of the diluted product or its components, needs to be evaluated frequently to enable optimized performance. (Byers, 2018)

4.3.5 NEUTRALIZATION NUMBERS

The neutralization number is an indicator of the alkalinity or acidity of an MWF. All lubricants and MWFs possess inherent alkalinity or acidity from the various components added to the formulations. These may also change over time during the operation as various degradation products form and contaminants enter the fluid. Hence, these neutralization numbers are often used as an indicator of the remaining useful life of the product. Additionally, the neutralization numbers (along with other parameters), can provide an idea of the concentration of additives in the formulation.

The neutralization numbers for metalworking fluids are usually represented as total acid number (TAN), total base number (TBN), and reserve alkalinity (RA). These are determined by colorimetric or potentiometric titration.

4.3.6 FLASH POINT

The flash point is a critical parameter concerning the safety of storing, transporting and handling the MWF. The flash point refers to the minimum temperature at which the vapours of the fluid can be ignited (or flashed) with the introduction of an ignition source. This is directly correlated with the flammability of the oil. The higher the flash point, the lower the flammability and the safer the fluid is for handling.

The fire point is a similar test which can be performed along with the flash point and provides additional safety information. It is the minimum temperature at which the vapours can catch and maintain a sustained flame with the introduction of an ignition source.

4.3.7 DENSITY

Density is the mass of a substance per unit volume. As the volume of a substance is dependent on the temperature, density is often reported along with the temperature. The specific gravity of a fluid is the ratio of the density of a substance to the density of water at the same temperature.

The density or specific gravity of a metalworking fluid is relevant as a property of a specific product, and as a quality control parameter during production. However, for water-extendable MWFs, density affects emulsion stability. The terminal velocity of the emulsified concentrate is affected by the difference in density between the concentrate and water. In this case, the closer the density of an MWF is to the water in which it is emulsified, the more stable the product.

4.3.8 pH

Water-based MWFs are usually slightly alkaline in nature. The higher pH helps in corrosion protection in ferrous metals and magnesium. It also helps in microbial control as it creates an inhospitable environment for microbes and the stability of the emulsion. However, exposure of the operator to high-pH fluids can cause dermatitis. Hence, the pH of a product is typically balanced between 8.5 and 10.5.

pH can also function as a concentration check technique. The pH of an emulsion is dependent on the concentration of the additives in the dispersed phase. The pH reduces over time as additives become depleted or the concentration of emulsion changes. Tracking the changes in pH can help determine the replacement interval for the fluid.

4.4 PERFORMANCE CHARACTERISTICS OF METALWORKING FLUIDS

4.4.1 FOAMING

Foaming refers to the surface phenomena where air is dispersed in the liquid. In metalworking fluids, this is a major concern as most of the additives added are surface active. The bubbles formed during the use of the MWF rise to the surface and agglomerate, but do not collapse due to the presence of the surface active additive. This phenomenon can be exacerbated by contaminants like water or other lubricants. The addition of anti-foam or defoamer additives beyond a critical concentration may also increase foaming.

Foaming causes many issues if circulated through the system. The presence of foam in the fluid can

- Reduce lubricity, leading to increasing tool wear
- Reduce cooling of the tool and workpiece
- Increase cavitation wear in the pumps used for pumping the MWF
- Increase oxidation
- Lead to the depiction of incorrect fluid levels in the system

A similar phenomenon, air entrainment refers to when the dispersion of air occurs in the bulk of the fluid. Air entrainment is also undesirable, and makes even clear emulsions and synthetic fluid emulsions look cloudy. Air entrainment is mainly dependent on the chemistry and viscosity of the base oil used, yet the problems it can cause are similar to that of foaming.

Another air-related phenomenon observed is misting when the fluid is forced out of the system in fine droplets. Misting can be beneficial as it helps settle fine metal particulates generated from machining operations. However, it can also lead to an increased risk of inhalation of metalworking fluid, dermal contact, and the loss of fluid from the system.

4.4.2 CORROSION PROTECTION

There is a propensity for corrosion in any metalworking operation, and the issue mostly arises from the MWF itself.

In neat MWFs, the sulphur containing EP additives may lead to corrosion or staining of yellow metal and related alloys such as copper and bronze, respectively. In these oils, additional metal deactivators are added to protect these yellow metals. However, as these additives are also polar and surface active, they may negatively affect the EP additive performance by competing with them for the surface. Hence, the formulation of the neat MWF must ensure a balance between the two.

In the case of water-extendable fluids, their usage as dilute emulsions means that ferrous metal and ferrous alloys come in contact with the water meant for cooling. This can lead to the rusting of not only the equipment and tool but the workpiece as well. Rust and corrosion inhibitors must be added to ensure rust inhibition in these metals. Water-extendable MWFs are also observed to cause staining or corrosion induced material failure in aluminium, magnesium, copper and titanium machining if improperly formulated.

4.4.3 TRAMP OIL REJECTION

Contamination of the emulsion in the sump is inevitable, especially by the lubricants used in the operation of the equipment – such as hydraulic oils and slideway oils. The contaminant oil or tramp oil contains several additives that when added to the emulsified MWF at significant concentrations can cause several issues, including

- Demulsifiers in the oil can destabilize the emulsion
- Anti-wear additives can compete with MWF's AW and EP additives and reduce lubricity
- The contaminant oil can reduce the cooling properties of the MWF
- Polar additives can increase foaming and mist
- Phosphorus- and nitrogen-containing additives such as AW additives can promote microbial growth causing malodour

Hence, to control the influence of tramp oil, the emulsion must be able to separate the oil or it must emulsify the oil into itself, and the MWF undiluted concentrate is formulated accordingly. The product can be formulated to have just sufficient emulsifiers so that the diluted form is stable for metalworking fluid. Thus, there will be insufficient emulsifiers to emulsify the tramp oil contaminating the sump. However, the product may not be able to withstand changes in the hardness of water, fine particulate suspension, etc. The product can alternatively be formulated to have an excess of emulsifiers to disperse any oil contaminants.

4.4.4 MATERIAL COMPATIBILITY

Aside from metallic components, the MWF may come in contact with a variety of other materials such as elastomer seals, plastics and even paint on the equipment, which may cause changes in their size, strength and durability.

4.4.5 LUBRICITY

The lubricity of the MWF determines how well the tool–workpiece point of contact is lubricated and protected. In most operations, due to the high load applied and the high speeds involved, lubrication regimes tend towards the boundary regime where the gap between tool and workpiece is minimal and direct metal–metal contact occurs. It is the lubricity that ensures the tool is protected from wear and excessive friction by preventing direct contact, and also protects the surface integrity of the workpiece.

Lubricity in an MWF is offered by an array of additives – AW, EP or FM additives. These additives are polar additives that adhere to the surface of the tool or workpiece forming low-shear films and protecting the surface underneath. The three types of additives differ in terms of their activation energy. FM additives require the lowest activation, while EP additives require the highest. EP additives are more suited to heavier loads, while AW and FM offer better lubricity at low to moderate loads.

The lubricity offered by an additive depends on the operating conditions and the metallurgy involved in a metalworking operation. The hardness of the metals involved and the load and speed applied, can vary the temperature at the point of contact between the tool and workpiece, which can affect the activity of the additive.

4.4.6 MISCELLANEOUS PROPERTIES

Several other parameters can also be evaluated, depending on the type of MWF and the application requirements, such as

* Residue content
* Cloud point
* Surface tension
* Disposal

TABLE 4.3

Summary of Major MWF Parameters and Their Recommended Performance Levels

Property	Neat Oils	Soluble Oils	Semi-Synthetic Oils	Synthetic Fluids
Lubricity		Optimized for application		
Viscosity		Optimized for application		
Product stability		Stable		
Emulsion stability	Not applicable		High	
Foaming tendency			Low	
Corrosion protection			High	
Tramp oil rejection	Not applicable		High	
Flash point	High			Not applicable
Material Compatibility		No effect on other materials		
pH of emulsion	Not applicable		8.5 to 10.5	

4.5 LABORATORY EVALUATION OF PHYSICO-CHEMICAL CHARACTERISTICS

The physical and chemical characteristics of an MWF need to be quantified for a multitude of reasons. Several of these properties can directly affect the performance of a lubricant, some for quality control and quality assurance during production, and others for condition monitoring of the fluid or emulsion during their use. This section covers the different methods usually employed to measure or evaluate these main physico-chemical properties.

4.5.1 VISCOSITY

The viscosity of a fluid is usually expressed as kinematic viscosity. Kinematic viscosity is the resistance of a fluid to flow under gravity. This is determined in calibrated glass capillary tubes, commonly called viscometers, by measuring the time taken for a specific volume of fluid to flow between two points on the tube. A fluid with higher kinematic viscosity will take a longer time to flow.

For petroleum products like lubricants, MWFs and undiluted concentrates, the ASTM D445 and D446 methods are employed to measure kinematic viscosity. ASTM D446 describes the standard specifications of the glass viscometers and their operating instructions. ASTM D445 describes the procedure for determining the kinematic viscosity of the fluid. The time taken for the fluid to flow between the points is multiplied with a constant (to be determined for each capillary viscometer) to obtain the kinematic viscosity. The results are expressed in centistokes (cSt) or millimetre^2 per second (mm^2/s), and the test temperature is also mentioned. For metalworking fluids, a kinematic viscosity at 40 °C is usually reported.

A related parameter is the viscosity index (VI). The VI is an arbitrary number representing the dependence of the viscosity of a fluid on the temperature. The

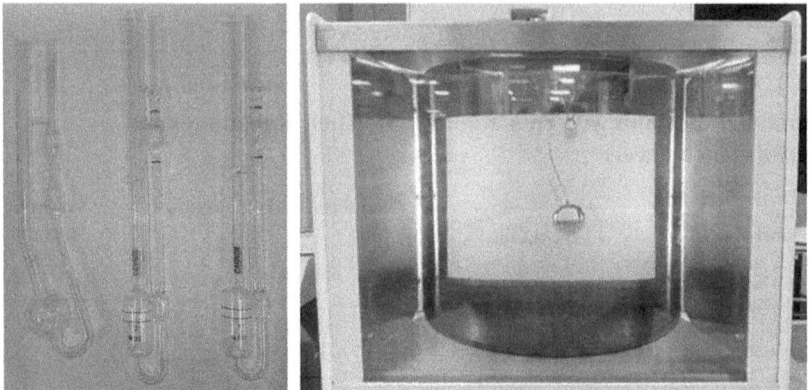

FIGURE 4.4 Image on left shows examples of glass capillary viscometers for measurement of kinematic viscosity. Image on right shows the testing of a lubricant in a Cannon-Fenske routine capillary viscometer.

kinematic viscosity in cSt is measured at 40 °C and 100 °C and then used to calculate using formulae from the ASTM D2270.

4.5.2 STABILITY

4.5.2.1 Neat Metalworking Fluids

As mentioned earlier, the stability of a neat MWF is essential to determine its shelf life. Additionally, the product must also be able to maintain its stability at different temperatures.

It is not feasible to determine the shelf life by waiting one or two years. Instead, the process can be accelerated by heating or cooling the MWF. The product is tested at room temperature, at elevated temperature, and at low temperatures, to ensure stability at a wide range of temperatures. Typically, the elevated temperature is 50 °C and the low temperature is 0 °C.

Alternatively, a modified cyclic stability test based on the homogeneity and miscibility studies used in engine oils can be adopted for neat MWFs (ASTM 6922, IS 13656, Annex A). The test involves the testing of the pour point of the sample, then heating the oil to over 232 °C, and storing the sample at its pour point for 18 to 24 hours. Visual measurements are taken at the end of every stage for any change in state, such as phase separation, clouding, etc. Instead of heating the sample to 232 °C as is for engine oils, the MWF sample may be heated to more moderate temperatures of 85 °C or 100 °C.

4.5.2.2 Product or Concentrate Stability

Similar to neat MWFs, it is necessary to evaluate the long-term stability and shelf life of the concentrate or undiluted water-extendable MWFs. As mentioned in the

FIGURE 4.5   Concentrate showing phase separation in a low temperature stability test.

previous section, this is a major concern because of the variety of polar additives present. Semi-synthetic and synthetic product concentrates are even more difficult as they are dispersions. To ensure that the key components such as emulsifiers do not separate, stability tests must be conducted.

For soluble cutting fluids, the IP 311 (IS 1448-100, IS 9611, Appendix B) thermal stability test is used. The method determines the thermal stability of emulsifiable cutting oils at 0 °C and 50 °C by examining phase separation.

4.5.2.3 Dilute Emulsion Stability

The typical practice is to evaluate the stability of emulsions of different concentrations (1%, 5% and 20%), and in water of different degrees of hardness (demineralized, tap, 200 ppm as $CaCO_3$, 400 ppm as $CaCO_3$, 600 ppm as $CaCO_3$, etc.). The range of concentrations and water hardness in the test ensures that the emulsion stability can be assured for any real variations at the end user's side. The majority of the standardized tests are variations of this methodology.

CNOMO D 655202 recommends the preparation of a 5% emulsion of the concentrate in water of 200 ppm $CaCO_3$ hardness. The number of inversions of the measuring cylinder or flask to mix the concentrate and water is noted (the fewer the inversions, the better the dispersibility). The stability is measured after 24 hours at room temperature, with the oil and cream layers reported. (Byers, 2018)

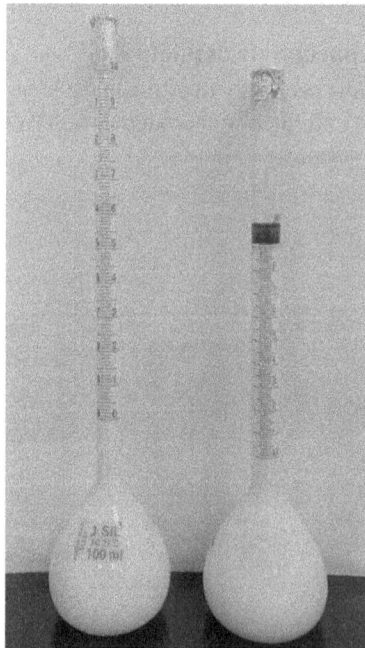

FIGURE 4.6 Typical emulsion stability test, showing a stable emulsion (left) and an unstable emulsion with clear oil phase separation (right).

ASTM D1479 describes a method where stability is tested by measuring the extent of oil depletion over time. The emulsion is prepared at various dilutions and stored in separatory funnels for 24 hours. One-fifth of the emulsion is drawn from the bottom. The emulsion is broken by the addition of concentrated sulphuric acid and is then centrifuged to measure the oil content. The change in oil content before and after the storage test is reported as percentage oil depletion. DIN 51367 is a similar test but reports the stability as percentage change in oil content.

In many cases, the emulsion may look homogeneous to the eye but have varied oil concentrations throughout. Some methodologies propose spectrophotometry or light scattering to examine the stability instead of visual examination.

Deluhery et al. employed a turbidimetric method using a spectrophotometer to detect changes in the wavelength over a shorter duration of 10 minutes. The method was found to be a more conservative test in comparison to the standard 24-hour phase separation tests.

The instrument Turbiscan is also used for evaluating the stability of MWF emulsions. The instrument uses a multiple light-scattering method. The test sample is scanned across its entire height, and based on the amount of light transmitted or back-scattered the turbidity of the sample is measured.

The measurements are performed over time, and the change in turbidity is recorded at different heights of the sample. This way, the extent of sedimentation, clarification and flocculation of the dispersed phase is tracked. The instrument can also be used to evaluate the stability of neat oils or product concentrates. In Figure 4.8, the instrument has been used to determine the stability of two emulsions of semi-synthetic MWFs. The change in back-scattering is measured at different heights, and at different times. The top image for emulsion in water of 200 ppm

FIGURE 4.7   Turbiscan™ instrument for turbidimetric analysis for stability.

hardness shows an increase in back-scattering at both the bottom and top of the sample over the test duration. This suggests that sedimentation is occurring at the bottom of the sample, and creaming at the top. In the bottom graph for the emulsion in water of 1000 ppm hardness, increase in back-scattering is observed evenly across the entire height of sample over time. This shows that the sample is evenly destabilizing throughout the sample, at a significantly higher rate than in the emulsion in 200 ppm hard water.

The fish tank or aquarium apparatus described by Smith et al. (1973) circulates metalworking fluid in a 5-gallon glass tank, with the MWF being pumped and passed through a filter.

4.5.3 COMPOSITION

In neat MWFs, the lubricity additives are usually natural esters, compounds containing sulphur, phosphorus, nitrogen or chlorine, or combinations thereof.

The ester or fat content of the oil can be estimated by the saponification number (ASTM D94). A known mass of the oil is heated with an alcoholic potassium hydroxide (KOH) solution. Part of the KOH will react with the fat and ester, saponifying them. The quantity of the remaining KOH is determined by titration. Based on the quantity remaining, the amount used for saponification is determined and used to calculate the saponification number. This number is reported as milligrams of KOH per gram of sample (mg.KOH/g).

The amount of sulphur, phosphorus, nitrogen and chlorine can be determined mainly by different analytical techniques such as XRD, ICP, etc. and is usually represented in percentage weight (wt%) or ppm.

Phosphorus content can be determined using the ASTM D1091 method (now withdrawn), where all the phosphorus in the oil is converted into phosphates by oxidation and then measured by the photometric or gravimetric method.

Chlorine content is measured by wavelength dispersive X-ray fluorescence spectroscopy (WD-XRF) (ASTM D6443) or by the now inactive sodium alcoholate titration method (ASTM D1317).

For nitrogen content, the syringe/inlet oxidative combustion chemiluminescence method (ASTM D4629), or the boat-inlet chemiluminescence method (ASTM D5762) or ASTM D5291 can be used.

Sulphur content can be determined using several test methods, including energy dispersive X-ray fluorescence spectroscopy (ED-XRF) (ASTM D4294), wavelength dispersive X-Ray fluorescence spectroscopy (WD-XRF) (ASTM D2622, D6443), UV fluorescence (ASTM D5453), the general bomb method (ASTM D129), and the lamp method (ASTM D1266).

ASTM D1662 can be used to determine the amount of active sulphur present in the oil. Active sulphur is the readily reactive sulphur additive present in the oil. It provides excellent EP properties but is highly corrosive to yellow metal. The method heats a known quantity of fine copper particles in a known volume of fluid. The sulphur content is measured before and after the test, and the difference represents the active sulphur concentration.

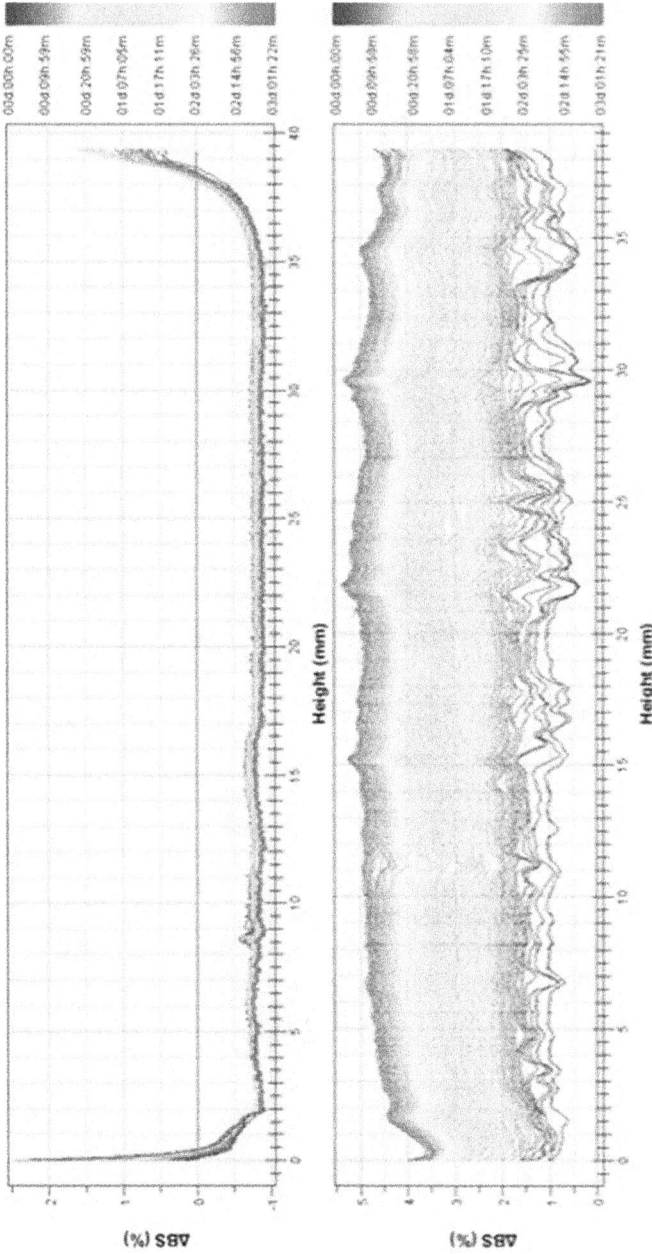

FIGURE 4.8 Change in back-scattering for emulsions of semi-synthetic type metalworking fluid. Top figure is for an emulsion in water of 200 ppm hardness, the bottom figure is for an emulsion in water of 1000 ppm hardness.

The ICP-OES method ASTM D4951 can also be used to determine the sulphur and phosphorus content, among other additive elements.

4.5.4 CONCENTRATION

For most metalworking applications using emulsified coolants, the dilution is predetermined or recommended by the MWF supplier. Straying away from this dilution can affect the overall performance and maintenance of the fluid. Several methods are commonly used to ensure that the concentration of the product in the emulsion lies within the recommended range.

Refractometers are used to determine the concentration. The refractometer can be either optical or digital. In optical refractometers, the oil sample is placed on a glass prism, exposed to a light source and observed through an eyepiece. A band of light falls on a numerical scale visible through the eyepiece. As the concentration increases, the optical density of the sample also increases, changing the position of the band on the scale. This value on the scale is reported. In a digital refractometer, the light emitted by an LED reaches the sample through a prism. Using a detector, the angle at which no refraction occurs is determined and then used to calculate the refractive index. In both cases, a correlation graph is first generated for a range of emulsions, which relates the refractometer reading with the concentration. During the evaluation, the refractometer reading and this graph are used to calculate the concentration. This is not an accurate measurement of concentration as the reading can be affected by the presence of contaminants, and does not correlate well at very high concentrations. However, it can be used as a quick indicator of concentration.

The acid breaking test is a method used to determine the concentration by evaluating the oil content. Similar to refractometers, a concentration-oil content correlation curve is generated and the oil content from the test sample is used to calculate the concentration. In one method, 10 ml of an emulsion is taken in a Babcock bottle and topped with 30% sulphuric acid. The bottle is centrifuged for 10 minutes at 1000 rpm. The volume of the oil phase that separates is used to calculate the oil content (Byers, 2018). Another technique requires 100 ml of a 10% emulsion in a Babcock bottle to be topped up with 0.1 N HCl. The bottle is then inverted a few times and placed in an oven at 50 °C for 16 hours. After cooling to room temperature, the volume of oil floating is used to calculate the oil content. A similar technique is described in DIN 51368.

Although the pH, alkalinity and RA of the MWF cannot directly provide the concentration of the product in the emulsion, they can be used as indicators for changes in concentration or for signifying the need for fluid top-up or maintenance. If any of these properties change significantly, the product or the additives contained in it become depleted. A number of methods can be used to determine the pH. The simplest method is to use a pH indicator paper, where the pH can be determined by the colour of the strip after dipping in the fluid. Other methods such as electrometric methods (ASTM E70 and DIN 51639), or titration can be performed to accurately determine the pH.

Anionic emulsifiers constitute a major portion of the additives used in water-extendable MWFs and play an important role in stability. Monitoring the concentration of the anionic emulsifiers can help track their depletion during use. Cationic titration can be used to determine the anionic surfactant content of the emulsion. The MWF sample is taken in a bottle or flask, along with a suitable colour indicator and a water-insoluble solvent like chloroform. Diluted quaternary ammonium chloride solution is added in steps. Between each step, the mixture is shaken and permitted to rest until the solvent phases out of the water layer. At one point the anionic surfactants are neutralized. The addition of the cationic quaternary ammonium chloride will turn the mixture more cationic, causing a change in colour in the solvent layer. The quantity of quaternary ammonium chloride solution required is representative of the concentration of anionic emulsifiers. Although more expensive, chromatographic methods like gas chromatography and high-pressure liquid chromatography can be used.

Biocide concentration can be used as another indicator for metalworking fluid condition in use. Different analytical techniques can be used to determine the biocide content based on their chemistry. To detect the depletion of biocides, a dipstick method to determine the biocidal efficacy can be used. The dipstick contains nutrients or media to promote the growth of microorganisms, and a growth indicator to show the presence of the microorganisms. It is incubated in the MWF for 24 hours. The biocidal efficacy is assessed based on the colour developing on the dipstick. If the efficacy is high, no colour will develop on the dipstick, showing that the MWF still contains biocides.

4.5.5 Neutralization Number

Total acid number (TAN) and total base number (TBN) are usually measured for neat MWFs to provide an idea of their neutralization capabilities and composition, with the former more common than the latter.

The TAN can be determined by either ASTM D664 or ASTM D974. Both methods are based on titration, using KOH as a base to neutralize the acidic components in the oil. The difference between the two lies in the method of determining the endpoint. ASTM D664 uses a potentiometric determination of the endpoint, while the ASTM D974 method is a colour indicator titration method.

Similarly, ASTM D2896 is a potentiometric titration method and ASTM D974 is the colour indicator titration method for the determination of the TBN of the MWF.

4.5.6 Flash Point

Different methods are used to determine the flash and fire points of a fluid, based on the volatile nature of the test fluid.

The most common method is the Cleveland Open Cup (COC) flash point test, described in ASTM D92. The fluid is taken in a metal cup and slowly heated, while a small flame is passed in intervals of 2°C. The temperature at which a small flash (momentary ignition of the vapour) occurs is reported as the flash point. For the fire point, the sample is heated further until a sustained burning for 5 seconds is obtained.

For fluids that may contain a significant concentration of low boiling point materials, closed cup tests such as the Pensky-Martens closed cup (PMCC) test can be used. The PMCC test is described by ASTM D93, and can only be used to determine the flash point. The fluid is stirred and heated in a closed cup. The ignition source is momentarily introduced in intervals until a flash is observed.

Other methods that are often used include the tag open cup apparatus for the flash point or fire point, (ASTM D1310), the tag closed cup apparatus (ASTM D56) and the Abel flash point apparatus (IP 170, IS 1449-20)

4.5.7 DENSITY

The density or specific gravity of an MWF is usually determined by one of two methods.

ASTM D1298 uses a hydrometer floating in the test fluid taken in a hydrometer cylinder at a particular temperature. The hydrometer scale reading is taken and then converted to density or API gravity using the petroleum measurement tables in ASTM D1250.

The ASTM D4052 method takes a small quantity of fluid in an oscillating tube. The change in the frequency of oscillation is correlated with the change in the mass of the tube from the addition of the fluid and together used to determine the density.

4.6 LABORATORY EVALUATION OF PERFORMANCE CHARACTERISTICS

The performance of a lubricant can be quantified by bench tests, metalworking simulation tests, or by field trials on the production line itself. Field trials are the most complex and carry the most risk as the MWF will be tested in the application. These trials will be most relevant, but failure involves significant costs and maintenance. Laboratory bench tests are comparatively easier to conduct, and they can evaluate a large number of parameters and have lower costs and risks involved. However, they cannot accurately simulate metalworking process and the forces involved. Simulation tests lie in between the two other methods. These tests miniaturize metalworking application to smaller rigs or bench tests, making it easier to study the fluid with a lower investment requirement.

Although bench tests only provide a narrow view of MWF performance, they play a vital role in formulating MWFs, as well as in fluid maintenance. They can work as a method to screen likely candidate formulations and can provide a measurable quantity for comparison with a benchmark fluid. However, it is imperative to select the methods which simulate some aspect of metalworking operation in focus, are repeatable, and can differentiate product characteristics in some way. Many of the popular evaluation methods are standard tests, where the test equipment design, the materials and the parameter are fixed. This means that the test is repeatable in terms of the results, but becomes limited in terms of scope. The test may not be able to accurately evaluate field performance with modifications in design or test parameters or both (Canter, 2018). On the other hand, custom tests offer flexibility in design and can closely simulate metalworking operating conditions, but there is limited information on the consistency of the results obtained.

This section describes the test methods, both standard and customized, that have been used to evaluate the lubricant characteristics and to determine field performance.

4.6.1   FOAMING

4.6.1.1   Neat Metalworking Fluids

Unlike emulsifiable MWFs, foaming is not as significant a concern as other properties when it comes to neat cutting fluids, but can still be an issue in the field when high-speed pumping or splash lubrication is involved.

Foam inhibition properties of the neat metalworking fluid can be determined by using the ASTM D892 method. The method involves the aeration of the sample through diffusers. The amount of foam generated at the end of aeration and the amount of foam after a 10-minute settling time are reported. The test is done in three sequences. The first sequence is performed at 24 °C for a fresh sample. For the second and third sequences, another fresh sample is first evaluated at 93.5 °C and then at 24 °C. The method provides an idea of the foaming tendency of an MWF when it is fresh and when it is in use.

4.6.1.2   Emulsion Foaming

Unlike neat metalworking fluids, foaming is a significant concern for water-extendable MWFs. Emulsifiers and other polar additives can stabilize foam in the emulsions during its use. Hence, several methods have been developed to evaluate the foaming tendency of emulsions.

A simple method is described in IP 312, where 100 ml of the emulsion is prepared in 200 ppm $CaCO_3$ water, shaken in a 250 ml measuring cylinder for 15 seconds and allowed to settle. The froth or foam levels are reported in the 0th, 5th, 10th and 15th minutes.

Another simple and popular test is the bottle test (ASTM D3601). The method involves agitation of 200 ml of an emulsion in a Boston round bottle. The initial height of the emulsion is recorded. The sample is then agitated for 10 seconds by hand, and the maximum foam height is recorded. The emulsion is allowed to rest for 5 minutes. If the foam settles to a height 10 mm higher than the initial height, the time taken to do so is noted. If foam does not settle to a height 10 mm higher than the initial height after 5 minutes, the final height is noted. The method can be tested at different concentrations of emulsion and different hard water levels.

The blender test (ASTM D3519) builds on the bottle test by including a high-speed agitation step. The emulsion is taken in a blender jar at 25 °C. The initial height is noted before blending at 4,000 to 13,000 rpm for 30 seconds. The maximum height of foam, time taken to collapse to 10 mm above initial height (if foam settles within 5 minutes), and final height of foam after 5 minutes (if the foam does not settle to 10 mm within 5 minutes) is reported.

Recirculation tests are considered to simulate real operating conditions most closely. The CNOMO D655212 test involves circulating 1 litre of emulsion with a pump through a 2-litre graduated cylinder. The emulsion is drained from the bottom of the cylinder and returned to the top at a flow rate of 250 L/h for a duration of five hours. The foaming tendency is evaluated based on the time taken to reach the 2-litre mark (if it occurs before 5 hours), the volume of foam above the 1-litre mark immediately after the pump is switched off at 5 hours and the volume of foam above the 1-litre mark 15 minutes after (Broze, 1999) (Byers, 2018).

4.6.2 Corrosion Protection

4.6.2.1 Neat Metalworking Fluids

Due to the usage of highly surface active sulphur additives in neat cutting or grinding oils, especially in high-load applications, yellow metal corrosion is a concern. ASTM D130 evaluates the tendency of an oil to corrode copper. The method involves heating a polished copper strip in oil at a constant temperature and fixed duration. After the test, the strip is washed and rated against a standard chart based on its colour and extent of tarnish. The temperature and duration of the test are dependent on the type of product being evaluated. The typical temperatures are 75 °C and 100 °C, and a duration of three hours is normally selected.

4.6.2.2 Water-Extendable Metalworking Fluids

Ferrous metal and ferrous alloy corrosion are most commonly evaluated using the cast iron chip corrosion test (ASTM D4627). In this test, cast iron chips are evenly distributed over filter paper in a petri dish. The chips are then submerged with the MWF emulsion of known dilution. This is allowed to stand for 24 hours. Afterwards, the area of filter paper stained by the rust on the chips is estimated. The breaking point is also calculated based on the lowest dilution of MWF at which no staining of filter paper is observed. IP 287 and DIN 51360-2 also describes a similar test for cast iron chip corrosion, but the quantity of cast iron chips taken is lower and test fluid only wets the chips for two hours. IP 125 and IS 1115 (Appendix A) describes a method where rusting characteristics are determined where steel chips are in contact with a cast iron plate. Also, a 0–5 rating for the area of corrosion on the plate and the intensity of staining, and the number of pits are reported.

A simple method to determine whether an emulsion will corrode or stain a metal is the metal immersion test or the coupon test. Coupons of metals and alloys are half submerged into the MWF emulsions at different dilutions. The coupons are left in the fluid at room temperature for 24 hours. The coupons are then visually evaluated

FIGURE 4.9   IP 125 cast iron corrosion test. Image on the left shows the wetting of the steel chips on the cast iron plate. The image on the right shows the staining and corrosion at the end of the test.

FIGURE 4.10   Results from ASTM D4627 cast iron chip corrosion test. Tested filter paper on the right shows a lower corrosion rating.

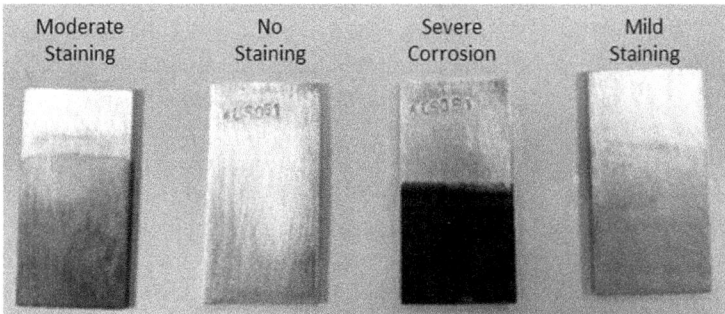

FIGURE 4.11   Metal immersion test performed for different soluble oils on aluminium coupons.

for staining, and the loss of weight of the coupon is reported. This method is commonly used for evaluating the MWF effect on aluminium, copper and titanium.

Aluminium corrosion can also be evaluated by the sandwich corrosion test (ASTM F1110). Although designed for evaluating fluids or dry materials used for aircraft maintenance, it can also be extended to evaluate MWFs. Here, aluminium coupons are sandwiched together with a layer of filter paper with the MWF in between two coupons. Three sets of coupons are evaluated – one for water, one for the product in concentrate and one for the product in dilution. These coupons are warmed using ambient air in an oven and humid air in a humidity cabinet in cycles for seven days and then visually examined for corrosion.

MWF's effect on copper is commonly evaluated by the metal immersion test described earlier. Sun et al. (2018) in their study tested corrosion on copper coupons of different MWF dilutions at elevated temperatures. The rate of copper loss was then determined. The surface was also examined by a scanning electron microscope (SEM) to understand the extent of corrosion.

4.6.3 MATERIAL COMPATIBILITY

Material compatibility of MWF is ascertained by treating the material specimen to the MWF and then evaluating the changes in the strength, hardness, volume, etc. In most cases, it is preferred to have minimal change in these characteristics.

ASTM D471 provides the procedure for testing compatibility with rubber. The test specimen is immersed in the fluid for a specified time and at a specified temperature. Following this, the changes in mass, volume, tensile strength, elongation, and hardness are reported.

ASTM D543 describes a method for evaluating plastic compatibility with the MWF. The test plastic specimen is immersed in the fluid for seven days at room temperature. The duration of the test can be modified, and the test can be performed at elevated temperatures of 50 °C or 70 °C. After the treatment, the plastic specimen is tested for changes in dimension, mass, tensile strength or flexural properties. The method also provides the procedure for treating the specimen while a mechanical strain is applied to it under standardized conditions.

ASTM D4289 is used for testing the elastomer seal compatibility of lubricating oils and greases. The method is mainly intended for automotive engine oils but can be used for industrial fluids. The standard method describes immersing coupons of the seal material in the oil for 70 hours at either 100 °C or 150 °C. For MWFs, the testing can be performed at a lower temperature of 50 °C. The changes in volume and hardness as measured by Durometer A are reported.

To evaluate compatibility with paint, two simple methods can be adopted (Byers, 2018). Both tests are performed on painted steel panels. In one method, the panel is completely immersed in the diluted MWF. Scratches may be made on the painted panel to permit fluid ingress under the paint layer. In another method, drops of the concentrate product are applied to the painted panel. Over a period of time, the panels are inspected for changes in appearance or colour.

4.6.4 TRAMP OIL REJECTION

The ability of the emulsion to either reject or emulsify tramp oil can be evaluated by a simple stability test. An amount of 90 ml of the emulsion (at the recommended dilution for the application) is taken in a measuring cylinder and 10 ml of hydraulic oil is added to the emulsion. The hydraulic oil is preferably sampled from the actual equipment itself. The mix is then inverted several times and allowed to rest at room temperature for 24 hours. The volume of oil not mixed with the emulsion is reported. Depending on the sump and equipment, a higher volume (or higher rejection of oil) may be preferred. A more robust evaluation is offered by the CNOMO D 655203, where the mixture of oil and emulsion is agitated at 10,000 rpm for 15 seconds.

4.6.5 LUBRICITY

Tribological bench tests and simulator test rigs can be used to evaluate the lubricity of an MWF. These tests can simulate a sliding contact, rolling contact or both, or can simulate a point, line or area contact. Although the tests cannot directly correlate with field performance, a combination of these tests may be employed to predict the effectiveness of the fluid in the application.

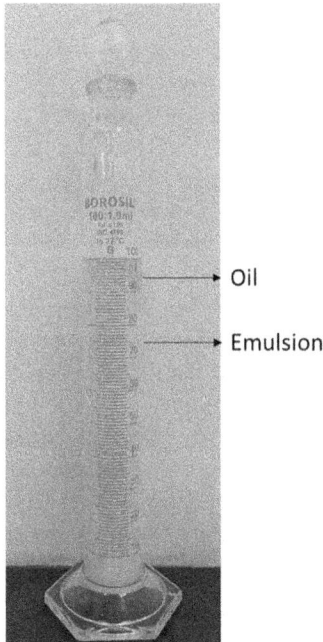

FIGURE 4.12 Tramp oil rejection test. The image shows the separation of cream and oil out of the emulsion after 24 hours.

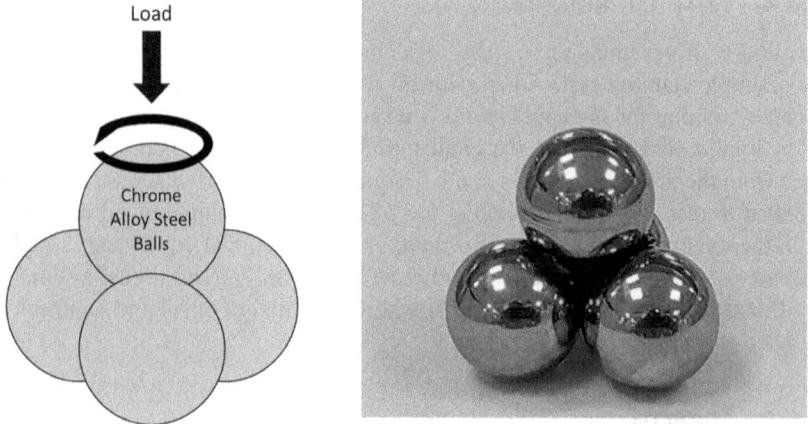

FIGURE 4.13 Image on left shows a representation of the 4-ball test. Image on the right shows the test balls welded together after reaching the weld load stage in the ASTM D2783 test.

The 4-ball tester is considered to be the industry standard for wear and EP tests, particularly for industrial fluids. The equipment is also used for evaluating the AW and EP properties of MWFs. In the 4-ball test, a chrome steel alloy ball is loaded onto three similar balls, held stationary in a cup containing the test fluid. The loaded ball rotates over the three stationary balls. However, depending on the parameter being evaluated different test conditions are used. In the ASTM D2783 test for EP properties of the lubricant, the ball rotates at 1,760 rpm for 10 seconds at specified increasing loads. The load at which the balls weld together is reported as the weld load in kg. The test method IP 239 is similar to this but has more test load stages. The higher the weld load stage, the better the EP performance.

In the ASTM D4172 test for AW properties of the lubricant, the oil is heated to 75 °C, and the ball rotates at a lower speed of 1,200 rpm with 15 kg or 40 kg of load applied. Depending on the load of metalworking operation, fluids can be evaluated at either load. The AW property is assessed by the diameter of the wear scars formed on the three stationary balls. The better the AW characteristics, the lower the wear scar diameter (WSD).

Similarly, the Pin and V-Block test rig can be used to determine the AW and EP properties of a metalworking fluid. A steel journal (or pin) rotates against two steel jaws with v-shaped notches in them (v-block). The v-block is loaded against the pin using a ratchet mechanism and is completely immersed in the test fluid. ASTM D2670 describes a method to evaluate the AW properties of the lubricant. As the pin rotates, the v-block experiences wear and the load applied on the pin reduces. So to maintain the load on the pin, the ratchet wheel is turned further. The wear is determined by the number of gear teeth on the ratchet wheel advanced to maintain the applied load on the pin. The ASTM D3233 method provides two procedures to evaluate the load-carrying capacity of the oil. In both procedures,

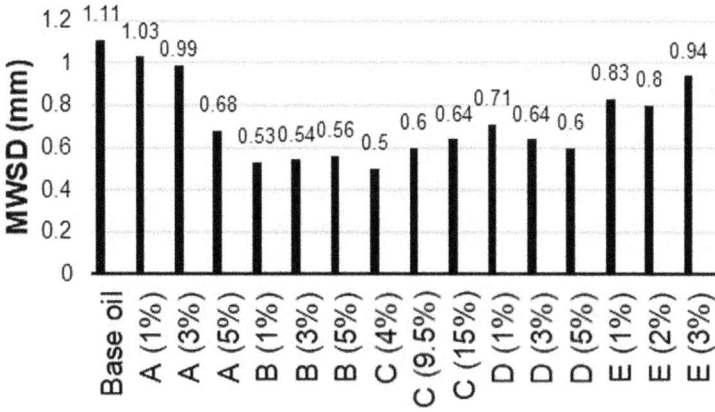

FIGURE 4.14   Effect of addition of additives of different chemistries and different dosages on the AW properties as measured on the 4-ball test. The figure plots the mean wear scar diameter (MWSD) against the type of additive and concentration (A – natural ester, B – sulphurized ester, C – chlorinated paraffins, D – phosphate ester, E – sulphurized olefins).

the load on the v-block is increased until seizing occurs. In Procedure A, the load is increased continuously. In Procedure B, the load is increased in increments.

Block-on-ring tests like the Timken test, Falex Block-on-Ring test and the Brugger test are regularly used to evaluate cutting and grinding fluids. A solid block is loaded against a rotating steel ring. The block material is often varied to simulate the workpiece metallurgy. In these tests, the dimensions of the wear scar on the block and the friction force or the COF are usually reported. The seizing load and the load for rupture of the lubricant film are reported in the Timken test (ASTM D2782). The weight of the block lost during the test is also measured in the Falex Block-on-Ring test (ASTM D2714). The Brugger test follows the DIN 51347 standard.

Similar to the Brugger test, the Reichert test uses a roller-on-ring configuration. Only the lower third of the ring is submerged in the oil. The test is very much suited to applications where the lubricant may be applied on the workpiece before it comes in contact with the tool. The wear on the roller is measured and related to the lubricity. The COF and the acoustic emissions can also be recorded for additional information.

The twist compression test (TCT) can help evaluate forming fluids like MWFs for drawing, deep drawing, wire drawing and stamping. The methodology involves the loading and rotation of a metal ring or cylinder on its flat annular or circular face into a flat block or sheet of metal. The metal test specimen can be customized to the application. The ring, which represents the tool, rotates slowly into the test metal. This creates a lubricant deprived environment at the point of contact. The COF is recorded for the duration of the test. The test can be performed with a constant load for a fixed duration or by applying a ramping load. In both cases, the COF and the time taken for tribofilm failure is determined.

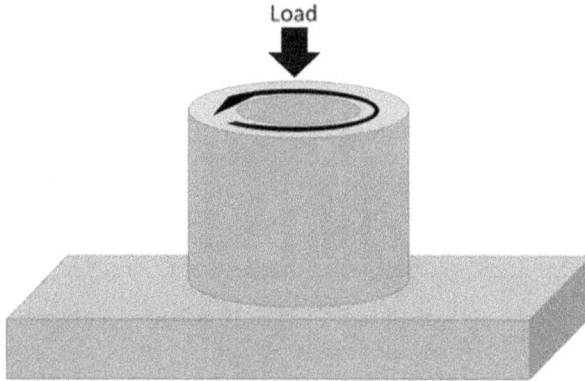

FIGURE 4.15   Representation of the twist compression test.

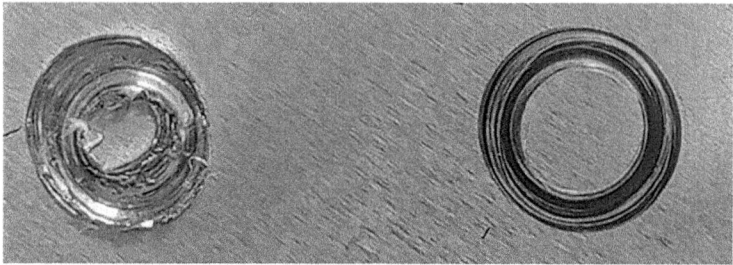

FIGURE 4.16   Results from the twist compression test. The result on the left is for a test with an oil with poorer lubricity, seen from the extensive wear, increased depth and irregular finish. The result on the right shows less wear, implying better lubricity.

The tapping torque test (TTT) is commonly used for evaluating the lubricity of both neat and water-extendable cutting fluids, but now is gaining traction in evaluating MWFs for forming as well. In this method, a tapping tool descends on a pre-drilled hole in the test specimen, cutting threads into it. The material of the test specimen and the tool can be changed to suit the application. The torque is measured during the tapping process. The torque from five separate tests is averaged and reported. Often, the ratio of the average torque for a reference fluid to the average torque of the test fluid is reported as percentage efficiency, as in the case of ASTM D5619. This method has been withdrawn and replaced by ASTM D8288. The temperature of the tool and acoustic emissions in the workpiece during the threading process can be measured to provide a better idea of the MWF's effect on cooling and tool protection. The results from TTT, in general, are highly dependent on the tool and workpiece material, the speed of the tool, the dimensions of the hole and the state of the tapping tool.

The SRV ball-on-disc test can be used to evaluate the lubricity of a neat cutting oil. Several ASTM and DIN test methods have been published to evaluate the AW

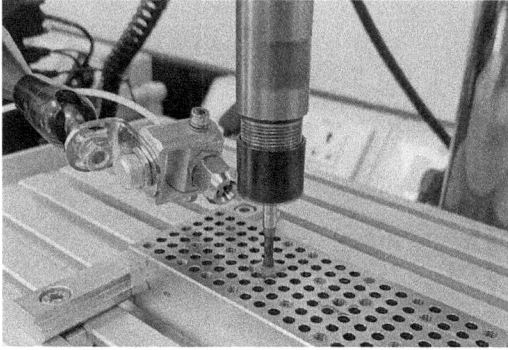

FIGURE 4.17 The tapping torque test for evaluating metalworking fluids. The image shows the testing of a soluble oil using a forming tool with a steel test bar.

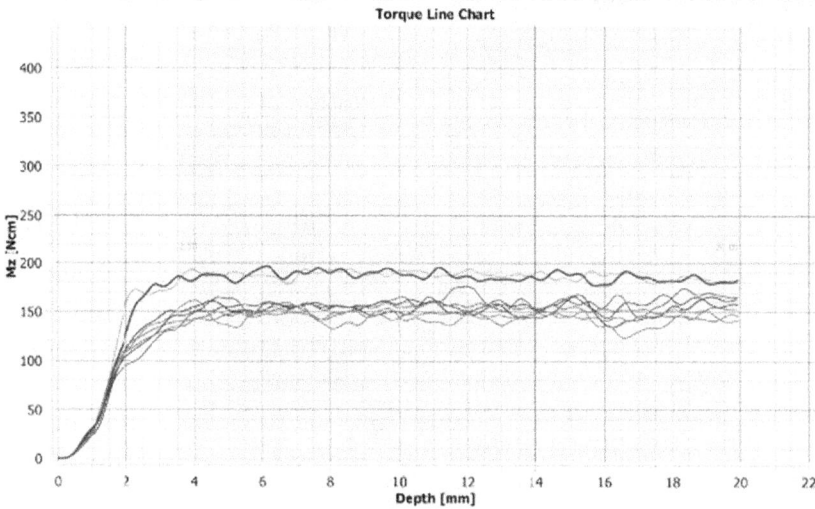

FIGURE 4.18 Comparison of torque developed on the tapping torque test for different soluble oils. The lower the torque, the better the lubricity.

and EP properties of lubricants. In this test, a steel ball is loaded on a steel disc and oscillates at a very high frequency in short amplitudes. The COF is recorded for the duration of the test. The wear on the ball and disc can also be studied. A modified version of the test using a ball of tungsten carbide YG8 against a disc of Inconel 718 was used to evaluate the efficacy of sulphurized additives in cutting fluids for machining nickel-based superalloys. (Yang et al., 2022)

Mini traction machine (MTM) is a ball-on-disc test that measures the traction coefficient of oils. However, in this test both the ball and disc rotate simultaneously at different speeds, creating a rolling-sliding point contact. The instrument

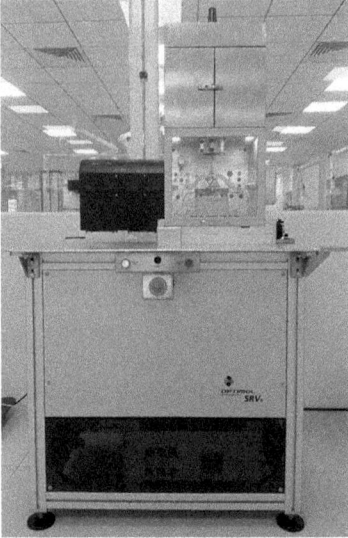

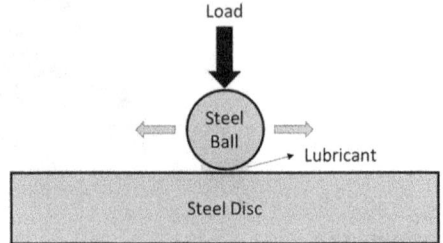

FIGURE 4.19   SRV test equipment (left); representation of the SRV linear reciprocating test (right).

permits the variation in the slide-to-roll ratio (SRR). By varying the load and speed, it is possible to evaluate the oil in all lubrication regimes (boundary, mixed, elasto-hydrodynamic and hydrodynamic). The instrument has been used for the evaluation of water-soluble PAG-based cold rolling oils for aluminium (Reichardt et al., 2013), as well as in a study evaluating the frictional properties of ethanol-amine and ethylamine corrosion inhibitors in a water-based metalworking fluid. (Tomala et al., 2014)

A ring compression test can be used to evaluate the efficacy of a lubricant in a bulk forming operation. A ring specimen is compressed between two dies in the test with the fluid in between, causing the height of the ring to reduce and the diameter to increase. However, if the friction between the ring and die is too high,

FIGURE 4.20   Ball-on-disc test on the mini traction machine.

FIGURE 4.21 Results from the ring compression test. Image on the left shows an untested specimen. The image in the centre shows the test done with lubricants, while the image on the right shows specimen tested in dry conditions. (Images are not to scale.)

the material flow will happen both inwards and outwards. This reduces the inner diameter significantly but does not increase the outer diameter as much. The reduction in internal diameter is plotted with the reduction in height to assess the lubricity of the oil (Bucur et al., 2017). Figure 4.20 shows the effect of a lubricant in ring compression test. The image in the centre shows the benefit of using a lubricant. The surface finish is largely unchanged, while the outer diameter has increased. The image on the right shows the detriment of not using lubricants, leading to poor surface finish or significant damage to the surface integrity.

Some customized tests can also be devised using real metalworking equipment to closely simulate the tribology of the process. These test facilities could be the prototypes of the actual field machinery like experimental rolling mills or actual machines with additional monitoring systems. These tests monitor parameters like cutting force, cutting tool temperature, efficiency, G-ratio, wear, etc. to evaluate the MWF.

Samykano et al. used a lathe machine to evaluate a proposed hybrid nano-coolant of cellulose nanocrystals in an ethylene glycol and water solution. The fluid was compared with a conventional metalworking fluid in the turning of mild steel under minimum quantity lubrication conditions. The study compared surface roughness on the workpiece, tool flank wear and chip formation, and commented on the superiority of the proposed fluid. (Samykano et al., 2021). An orthogonal cutting machine was used in a study to evaluate the lubricity of chemically modified jatropha oil with boron nitride nanoparticles (Talib et al., 2019) and copper oxide or activated carbon (Talib et al., 2022) as a metalworking fluid, reporting the effect on cutting tool temperature, and cutting forces. Somarajan et al. devised a tribometer that can measure in situ the lubricity of an MWF, by measuring the COF between a pin and the machined surface during the machining process (Somarajan et al., 2019).

The G-ratio, grinding efficiency and surface finish can be used to evaluate MWFs when tested on grinding machines. A comparative study of the effect of the grinding wheel material, metalworking fluid and material removal rate on reducing tool wear and increasing productivity showed that the right MWFs can significantly improve productivity (Krueger et al., 2000). Another study measured COF using grinding force, surface roughness and chip formation to evaluate aluminium oxide (Al_2O_3) and cupric oxide (CuO) nanoparticles containing MWFs for the machining of titanium alloys (Setti et al., 2015).

4.6.6 Methods for Other Miscellaneous Properties

Measurement of residue content of MWF can indicate the likelihood of it leaving a solid or sticky residue on the workpiece or metalworking equipment. In most cases, such a film is undesirable. The property is evaluated by heating a measured quantity of the MWF in a petri dish. After heating, the amount of residue left is reported as a percentage of the initial weight. In the case of water-based MWF, the residue may be washed with demineralized, tap and hard water to check whether the residue can recombine for an emulsion again. This can indicate the washability of the fluid.

Cloud point is often used as an indicator of the instability of synthetic or semi-synthetic MWFs. These formulations become cloudy at elevated temperatures due to the destabilization of the emulsifiers and couplers. EN 1890 provides a method to determine the cloud point of surfactant containing solutions.

Surface tension provides a measure of the wetting and spreading capabilities of the MWF. If the fluid has low surface tensions, it spreads better but may cause more foaming and misting. Lower surface tensions also allow quicker settling of fine metal particulates. Surface tension can be measured by standard tensiometers or methods such as ASTM D1331 (du Noüy ring and Wilhelmy plate method), or ASTM D3825 (fast bubble technique). A related parameter is the contact angle. It is dependent on the surface tension and the surface in question and can be used to assess the wettability of the fluid. If the contact angle is too high, its wettability is low. This property can be measured by goniometers (Byers, 2018).

The compatibility of MWF to disposal strategies cannot be easily evaluated, as different types of fluids require different methods for waste treatment, and different end users employ different processes for waste treatment. Nonetheless, the ability to safely dispose of MWF is usually the first property scrutinized by the production or maintenance facility. Laboratory evaluation is not possible to evaluate the suitability of all possible treatment processes. Instead, the formulator and end users can use the other properties of MWF to gauge suitability to the plant's treatment system.

4.7 PERFORMANCE CHARACTERISTICS AND SUSTAINABILITY

The US EPA has defined sustainable manufacturing as 'the creation of manufactured products through economically-sound processes that minimize negative environmental impacts while conserving energy and natural resources. Sustainable manufacturing also enhances employee, community and product safety.' (US EPA, 2023) (Gholami et al., 2020). While adding economic value during the transformation of raw material into a product, sustainable manufacturing aims at ensuring the quality of the environment, natural resources and energy sources for the coming generations. This must also not compromise the quality of the product, or organizational requirements (Khan et al., 2022). Sustainability must not be achieved at the sacrifice of performance and quality.

Sustainable metalworking operations are critical to making the entire manufacturing value chain sustainable. These operations are dependent on multiple operational

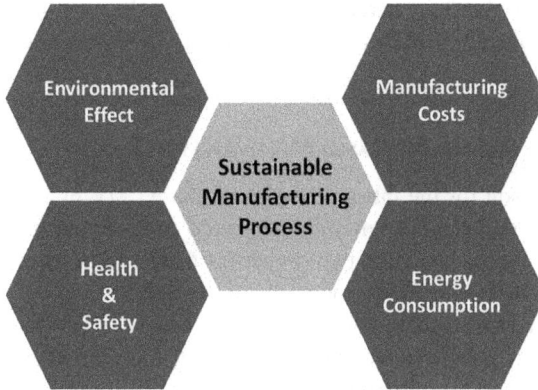

FIGURE 4.22   Factors that contribute to a sustainable manufacturing process.

elements or indicators, each of which can be optimized for sustainability. In most cases, optimizations in the product quality or design are not possible as they are specified for the end user. Hence, the focus is on other factors such as waste management, health, safety and environment (HSE), equipment, tools, lubricants and coolants, etc. (Gholami et al., 2020).

Metalworking fluids are one of the aspects which can improve the sustainability of the manufacturing process. There is tremendous scope to improve MWF formulations to further enhance sustainability. Nearly all metalworking fluids in use today are of petroleum origin. A significant number of studies have shown the environmental impact their disposal poses, along with the effect on the health of the operators. Several additives used in conventional MWFs are potentially carcinogenic, including alkanol amines, nitrosamines, sulphur, and formaldehyde-releasing biocides. The products may also contain polyaromatic hydrocarbons which are known carcinogens. Due to misting from metalworking process, alkaline or acidic emulsion-based MWFs can cause dermatitis in operators. Inhalation of the mist can also cause respiratory and digestive issues (Khan et al., 2022). If microbes have proliferated in the emulsion MWFs, the mists generated can cause infections. Additionally, a majority of the components in MWFs are toxic to animals, non-biodegradable or both. Disposing of MWFs without appropriate treatment has significant environmental impact (Skerlos et al., 2008). These long-standing concerns have initiated a slow shift towards green and sustainable lubrication.

At the same time, enhancing the performance of MWF aids sustainable manufacturing. Longer life from MWF can reduce waste and conserve the resources that have gone into the product. Better tool protection can improve tool life and reduce consumption. Compatibility with the workpiece metallurgy can reduce the wastage of raw materials.

Several studies even show that it is possible to improve both metalworking performance and the health and environmental safety of metalworking by using bio-based components in metalworking fluid. Using vegetable oils and natural esters can improve lubricity and tool protection while being non-toxic and biodegradable. There are even studies on the complete replacement of mineral oils with vegetable oils.

4.7.1 Properties for Bio-based Metalworking Fluids

The primary reason for the introduction of bio-based MWFs is the use of components from a biological or renewable resource and the ease of disposal due to their biodegradation. Hence the two properties that need to be evaluated are the bio-based content and the biodegradability of the MWF.

In petroleum derivatives and oil, the organic carbon content originates from the transformation of fossilized organic carbon from millions of years ago and is called old carbon or fossil carbon. However, if organic carbon is from recent sources such as vegetable oils, solid wastes, etc. it is called newer organic carbon. It is possible to differentiate the two types by the presence of radioactive carbon (C_{14}). Any radioactive carbon that would be present in petroleum or coal, would have decayed to non-detectable amounts during their fossilization. Hence, only newer carbon has radioactive carbon. Bio-based content is a measure of the amount of newer organic carbon in the entire sample.

The bio-based content of the oil can be determined by ASTM D6866, which evaluates the concentration of the components of natural or renewable origin. The method offers two procedures (Methods B and C) for measuring the amount of C_{14} in the sample. The ratio of the amount of C_{14} in the sample to that in a standard is represented as percentage modern carbon (pMC). The value of pMC between 0 (representing value for pure petroleum oils) and 105 (representing value for oils of purely biological origin) can provide the proportion of bio-based content in the sample.

Biodegradability is a measure of the ability of the fluid to be degraded by microorganisms. This directly relates to the life of MWF after disposal. A high biodegradability is preferred as it would reduce concerns over untreated disposal or spillage. Several terms are used in association with this, denoting different types or extents of biodegradability

- Primary biodegradability refers to the loss of activity of molecules by microbial action. Essentially it evaluates the extent to which a molecule is changed by degradation that the original molecule is no longer detectable
- Ultimate biodegradability refers to the complete breakdown of the substance until it is completely removed from the environment

A substance is said to be readily biodegradable if biodegradability greater than 60% is observed in the OECD 301 tests (B, C, D, F Methods). If only 20 to 60% biodegradability is obtained, the substance is said to be inherently biodegradable.

Bio-based content is a necessary evaluation for an MWF to be labelled in the USDA Biopreferred program (USDA, 2013). Similarly, biodegradability must be established for an MWF for ecolabelling schemes such as EU Ecolabel (European Commission, 2020).

The extent of biodegradation can be determined by numerous methods, such as evaluating the amount of carbon dioxide (CO_2) evolved, the chemical oxygen demand (COD), the dissolved organic carbon (DOC) of the mixture, or the oxygen uptake. ASTM D5864 and ASTM D6731 describe similar procedures for aerobic biodegradability. The biodegradability of metalworking fluid can also be evaluated by the OECD 301 and 302 test methods for ready biodegradability and inherent biodegradability of metalworking fluid. The methods involve the mixing of the evaluated component with an inoculum of activated sludge and a medium of mineral nutrients, with aeration in dark or diffused light for 28 days. The microorganisms in the sludge consume the components of the sludge, converting them into CO_2.

A summary of the impact the properties and performance characteristics of an MWF can have on sustainability is described in Table 4.4 below.

TABLE 4.4
Performance Characteristics of Metalworking Fluids and Their Contribution to the Sustainable Metalworking Process

Sustainability Factor	Property	Contribution
Manufacturing costs	Lubricity	Reduces tool replacement frequency by reducing tool wear
		Reduces workpiece rejection by improving surface finish
	Corrosion protection	Reduce workpiece rejection by protecting the workpiece from corrosion
	Product stability	Increases the life of MWF in the application and reduces the frequency of fluid replacement
	Emulsion stability	Increases the life of MWF in the application and reduces the frequency of fluid replacement
	Tramp oil rejection	Increases the life of MWF in the application and reduces the frequency of fluid replacement
	Microbe control	Increases the life of MWF in the application and reduces the frequency of fluid replacement
	Concentration	Reduces MWF consumption if low dilution emulsions can be used
Energy consumption	Lubricity	Reduces the machining force required for the operation and improves efficiency
	Viscosity	Reduces fluid friction if a low-viscosity fluid is used
Environmental effect	Biodegradability	Completely biodegrades in the environment, reducing environmental stress when disposed of or spilt
	Bio-content	Reduces the use of non-sustainable resources by replacing with renewable sources
	Composition	Reduces resource consumption if low additive treat rates are used
		Reduces toxicity and environmental effects by careful selection of additives
	Concentration	Reduces resource consumption if low dilution MWFs can be used
Health and safety	Flash point and fire point	Improves safety during high-temperature operations
	Microbe control	Reduces risk of infections when personnel come into contact with MWF mists
	pH	Reduces risk of dermatitis when personnel come into contact with MWF mists if pH is optimized

4.7.2 Minimum Quantity Lubrication

Minimum quantity lubrication (MQL) is an alternate advanced technique of providing lubrication to the worksite, compared to the conventional flooding method. The motivation for the innovation of the method was the occupational hazard of conventional fluids, and the concerns about their disposal (Sharma et al., 2015). The topic will be expounded upon in Chapter 6 in the section on minimum quantity lubrication. However, a brief mention is made here to show the evolution of the metalworking process, and the changes in stresses the lubricant experiences.

An MWF is applied to the site of machining via misting. A fine layer of mist covers the tool and workpiece, applied by either an internal or external nozzle. The technique can be used for both neat oils and water-soluble oils, but the fluid is never recirculated as MQL uses a fraction of the amount of MWF required in the conventional machining process (10–100 ml/h). A study reports that the quantity of lubricant required can be reduced by 10,000 times (Lawal et al., 2012; Li et al., 2022). This reduces the space required for an MWF sump, the pumping and energy needs, and frequent fluid replacement, disposal and cost. Other studies have shown that MQL can significantly boost performance, reduce tool wear and improve surface integrity. Vegetable-based metalworking fluids are found to be very compatible with MQL systems. The bio-based and renewable nature of these fluids and the minimalistic approach of MQL, have led to significant research on the subject.

4.8 CONCLUSIONS AND FUTURE SCOPE

Metalworking fluids are as multifarious as the manufacturing operations used in the industry. It is reiterated that the performance of the lubricant in real-life applications is not easy to predict, but identifying some of the characteristics of the lubricant can help. This chapter is an attempt to provide an overview of the several properties that can help develop and use an MWF, and the various methods to evaluate them. The list provided in the chapter is not exhaustive and does not imply that these are the best methods for evaluation. The list, however, can provide a guide to metalworking fluid formulators, machine operators and manufacturing technologists to help improve the efficiency of the manufacturing value chain as a whole.

Significant developments are occurring in the machining application and metalworking fluid formulations. The properties that are significant today may no longer be relevant in the coming years. New methods to evaluate performance are constantly being created to help simulate working conditions better or to accommodate new concerns. As the drive for sustainability pushes for newer technologies and solutions, there will be a shift in the performance requirements. The constant innovation in these fields needs to be reviewed and updated frequently.

REFERENCES

Black, J. T., Kohser, A. R. (2008). *DeGarmo's Material and Processes in Manufacturing*. Tenth Edition. John Wiley & Sons.

Broze, G. (1999). *Handbook of Detergents, Part A: Properties.* CRC Press, Taylor & Francis Group.

Bucur, A., Achimaş, G., Lăzărescu, L. (2017). Evaluation of Environmentally Friendly Lubricants By Ring Compression Test. *Academic Journal of Manufacturing Engineering*, 15(2), 37–42.

Byers, J. P. (2018). *Metalworking Fluids.* Third Edition. CRC Press, Taylor & Francis Group.

Canter, N. (2018, March). Bridging the Gap between Metalworking Fluid Bench Tests and Machining Operations. *Tribology & Lubrication Technology*, Society of Tribologists and Lubrication Engineers.

Deluhery, J., Rajagopalan, N. (2005). A Turbidimetric Method for the Rapid Evaluation of MWF Emulsion stability. *Colloids and Surfaces A: Physicochemical and Engineering Aspects*, 256(2–3), 145–149.

European Commission. (2020). Commission Decision (EU) 2018/1702 establishing the EU Ecolabel criteria for lubricants.

Felix, C. (2019). The Real Cost of Metalworking Lubricant. https://www.productionmachining.com/articles/the-real-cost-of-metalworking-lubricant

Gholami, H., Saman, M. Z. M., Sharif, S., Khudzari, J. M., Zakuan, N., Streimikiene, D., Streimikis, J. (2020). A General Framework for Sustainability Assessment of Sheet Metalworking Processes. *Sustainability*, 12(4957), 1–18.

Khan, M. A. A., Hussain, M., Lodhi, S. K., Zazoum, B., Asad, M., Afzal, A. (2022). Green Metalworking Fluids for Sustainable Machining Operations and Other Sustainable Systems: A Review. *Metals*, 12(1466), 1–22.

Krueger, M. K., Yoon, S. C., Gong, D. (2000). New Technology in Metalworking Fluids and Grinding Wheels Achieves Tenfold Improvement in Grinding Performance. *Technical paper presented at the Coolants/Lubricants for Metal Cutting and Grinding Conference.*

Lawal, S. A., Choudhury, I. A., Nukman, Y. (2012). A Critical Assessment of Lubrication Techniques in Machining Processes: A Case for Minimum Quantity Lubrication using Vegetable Oil-Based Lubricant. *Journal of Cleaner Production*, 41, 210–221.

Li, C., Zhang, Y., Xu, W., Ali, H. F., Sharma, S., Li, R., Yang, M., Gao, T., Liu, M., Wang, X., Said, Z., Liu, X., Zhou, Z. (2022). Electrostatic Atomization Minimum Quantity Lubrication Machining: From Mechanism to Application. *International Journal of Extreme Manufacturing*, 4(042003), 1–43.

Mang, T., Dresel, W (2007). *Lubricants and Lubrication.* 2nd Edition. WILEY-VCH Verlag GmbH & Co. KGaA.

Reichardt, T., Duchaczek, H., Chtaib, M., Raulf, M., Kudermann, G., Mömming, C., Espinosa, D., Persson, K. (2013). Increase of Cold Rolling Performance by New Lubricant and Innovative Work Roll Concepts, Report for the Directorate-General for Research and Innovation, European Commission.

Samykano, M., Kananathan, J., Kadirgama, K., Amirruddin, A. K., Ramasamy, D., Samylingam, L. (2021). Characterisation, Performance and Optimisation of Nanocellulose Metalworking Fluid (MWF) for Green Machining Process. *International Journal of Automotive and Mechanical Engineering (IJAME)*, 18(4), 9188–9207.

Setti, D., Sinha, M. K., Ghosh, S., Rao, P. V. (2015). Performance Evaluation of Ti–6Al–4V Grinding Using Chip Formation and Coefficient of Friction Under the Influence of Nanofluids. *International Journal of Machine Tools & Manufacture*, 88, 237–248.

Sharma, V. S., Singh, G., Sørby, K. (2015). A Review on Minimum Quantity Lubrication for Machining Processes. *Materials and Manufacturing Processes*, 30, 935–953.

Skerlos, S. J., Hayes, K. F., Clarens A. F., Zhao, F. (2008). Current Advances in Sustainable Metalworking Fluids Research. *International Journal of Sustainable Manufacturing*, 1(1/2), 180–202.

Somarajan, S. P., Horng, J., Kailas, S. V. (2019). Development and Testing of a Modular Lathe Tribometer Tool to Evaluate the Lubricity Aspect of Cutting Fluids on Freshly Cut Surfaces. *Tribology Online*, 14(5), 417–421.

Sun, J., Yan, X., Meng, Y. (2018). Experimental Insight into the Chemical Corrosion Mechanism of Copper with an Oil-in-Water Emulsion Solution. *RSC Advances*, 8, 9833–9840.

Talib, N., Sabri, A. M., Sani, A. S. A., Jamaluddin, N. A., Suhimi, M. H. A. (2022). Effect of Vegetable-Based Nanofluid Enriched with Nanoparticles as Metalworking Fluids During Orthogonal Cutting Process. *Jurnal Tribologi*, 32, 29–39.

Talib, N., Sani, A. S. A., Hamzah, N. A. (2019). Modified Jatropha Nano-Lubricant as Metalworking Fluid for Machining Process. *Jurnal Tribologi*, 23, 90–96.

Tomala, A., Suarez, A. N., Ripoll, M. R. (2014). Tribological Behaviour of Corrosion Inhibitors in Metal Working Fluids under Different Contact Conditions. *Advanced Materials Research*, 966–967, 347–356.

United Nations Industrial Development Organization (UNIDO). (2022). International Yearbook of Industrial Statistics, Edition 2022.

United States Department of Agriculture (2013). Biopreferred Program Overview. https://www.biopreferred.gov/BioPreferred/

United States Environmental Protection Agency. (2023). https://www.epa.gov/sustainability/sustainable-manufacturing

Vasudevan, S. A., Saravanan, V., Anandakumar, D. (2019). Experimental Investigation on Tribological Behavior of Bio Based Lubricant as Metal Working Fluid in Machining Process. *International Journal of Research and Scientific Innovation*, 6(3), 59–62.

Yang, Y., Luan, H., Guo, S., Liu, F., Dai, Y., Zhang, C., Zhang, D., Zhou, G. (2022). Tribological Behaviors of Inconel 718–Tungsten CarbideFriction Pair with Sulfur Additive Lubrication. *Metals*, 12(1841), 1–13.

Youssef, H. A., El-Hofy, H. (2008). *Machining Technology: Machine Tools and Operations*. CRC Press, Taylor & Francis Group.

5 Green Lubrication in Manufacturing

C. T. Chidambaram, Sanjay Kumar,
Vincent Martin, P. V. Joseph, S. Vignesh, and
S. Vinith Kumar
Gulf Oil International Global R&D Centre, Chennai, India

5.1 INTRODUCTION

In the early 1960s, there was a large rise in the reported breakdowns of equipment and machinery due to wear and tear, resulting in huge economic losses. Although technology was advancing, the expense of running a factory was simultaneously rising. The continuous process made equipment failures more frequent, costly and serious than before. This tendency was recognized by experts in the fields of friction, wear, and lubrication. As a result, the British government appointed a working group under Peter Jost to investigate this situation. This group coined the term "tribology" to highlight the scientific aspects of understanding the interactions of solid contacting surfaces in relative motion, covering three disciplines, viz. friction, wear, and lubrication as shown in Figure 5.1. Friction is the resistance force to motion during sliding or rolling that is experienced by a solid body when it moves over another. Wear is the surface damage or removal of material from one or both of two solid surfaces from sliding, rolling, or impact motion relative to one another. They also suggested that friction is the property of the material, while wear is the consequence of the load and the relative motion of the two contacting surfaces. To mitigate this friction and wear, an intermediate layer of a material is introduced between the interacting surfaces to minimize friction and wear; that material is called a lubricant and the process of applying it is called lubrication (Julieb et al., 2003, Abdalla et al., 2007).

5.2 LUBRICANTS AND LUBRICATION

Lubricants are used to reduce wear and friction between interacting surfaces. A lubricant performs a variety of tasks, including removing heat, preventing corrosion, transferring power, and forming a liquid seal at moving contacts, in addition to decreasing friction and wear. Lubricants are utilized in a variety of machinery types, ranging from large metal-rolling mills to tiny equipment such as computer hard drives. Lubricants are classified into liquid, semi-solid and solid based on its state of matter. The majority of lubricants are liquids and these are classified based

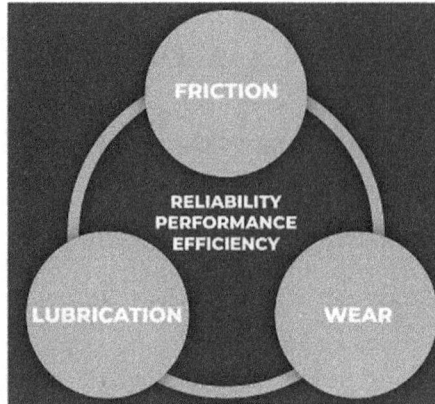

FIGURE 5.1   Tribology and its components.

on their composition, such as mineral oil-based lubricants, synthetic oils and bio-based base oils. Mineral oils are refined crude oil products created in a refinery; synthetic oils, such as poly alpha olefins, esters, and di-esters are produced in a chemical or petrochemical plant; and bio-based oils such as castor oil, rapeseed oil, sunflower oil, coconut oil are extracted from plants. Vegetable oils and vegetable oil-based esters are usually regarded as biodegradable oils due to their rapid degradability and eco-friendly nature (Bennett, 1983, Ozcelik et al., 2011).

In manufacturing industries, the machining process plays a major role. It is a process in which the cutting tool is utilized to remove excess or unwanted material from a work piece to obtain a desired size, shape, or dimension. During this process, a large amount of heat is generated due to the shearing and friction between tool and work piece. This heat affects the life of the tool tip and surface quality of the work piece after machining. Metalworking fluids or cutting fluid, also known as coolant, reduce the interfacial temperature between the tip of the cutting tool and work piece and lubricates the contact area between the machining zones where the friction is generated. Additionally, it promotes the generation of discontinuous chips which avoids the forming of built-up edges (BUE) on the tool, removes the chips from the machining zone, and reduces the frictional torque required for machining, thus extending the tool's life (Ozcelik et al., 2011, Zeman et al., 1995).

Petroil systems, bath or splash lubrication, flooded-type lubrication, pressure systems, and dry sump systems are a few commonly used lubrication systems in mechanical equipment, such as engines, gears, and bearings. In other manufacturing process like metalworking and metal cutting, the most commonly used lubrication systems are flooded lubrication, air blow lubrication, minimum quantity lubrication and lubrication with cryogenic gases (Sreejith and Ngoi, 2000).

5.3   GREEN TRIBOLOGY

The new concept of green tribology was defined by H.P. Jost (1992) as the science and technology of the tribological aspects of ecological balance and of environmental and biological impacts. Previously it had also been known as eco-tribology that emphasizes the ways in which contact systems interact with the environment. Green tribology aims to save natural resources, promote environmental protection, and enhance human well-being. It will have a direct impact on the economy by minimizing waste and extending the life of equipment; attaining a better environmental, technological, and economical balance; minimizing greenhouse gas emissions; enhancing safety in human society; and promoting sustainability (Jagadeesh et al., 2018).

Green tribology emphasizes the following points (Lee and Choong, 2011, Matthew et al., 2007):

- In order to make tribo systems more environmentally acceptable, surface texturing offers a technique to manage a variety of surface characteristics
- The environmental implication of coating should be taken in account
- Surfaces, coatings, and tribological components should be designed for maximum efficiency
- Biomimetic and bio-inspired approaches and eco-friendly green lubrication, e.g., vegetable oil, should be employed wherever feasible
- Controlling and reducing friction, which results in energy saving and the reduction of environment degradation from heat pollution

5.4   GREEN LUBRICANTS FOR MANUFACTURING

Lubricants made from plant-based oils or other natural sources are typically referred to as "green lubricants". They are based either on instinctive vegetable oils, genetically modified versions or on chemically modified vegetable oils. The other common names for green lubricants are, "environmentally friendly", "environmentally acceptable" and "environmentally benign" lubricants (Goyan et al., 1998). Green lubricants are safe and biodegradable, and because of these characteristics have good scope for use in manufacturing sector.

There are also some other factors that play an important role in using vegetable oil-based oils as green lubricants (Petlyuk and Adams, 2004, Lea, 2002)

- Physio-chemical characteristics
- Technical performance and economical compatibility
- Safety and health concerns
- Sustainability, disposal, and recycling of the used lubricants

Lubricant researchers are currently more focused on the development of sustainable lubricants to replace petroleum-based and mineral oil-based products. Alternative solutions, including synthetic, solid, and green lubricants are being considered as approaches. Due to environmental friendliness, renewability, lack of toxicity, operator ergonomics, and ease of biodegradation, vegetable oil-based lubricants are often compelling alternatives for mineral oil and petroleum-based lubricants. Numerous studies are being conducted to create new bio-based green cutting fluids that are based on different edible and non-edible vegetable oils, genetically modified vegetable oils, chemically modified vegetable oils, green ionic liquids, etc. (Srikant et al., 2009, Abdalla and Patel, 2006).

History and Evolution of Green Lubricants

History studies have shown that lubricants were used since ancient times where olive oils, rapeseed oils, and animal fats like tallow, sperm oil were used as lubricants before the invention of crude oil (Grushcow, 2005). These were first replaced by mineral oil and later by synthetic oils because of its high cost, raw material availability, low oxidation stability and poor low-temperature performance. Yet green lubricants have many benefits, including good dispersal capability, detergent quality, higher shear stability, lower volatility, higher lubricity, higher viscosity index and, most importantly, they are eco-friendly in nature. Currently, green lubricants such as glycols, vegetable-based oils, and natural esters are widely used in manufacturing industries as lubricants, coolants, cutting fluids, processing fluids, etc. Furthermore, studies have found that green lubricants have a wide scope in applications such as hydraulic, gear, turbine and compressor oils.

Kato and Ito (Koji & Kosuke, 2005) presented an approach named life cycle tribology (LCT) where the life cycle assessment (LCA) approach was applied to tribological systems. They found that improving the properties of lubricants used for friction reduction could affect climate change caused by fossil fuels by decreasing energy consumption positively. Later on, the following topics were mentioned by Sasaki (2010): tribo-materials to improve recyclability, friction and wear characteristics, and environmental safety; diamond-like materials as eco-tribological components; water-based lubricants, whose operational life should be prolonged to lower their environmental impact; and carbon coatings to reduce friction. The development of these and other tribo-elements may lower the need for energy in, for instance, machinery and other sectors.

Fox and Stachowiak (2007) approached innovation from a different angle, first considering the commercial issues related to friction and wear. or suspension Challenges like this give rise to new research areas, like bio-tribology, environmental tribology, and nano tribology. Bio-tribology emerged when more implants and prostheses were developed to fulfill the human desire for eternal life. The solution to issues of energy production and consumption, environmental deterioration, and climate change is environmental tribology. Advancement in the field of nano-technology naturally leads to the creation of nano tribology.

Green lubricants are produced from vegetable oils or modified vegetable oils (synthetic esters).

Vegetable oils are mainly classified into two categories based on sources:

1. Edible vegetable oils: These oils are widely produced across the globe and mostly consumed in food applications. Examples of edible oils include those of mustard, peanut/groundnut, olive, sesame, safflower, rice bran, soybean, coconut, palm, sunflower, corn, cottonseed oils, etc.
2. Non-edible oils: These oils are mainly used for commercial purposes other than food consumption. Commonly available non-edible oils include castor, jojoba, jatropha oils, etc.

5.4.1 Vegetable Oil Chemistry

Glycerides of fatty acids are the main component of vegetable oils, which are either liquid or semi-solid plant products whose composition is described in terms of plant-specific "fatty acids" which function as its building blocks. With regard to the area, environmental circumstances, and availability of resources for growth, it is likely that the fatty acid content for a specific oil can differ. Fatty acids are composed of straight chain carbon molecules ranging from 8 to 24 carbon atoms with a carboxylic acid functional group (Debnath et al., 2014).

Various types of fatty acids are present in vegetable oil:

Saturated Fatty Acids
The basic structure of these fatty acids does not have a carbon–carbon double bond: e.g., palmitic acid, stearic acid and lauric acid. They exhibit a high pour point, a relatively high melting point, and are chemically stable because of their molecule conformation also offer good resistance to oxidation.

Monounsaturated Fatty Acid
The basic structure of these fatty acids is only one carbon–carbon double bond, like oleic acid. They offer a reactive site for chemical modification, exhibit a relatively low pour point but are less stable to oxidation than saturated fatty acids.

Polyunsaturated Fatty Acid
The basic structure of these fatty acids have more than one carbon–carbon double bond, e.g. linoleic and linolenic acids. They offer two or more reactive sites for chemical modification, exhibit relatively low pour point, but are vulnerable to oxidation.

Hydroxy Fatty Acid
The basic structure of these fatty acids has one hydroxyl group (OH), e.g., Ricinoleic acid in castor oil is hydroxy fatty acid. They offer multiple reactive sites for chemical modification and exhibit a relatively low pour point. Vernolic acid is a special kind of fatty acid that is found in the seed oil of Vernonia anthelmintic.

Mineral oils and plant oils are entirely chemically dissimilar. The plant oil's major components are triacyl glycerols (TAGs) (98%), and the minor components are diacyl glycerols (DAGs) (0.5%), sterols (0.3%), monoacylglycerols (MAGs) (0.2%), fatty acids (FAs) (0.1%), and tocopherols (0.1%) (Neff et al., 1994, Loh and May, 2012, Dian et al., 2017, Rani et al., 2015).

5.4.2 Use of Vegetable Oils as Green Lubricants

Vegetable oils are good in lubricity but have poor low-temperature property and thermo-oxidation stability (Liu et al., 2015, Khan and Dhar, 2006). The drawbacks of vegetable oils are rectified by modifying it to synthetic esters and other molecules and also by modifying genetically.

Some of the approaches used to enhance vegetable oil properties are (Sharma et al., 2006)

- Chemical modifications:
  - Inter-esterification with other potential vegetable oils such as high oleic canola oil
  - Trans-esterification with various polyols
  - Epoxidation
- Genetic engineering to reduce saturates and increase monounsaturates
- Blending with synthetic esters such as trimethylol propane trioleate or pentaerythritol tetraoleate

5.4.2.1 Chemical Modification of Vegetable Oils

For production of biolubricants, some polyols like trimethylolpropane (TMP) are used to modify vegetable oils. Chemical modifications of seed oils were done by transesterification using TMP (Shirani et al., 2020) and the possible chemical modifications are shown in Table 5.1.

Palm olein extract from palm oil has moderate thermo-oxidative and low-temperature properties, and contains equal amounts of saturated and unsaturated fatty acids, allowing it to be readily converted into required biolubricants. The inclusion of methyl ester of palm oil up to 40%, shows good emulsion and foam inhibition characteristics, low air release value and good corrosion inhibition properties, along with high viscosity index compared with pure palm olein (Jagadeesh and Kailas, 2012).

TABLE 5.1

Vegetable Oils Modified Chemically Using TMP

Source of Oil	Oil and TMP Ratio	Catalyst	Conditions of Reaction	Yield in %	Properties
Jatropha seed	3.9:1	1% NaOCH₃	150 °C; 3h	46	PP(−3°C); VI(1785180
Jatropha seed	4:1	2% HClO₄	150 °C; 3h	70.00	PP(−22°C); FP(>30°C); VI(140)
Palm kernel	3.9:1	0.9% NaOCH₃	130 °C	98.00	V@40 °C(39.70 cSt); V@100 °C(7.700 cSt); FP(310°C); PP(2°C)
Castor seed	4:1	0.8% Phosphoric acid	120 °C	96.5	V@40 °C(45.3 cSt); V@100 °C(9.1 cSt); VI (190) FP(220°C); PP(−7 °C)

5.4.2.2 Genetic Modification of Vegetable Oils

Genetic modification is a technique to modify the genetic properties of plants to obtain oils with improved fatty acid distribution that meet lubricants' requirements, for an example, soybean oil genetically modified with oleic acid is widely used as a green lubricant as it provides better cooling and lubricity.

Due to the effectiveness in viscosity index, anti-foaming and anti-corrosion properties of oil formulations, genetically modified jojoba oil has received a lot of attention in the lubrication industry (Winter et al., 2012). It is entirely made of wax esters, which is responsible for its better lubricating properties. Currently, transgenic crambe plantseeds possess various levels of wax esters, for better performance combined with a higher percentage of conventional triacylglycerol (Kupongsak and Lucharit, 2013). The test evaluations show that crambe oil with 15% wax esters as a lubricant gives tribological characteristics and higher thermal stability up to a temperature of 200 °C than conventional crambe oil (Bhatia et al., 1990).

5.4.2.3 Modification of Vegetable Oils by Additivation

Vegetable oils' double bond (C=C) in their fatty acid causes them to have limited oxidative stability. As the unsaturated fatty acid content increases, so does the tendency to oxidation. The effects of lubricant oxidation include a rise in viscosity, the creation of silt and sludge, varnish, filter plugging, a loss of foam control, corrosion, and the formation of rust. Table 5.2 (Soni and Agarwal, 2014) lists the percentages of saturated and unsaturated fatty acids in various plant oils. According to the data, the majority of plant oils include a high percentage of unsaturated fatty acids, making them more likely to oxidize.

Lipid oxidation follows the same mechanism as hydrocarbon oxidation. When exposed to an oxygen-rich environment, oils and fats oxidatively age and form peroxides as a result of oxygen's interactions with the conjugated fatty acid chains of the lipid molecules. According to recent research (Gajrani and Sankar, 2017),

TABLE 5.2
Composition of Fatty Acids with the Source

Source	Fatty Acids (%)	
	Saturated	Unsaturated
Sunflower	12.3	87.3
Soybean	15.1	84.4
Canola	11.5	88.5
Coconut	92.1	7.8
Palm	49.9	49.7
Rapeseed	7.5	92.5
Olive	19.4	80.6
Linseed	9.4	90.6
Mahua	49.3	49.7

free-radical chain mechanism is commonly thought to be how lipid auto oxidation occurs at relatively high activation energy.

Step 1: Initiation (formation of free radicals)
   In. $+$ RH $\rightarrow$ InH $+$ R.
Step 2: Propagation (free-radical chain reactions)
   R. $+ O_2 \rightarrow$ ROO.
   ROO. $+$ RH $\rightarrow$ ROOH $+$ R.
Step 3: Termination (formation of non-radical products)
   ROO. $+$ ROO. $\rightarrow$ ROOR $+ O_2$
   R. $+$ ROO. $\rightarrow$ ROOR
   R. $+$ R. $\rightarrow$ RR

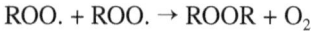

In this case, 'R' is fatty alkyl chain, 'In' is initiator, 'O' is oxygen atom, 'H' is hydrogen atom

The fact that vegetable oils have poor low-temperature characteristics is another problem that limits their use as a base stock for green lubricants. At room temperature (24 °C), this may result in exhibiting cloudiness, precipitation, poor flow, and solidification. Vegetable oils' poor low-temperature qualities are caused by the molecular structure of saturated fatty acids (Okullo et al., 2010). This is because saturated fatty acid chains tend to bundle their carbon atoms more quickly than unsaturated fatty acid chains do at low temperatures, resulting in a crystalline form. Some techniques to overcome oxidation issues are listed in Table 5.3.

Heshmat (1991) suggested that the introduction of pour point depressant (PPD) additives may improve the low-temperature performance of vegetable oils and investigated this. The improvement was shown to be dependent on the fatty acid composition of the vegetable oils. The sunflower (SO)/PPD blend had the most notable results in this regard, with a pour point temperature of 36 °C compared to

---

TABLE 5.3
Techniques to Overcome Oxidation Issues

Techniques	Procedure
Chemical alteration of molecular structures	Acetylation, epoxidation, esterification, etc.
Genetically modifying the fatty acid components	Plant breeding, hydrogenation, blending, etc.
Fractionalization	Extracting crystallized fats
Adding anti-oxidants	- Butylated hydroxyanisole – BHA
	- Butylated hydroxytoluene – BHT
	- propyl gallate – PG
	- Hydroquinone
	- Tetra-butylhydroquinone – TBHQ
	- Polyhydroxybenzenes
	- Aromatic amines, tocopherols, carotenoids
	- N-Phenyl-1-Naphthyamine – PANA

18 °C for plain oil. The high oleic sunflower oil (HOSO)/PPD blend, on the other hand, performed the poorest (21 °C) compared to HOSO (18 °C). However, it was shown that in addition to enhancing viscosity, ethyl cellulose (EC), a viscosity modifier, also causes a delay in HOSO crystallization, producing a similar effect to the PPD studied.

The above approaches are found to be viable for conversion of raw vegetable oils into suitable lubricants for manufacturing.

5.4.3 MERITS AND DEMERITS OF VEGETABLE OILS AS GREEN LUBRICANTS

Ruggiero et al. (2017) has studied that, compared to mineral oils, one or more ester groups found in vegetable oils and their modified forms exhibit better lubricating properties. Furthermore, they have very low volatility due to high molecular weight, good miscibility with other fluids, and a higher viscosity index, properties which make them the best mineral oil alternative; in addition, these green lubricants have high solubilizing quality for polar contaminants and additive molecules. The presence of polar groups in these oils makes it amphiphilic, which makes it stick to metal surfaces with good lubricity.

Vegetable oils have advantageous lubricating characteristics due to their triglyceride composition and structure. They have stronger lubricant films that reduce friction and wear significantly when interacting with metallic surfaces, which is attributed to its long, polar fatty acid chains (Heshmat, 1995).

Kolawole and Odusote (2013) suggested that vegetable oils as green lubricants are found to perform better than other oils in the following aspects:

- The lubricity qualities of vegetable oils are excellent. Vegetable oil exerts a good lubricating property because of its chemical structure and basic composition. Mineral oil molecules are charge-less, but vegetable oil molecules have a little polar charge so it sticks to a metallic surface more firmly than mineral oil and provides better lubricity. Vegetable oil molecules organize themselves in a dense and uniform manner to form a thick, robust, and long-lasting lubricant film.
- Mineral oil molecules vary in size, but vegetable oil molecules are alike in size. As a result, the characteristics of mineral oil, such as viscosity and boiling temperature, are more uncertain when compared to vegetable oils (Ruggiero et al., 2016).
- Vegetable oils have a greater flash point, which minimizes the risk of fire and the emission of smoke (Paswan et al., 2016). Can be used as cutting fluid in high temperature conditions because of its higher flash point.
- Good Viscosity Index: Vegetable oil is able to maintain a higher viscosity even when the temperature of the machining process is rising. Across the operating temperature range vegetable oils have a high viscosity index, which gives consistent lubricity.
- Good solvents – Vegetable oils and their esters are extremely effective solvents because of their polar character which removes dirt and wear debris.

- Good lubricity properties – They have a stronger affinity towards metal surfaces than non-polar mineral oils because of their polar composition. Compared to mineral oils, these oils exhibit good boundary lubrication.
- Load-carrying capacity – Vegetable oils have a better load-bearing capability compared to mineral-based oils because of the stronger adsorption of long chain fatty acids on metallic surfaces.
- Low emission of CO and hydrocarbons – Lubricants made of vegetable oil are more efficient at lowering the amounts of carbon monoxide and hydrocarbon emissions (Tanilgan et al., 2007).
- Biodegradability – Vegetable oils and modified vegetable oil esters are completely biodegradable into carbon dioxide and water molecules (Syahrullail et al., 2014, Zimmerman et al., 2003). Table 5.4 lists the standards followed and Table 5.5 shows the typical biodegradability values of some vegetable oils.
- Toxicity – Green lubricants are generally non-toxic to aquatic and terrestrial environments because vegetable oils and vegetable oil-based esters are made of fatty acid blocks. Table 5.6 shows the comparative data of toxicity for vegetable oils with mineral oils (Srikant and Rao, 2017, Abdalla and Patel, 2006).

TABLE 5.4
Biodegradable Standard Tests Usually Followed for Lubricants

Type of Test	Test Procedure
CEC-L-33-A-93 primary biodegradability test	Evaluates the degree to which C-H stretching vibrations in the test medium's per halogenated solvent extract disappear from the IR spectra over a 21-day period.
OECD 301B	Compares the actual amount of carbon dioxide evolved from a flask holding the test substance to a theoretical amount based on the sample's total organic carbon content over the course of 28 days.

TABLE 5.5
Biodegradability Test Values for Different Base Oils

Type of Lubricant	CEC-L-33-A-93 Primary Biodegradability Test (21 Days)	OECD 301B Ready Biodegradability Test (28 Days)
Mineral oil	10–45	10–40
PAO	20–80	5–60
Polyol ester	0–100	0–85
Vegetable oil	90–100	75–95
Alkyl benzene	5–20	0–20

TABLE 5.6
Comparison of Toxicity of Different Lubricant Oils

Base Oil (Lubricant)	Source of Base Oil	Level of Toxicity
Petroleum-based mineral oils	Fossil fuel	High
Glycols – PAG, PEGs	Synthesized hydrocarbons	Medium
Esters	Biological resources	Low
Vegetable and plant oils	Plant and seed extracts	Low

Some of the demerits of vegetable oils are found to be

- Poor low-temperature fluidity and higher pour point – Vegetable oils crystallize at higher temperatures than mineral oils because of long chain and saturated fatty acids. The monounsaturated and polyunsaturated fatty acids have double bonds, which form a link structure that causes them to solidify at a lower temperature than saturated fatty acids (Siniawski et al., 2007). However, the trans-esterification and other chemical modification routes are being adopted to manufacture lubricants with highly improved pour points.
- Hydrolytic stability – Compared to mineral oil, vegetable and trans-esterified oils are more likely to hydrolyze in the presence of water and produce corrosive acidic breakdown products. The hydrolysis-produced carboxylic acids have the potential to reduce life, erode metal bearings, and damage seals (Fox and Stachowiak, 2007).
- Poor resistance to foaming.
- Tendency to clog the filter – High viscosity vegetable oils and oxidized vegetable oils clog filters due to the presence of polymerized products.
- Narrow range of viscosities – Limited viscosity range, which is one of vegetable oils main drawback as potential lubricants.

To conclude, the triacyl glycerols of vegetable oils with three long chain fatty acids makes them an excellent lubricant candidate as a green alternative. Further, they have a high viscosity index, fluid miscibility, and low volatility due to high molecular weight. Vegetable oil is amphiphilic due to the polar groups it contains, which enables it to attach to metal surfaces with its strong lubricity (Belluco and De Chiffre, (2000). Therefore, employing vegetable oils as a base stock for green lubricants offers several benefits, but their use has been constrained because of their poor low-temperature and low oxidation qualities. These drawbacks can be overcome through different routes such as chemical modifications, genetic modification and blending.

5.5   GREEN TRIBOLOGY PRINCIPLES

De Chiffre and Nosonovsky (2010) have reported the following 12 principles of green tribology which are found to be essential:

(1) Heat minimization and dissipation of energy

   The amount of energy saved should be examined with the additives' or technology's environmental impact, taking into account their production, longevity, and end-life, in order to fully assess the sustainability of the techniques for dealing with energy and heat dissipation that have been provided. The evaluation is made more effective considering its lubricated component's wear because it affects maintenance and replacement costs. With regard to all of the information on frictional energy consumption and the associated issue of heat dissipation, accurate tribological study can help save a significant amount of lost energy and lessen its negative effects on the environment.

(2) Wear minimization

   As reducing wear is one of tribology's primary goals, it is the foundation of green tribology. Material loss while sliding between two surfaces is the wear phenomenon. With the exception of friction-required operations, for example, welding, it is best to avoid this harmful interaction. The importance of lubricants in dealing with wear has grown, and their massive use puts them at the forefront of discussions about tribo system lifetime preservation and environmental effect.

(3) Self-lubrication and complete elimination or reduction of lubrication

   The reduction, precise calibration, or elimination of lubricants opens the door for new research initiatives with the intriguing goal of determining the conditions that are optimal for the specific applications.

(4) Ntural lubrication

   Oil, primarily derived from biomass and vegetable sources, or water, are the two main components of natural lubrication. The presence of fatty acids, whose type, chain length, and polarity are the main influencing elements, is the common and essential component.

(5) Biodegradable lubrication

   Biodegradable lubricants (BLs) are made from plant- or animal-based, recyclable basic materials. Natural substances are those that are derived from plants or animals, while synthetic substances are those that have undergone various catalytic processes or have been chemically altered (Andres, 2004). Since they share similar compositions, traits, and production processes with natural lubricants, BLs are in fact closely related to them. They are primarily made up of bio-olefins, glycols, or saturated esters (produced from fatty acids).

(6) Sustainability chemistry in tribology

   Green tribology can be enhanced and value-added by the concept and principle of sustainability approaches like renewable energy production.

(7) Approach by biomimetic method

This is the concept of introducing biological approach into engineering. Micro textures and ablation are some of the processes that can be utilized in green tribology.

(8) Texturing of surface

Recreating the self-slippery liquid-infused porous surface (SLIPS) of lotus leaf surface over the material surfaces that can retain lubricity by creating a non-wetting layer due to capillarity which can lead to green tribology.

(9) Coatings and its environmental effects

Coatings are prone to quick deterioration and use more energy, resulting in the need to deploy new materials, incurring costs for maintenance and replacement. Green methods such as diamond-like carbon and self-healing coatings are some of the solutions that can overcome the above issues and improve the life of tribo systems.

(10) Degradability design

Green tribology also depends on a closed-loop supply chain concept. Waste oil management, collection and recycling, disposal of waste into environment are a few of the factors that are considered in degradability design.

(11) Real-time monitoring

Early fault detection, proper selection and monitoring of lubricants on site are some strategies in implementing green tribology. Contaminants from wear can be identified and removed for more efficient lubrication.

(12) Sustainable energy applications

When compared to conventional fossil fuel-based technologies, renewable energy usage can be more efficient, dissipate less energy, and be more cost effective with the help of a sustainable tribology study paving the way to green tribology.

Cetin et al. (2011) suggest that the interdisciplinary aspect of green tribology may define a potential useful contribution to waste and pollution reduction, material conservation, and energy efficiency. Each principle proposed by Nosonovsky and Bhushan (2010), has presented a useful plan to advance sustainability from various angles, including those pertaining to the environment, the economy, health, and security.

5.6 PERFORMANCE OF GREEN LUBRICANTS IN DIFFERENT METALWORKING APPLICATIONS

Green lubricants, especially cutting fluids, can be utilized for lubrication and cooling purposes in metalworking processes. The main advantages of these types of fluids is that along with better lubrication in machining processes, the release of harmful gases during machining operations is reduced and it is also biodegradable and non-toxic. Some of the commonly used green lubricants are groundnut, coconut, cottonseed, sunflower and soybean oils. The performance of green lubricants for different metalworking applications is discussed next.

5.6.1 PERFORMANCE IN TURNING OPERATIONS

Comparative studies were undertaken by Ojolo et al. (2008), to determine the effects of various cutting fluids on tool wear and surface roughness when AISI 304 austenitic stainless steel was turned with a carbide tool. Contrary to conventional mineral oil-based formulations, it was discovered that coconut oil was relatively successful in lowering tool wear and increasing the surface quality. This study also demonstrated an increase in surface quality and a decrease in cutting tool temperature when using coconut oil as a lubricant while turning AISI 1040 steel as shown in Figure 5.2. The findings further demonstrated that these oils are appropriate for metalworking operations, however, their impacts on cutting force varied depending on the work piece material due to their chemical compatibility and surface interaction.

Ozcelik et al. (2011) and Aggarwal et al. (2008) studied the performance of green lubricants on alloys like AISI 304L, Haynes 25 and Ti_6Al_4V. The primary goal of this study was to evaluate the performance of different lubricants in relation to surface roughness and machining capability at high cutting speeds. Sustainable neat oil-based machining fluids were made using sunflower, rapeseed, palm and high erucic rapeseed oils. Physical properties like kinematic viscosity, oxidation stability, and lubricity were investigated and low frictional forces were reported when compared to conventional cutting fluids. Also, some works using water-miscible green lubricants based on vegetable oil have demonstrated an improved tool life and decreased cutting temperature for the turning of SS 304. Vegetable oil derivatives from canola oil and coconut oil were used in this study

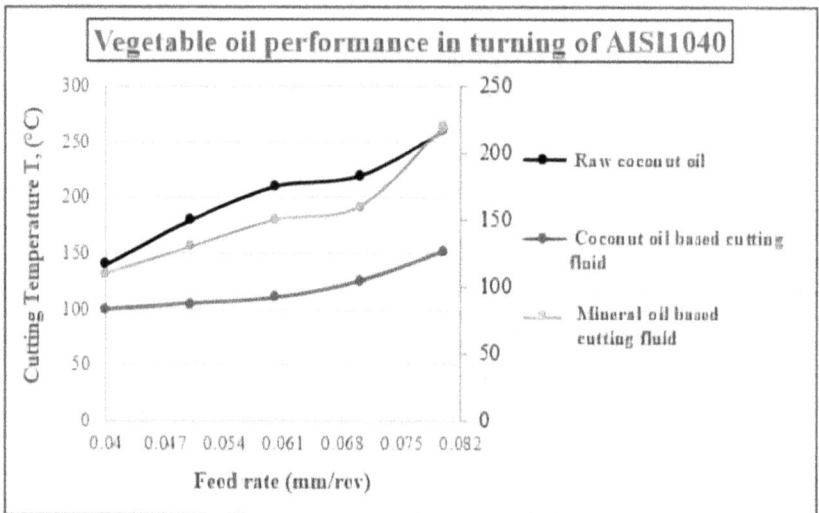

FIGURE 5.2   Vegetable oil performance in turning of AISI 1040.

and it was found to be superior when compared to neat cutting oils and conventional soluble oils (Shashidhara and Jayaram, 2010).

In another study by Zhang et al. (2015), some neat vegetable oils, namely groundnut, coconut, palm kernel, and shear butter oil showed better performance on cutting forces and other parameters during turning of materials like mild steel, aluminium, and copper using a tungsten carbide cutting tool. The experiments were conducted at spindle speeds 245, 330, 450, and 550 rpm with a constant feed rate of 0.15 mm/rev and 2 mm depth. The results as shown in Figure 5.3 suggest that the coefficient of friction has drastically reduced by using green lubricants. At high speeds, coconut oil and groundnut oil worked well in minimizing coefficient of friction which is attributed to their high oleic content (Wu et al., 2000).

In turning operations, green lubricants derived from vegetable oils have been found to perform better when compared to commercially neat oils in a few cases. Also, it has been identified that the performance of vegetable oils in the turning process is dependent on the work piece material. Surprisingly, emulsifiable vegetable oils were found to be good in cutting temperature reduction while turning stainless steel alloys.

5.6.2 Performance in Milling Operations

Palm oil-based green lubricants were evaluated in end milling of stainless steel AISI 420 using a carbide tool by Sharif et al. (2019). The machining conditions were at speeds of 100, 130, and 160 m/min; feed 0.05 mm/tooth; axial depth 12 mm and flute diameter 10 mm – 4 flutes. It was observed that the tool wear

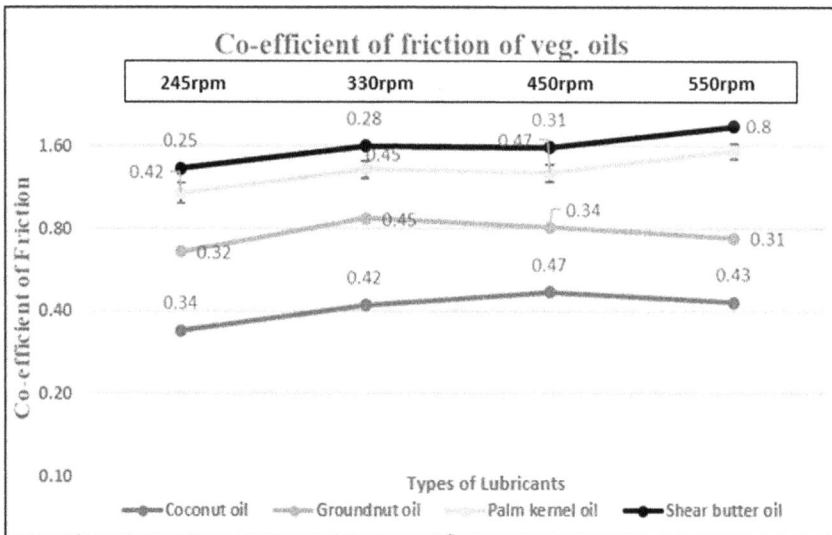

FIGURE 5.3 Coefficient of friction using vegetable oil-based lubricants.

gradually progressed for palm oil and fatty alcohols, whereas on the contrary it progressed rapidly for mineral oil-based commercial cutting fluids which is attributed to the better lubricity of vegetable oils. Furthermore, the surface roughness was measured at around 0.91 μm, 0.69 μm, and 0.73 μm for mineral oil-based cutting fluids, palm oil, and fatty alcohol, respectively. Burton et al. (2014) studied vegetable-based machining fluids on the end milling of ferrous alloys; the results suggested that the surface irregularity lowered, along with the cutting forces.

Shankar Mohanraj et al. (2020) and Zeman et al. (1995) discussed their findings on end milling process using edible oils (sunflower, coconut, palm and soyabean oils) and non-edible oils (cottonseed and canola oils), respectively. Their findings suggest that these oils provide better cooling and lubrication during high speeds and feed rates. They also provide minimum vibrations by lowering the cutting forces when compared to conventional coolants.

Shreeshail et al. (2021) studied vegetable oil's (groundnut and soybean) performance on end milling of aluminum alloy and mild steel. The influence on parameters, such as surface roughness, surface hardness and cutting forces involved were studied in comparison with conventional mineral oils and found to be better with these vegetable oils. The effect of vegetable oils in providing minimum surface roughness was found to be significant as shown in Figure 5.4.

In milling operations, vegetable oils were found to outperform commercially available mineral oil-based cutting fluids in reducing surface roughness and minimizing chatter. The better lubricity of vegetable oil-based green lubricants is credited to this performance.

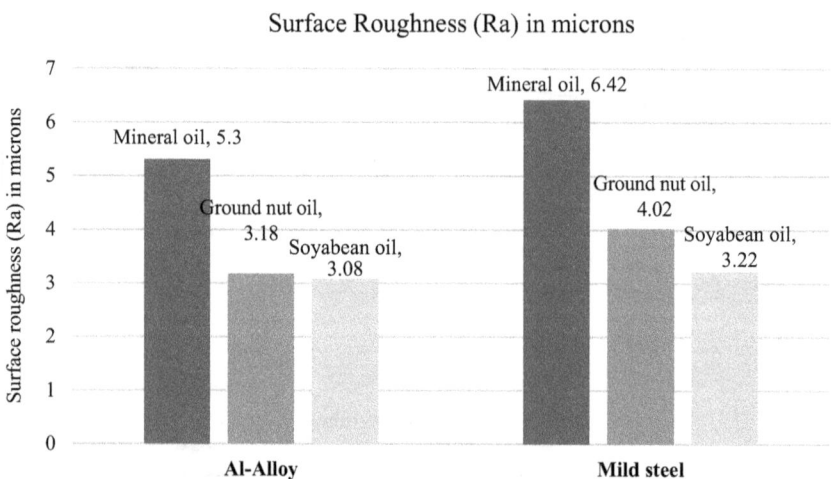

FIGURE 5.4 Surface roughness study with vegetable oil lubricants for milling.

5.6.3 PERFORMANCE IN DRILLING OPERATIONS

Lee and Nam (2018) used green lubricants as drilling fluids in micro drilling of Ti6Al4V and the results show that using vegetable oil-based green lubricants reduced thrust force and torque which thereby reduced the burr formation in drilled holes compared to mineral oil-based neat drilling fluids. Puttaswamy and Ramachandra (2018) researched Neem and Mahua oil-based green lubricants as drilling fluids and the results suggested that mahua oil outperformed neem oil in terms of thrust force and cutting temperature reduction, while both vegetable oils performed better compared with mineral oil-based lubricants.

The effectiveness of vegetable-based lubricants in drilling AISI 316L austenitic stainless steel using traditional HSS-Co tools was investigated by Belluco and de Chiffre (2004). They measured tool life, cutting forces, and chip formation using a commercial mineral-based oil as the reference product and a few vegetable oils with added blends at various concentrations as the test fluids. The measurements made at the beginning of the tool's life span and the tool's whole lifespan revealed that the thrust force reduction was significant as shown in Figure 5.5. They concluded that all vegetable-based fluids outperformed the commercial mineral oil used as the reference product as there were relative gains in tool life and the drop in cutting force was found to be good.

The effectiveness of MQL palm oil (MQLPO) and MQL synthetic ester (MQLSE) during the drilling of titanium alloys with a carbide drill coated with AlTiN insert was investigated by Ismail et al. (2019). Under the three different conditions, viz. external air blow, MQL, and flood coolant conditions (water soluble type), holes were drilled. Using cutting speeds of 60, 80, and 100 m/min, together with two feed rates of 0.1 and 0.2 mm/rev, the thrust force, torque, and

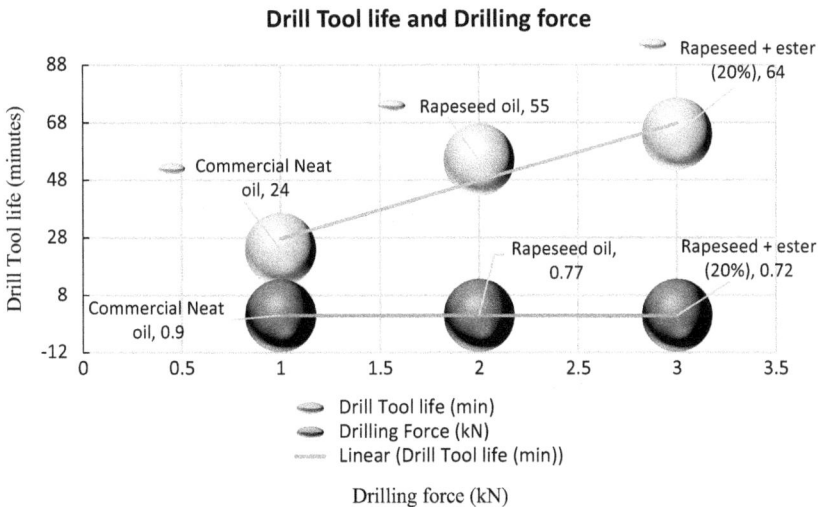

FIGURE 5.5   Drill tool life and drilling force.

work piece temperature were measured and compared in the first part of the experiment. In the second stage, a hole with a depth of 10 mm was drilled with a continuous cutting speed of 60 m/min and a feed rate of 0.1 mm/rev while maintaining the following tool life conditions: Vb (avg) = 0.2 mm for average flank wear; Vb (max) = 0.3 mm for maximum flank wear; Vb (avg) = 0.3 mm for corner wear. The thrust force was lowered by 30% and 6.5% relative to air blow cutting for the MQLPO and MQLSE conditions, respectively, resulting in similar and lowest thrust forces compared to other circumstances evaluated.

In drilling operations, vegetable oil-based green lubricants were found to perform better in terms of burr reduction, cutting force reduction and improvement of tool life. Some advanced techniques such as MQL and emulsifiable water-based drilling fluids made from vegetable oils tended to perform better than commercially available drilling fluids.

5.6.4 Performance in Grinding Operations

Herrmann et al. (2007) evaluated the technical, ecological, and performance aspects of natural ester-based green coolants in grinding operations. The research involved the study of various coolants, involving physical characteristics and grinding performance tests on hardened bearing steel (100Cr$_6$, 62 HRC) work piece material and a CBN grinding wheel. Cutting speed –(Vc) = 60 m/s, wheel diameter = 40 mm, width of cut = 10 mm, work piece diameter = 110 mm, and specific material removal rate = 2 mm^3/mm – were the parameters employed in the grinding process. In the study, the performance evaluation, life cycle assessment, and life cycle pricing were considered. While the grinding force ratios remained constant, the grinding forces achieved as shown in Figure 5.6 suggest that a steady

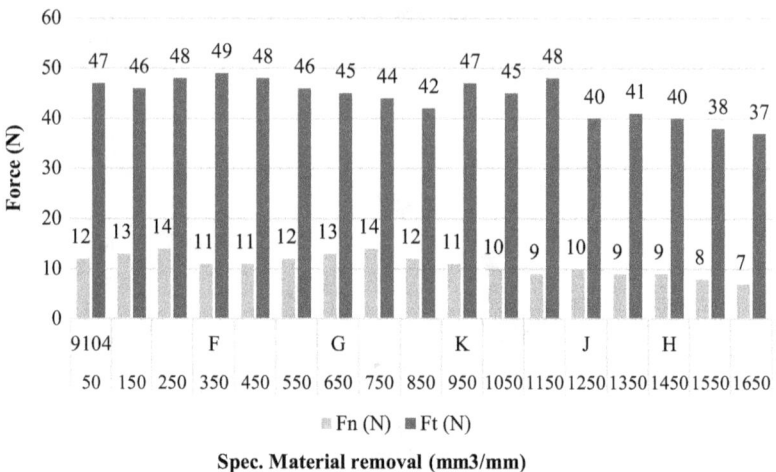

FIGURE 5.6   Grinding forces using methyl esters as grinding fluid: (9104 reference product – vegetable), F (animal fat methyl ester), G (used cooking oil methyl ester), K (methyl ester extracted from animal tallow), J (lard methyl ester), H (oleic fraction methyl ester).

grinding process was achievable with all tested ester oils. It is noted that all the green coolants performed better compared to conventional grinding fluids.

Emulsion-based vegetable oils are gaining popularity due to better cooling ability. Castor oil for grinding operations was investigated by Alves and de Oliveira (2006), who found that the newly formulated vegetable-based (castor oil) cutting fluid was easily biodegradable. When this cutting fluid was employed, its 5% dilution lowered wheel wear, grinding forces, and surface roughness. To compare the performance, two reference fluids, viz. soluble cutting oil and a semi-synthetic fluid, were studied with the novel emulsion-based castor oil cutting fluid. Radial wheel wear and work piece roughness were measured. In addition, the novel fluid was tested for biodegradability and toxicity and found to be easily biodegradable and non-toxic. The experiments were carried out under the following grinding conditions: cutting speed (Vs) = 3 m/s, work piece speed (Vf) = 11.5 mm/s, grinding width (b) = 6.5 mm, grinding wheel penetration (a) = 25 m, peripheral disc dresser velocity (Vr) = 38 m/s, dressing depth of cut = 10 m, and subsequent dressing strokes of 10 m in diameter until uniform profile was obtained. According to the study, the semi-synthetic cutting fluid generated an increased wheel wear of around 8 m in radius, whereas the new green cutting oil reduced wheel wear (high grinding and higher cooling ability, providing better lubricity.

In grinding operations, ester-based green lubricants were found to perform better when compared to conventional neat grinding fluids. Moreover, the tool life was found to be improved while employing emulsifiable castor oil-based green lubricants at 5% vol. concentration.

5.6.5 OTHER GREEN LUBRICANTS

Water-based lubricants have been used since long time and they are still widely used today for things like metalworking fluids and hydraulic fluids. Water-based fluids have the advantage of fire resistance, in addition to the advantages of eco-friendliness and non-toxicity. The need for water-compatible additives to enhance the performance of water-based lubricant compositions is ever increasing as suggested by Erhan et al. (2009).

Alkyl polyglucosides were investigated as additives for water-based fluids. They showed enhanced performance in terms of anti-wear qualities at lower concentration levels of 0.1% (Eziwhuo et al., 2019). Another work by Shao et al. (2004) used the process of conventional free radical polymerization, followed by sulfonation, to create a new fullerene-styrene sulfonic acid copolymer in the base stock of a triethanolamine aqueous solution. The performance of the water-soluble additive was assessed and found to improve the wear resistance, load-bearing capacity, and anti-friction qualities.

Improved initial seizure loads and weld loads were seen in studies using O, O-poly ethoxy glycol diester of dithiophosphoric acid as an addition to water in various concentrations. A similar result was shown using S-(carboxylpropyl)-N-dialkyldithiocarbamic acid as an additive in water (Vignesh and Mohammed Iqbal, 2022).

5.6.5.1   Green Powder Lubricants

In recent times, green powder lubricants derived from natural resources have been developed and used. According to research (Lancaster, 1966) solid lubricants like hexagonal boron nitride (h-BN) and boric acid, which carries outstanding lubricating qualities, are being tried under this category. These additives are easily accessible and safe for the environment and are not considered as contaminants, as per the Environmental Protection Agency (EPA, USA). In order to test the tribological performance of green powder lubricants, several studies were carried out. Boric acid (BA) powder varying in different sizes dispersed in canola oil, was subjected to tribological examination (Sikdar et al., 2022). Pin-on-disk experiments with a copper pin and an aluminium disc were carried out and the results are shown in Figure 5.7.

5.6.5.2   Green Ionic Liquids (GILs)

Tzani et al. (2022) and Katna et al. (2017) suggested that green ionic liquids (GILs), a "new class" of liquid lubricant, provide a better possible alternative to both conventional and natural oil-based lubricants. They generally feature a large, asymmetric organic cation and a suitable organic anion, and they display a variety of special qualities that make them potential lubricants. Ethyl lactates, salicylates, and saccharinates are some examples of GILs (Bhuyan et al., 2006). These are less toxic, biodegradable, and derived from natural resources.

In general, from Figure 5.8 it is observed that GILs have a promising tribological performance. Zhang et al. (2000) found that liquids' higher boundary lubrication ability is a result of the anion's ability to create a monolayer film that adhered to the positively charged metal surface, providing anti-wear properties. In the near

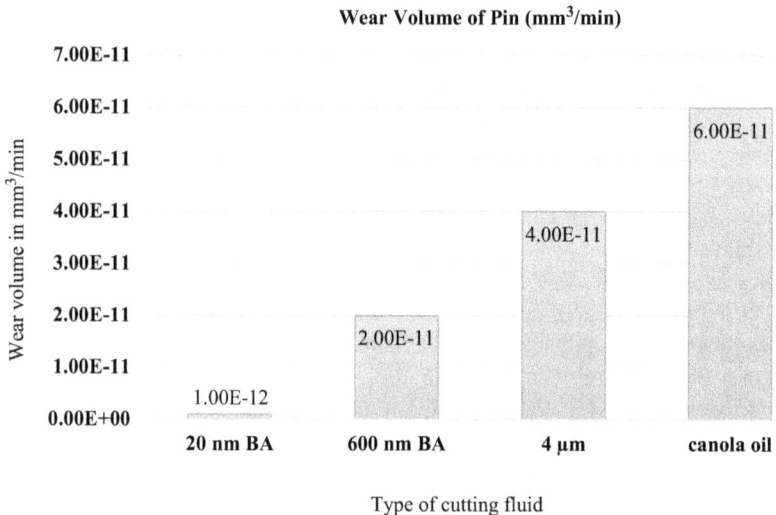

FIGURE 5.7   Tribological examination of canola oil with green additives.

Co-efficient of Friction

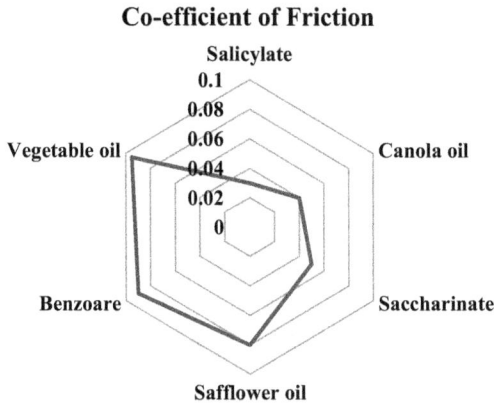

FIGURE 5.8   Coefficient of friction for different lubricants.

future, it is projected that the new class of green ionic liquid lubricants will play a key role in a variety of industries, including magnetic storage devices, mining and mineral processing, maritime equipment, and aerospace (Verma et al., 2022).

5.7 SCOPE OF GREEN LUBRICATION

For any lubricant applications that might pose threats to the environment, green lubricants are essential (Joseph et al., 2007). These situations may be where:

- Oils that are lost to environment during usage (chainsaw oils, 2-stroke engine oils, formwork release agents (forming process), and greases)
- Oils that are unintentionally leaked (hydraulic oils, oils for engines, gearboxes, axles, etc.)

When environmental preservation is a continual issue, such as in aquatic, mountainous, agricultural, and forest areas, or deep quarries, the use of green lubricants is especially advised (Katna et al., 2019). As a substitute lubricant for conventional oils, green lubricants have enormous potential for economic growth and environmental preservation.

5.7.1 MARKET POTENTIAL AND PRESENT SCENARIO

The market for green lubricants was valued at USD2.2 billion in 2019 and is anticipated to increase at a compound annual growth rate (CAGR) of 4.1% from 2020 to 2025, reaching USD2.46 billion (Market Research Report, 2020). The market for green lubricants is anticipated to be driven by strict regulations and increasing end-user acceptance. The tight rules implemented by nations throughout the world have been a major factor in the expansion of the international green lubricants sector. Due to ongoing technical advancements, there has been a rising acceptability among end-use industries to support the usage of green lubricants. The green lubricant market's hydraulic oil segment is anticipated to have the largest increase in the near future, both in terms of value and volume. Hydraulic elevators, sweepers, garage trucks, forklifts, motor graders, and end loaders—which are utilized in a variety of industries—are primarily responsible for this segment's rise. The use of green lubricants from 2020 to 2025 is projected to improve in a variety of segments that include marine, agriculture, and construction industries (Perera et al., 2015). Overall, a small worldwide development is predicted, taking into account the development potential in Asia where per capita consumption in certain regions is still quite low, and a persistent decline in volumes or stagnation in Western developed nations. The rising globalization of technology will encourage high-value goods even in developing and emerging lubricant markets like India, which will result in a more prominent rise in value. These nations will employ equipment and lubricants that are comparable or equivalent to those used in developed countries (Singh et al., 2019).

5.8 CONCLUSIONS

Vegetable oil-based green lubricants have gained significant importance in the current environment because they are derived from renewable resources. Environmentally friendly lubricants are fully biodegradable and substantially less polluting than petroleum-based oils, thus their production and use will have a big influence on society and the environment. The various categories of green lubricants in solid and liquid forms have been found to be useful in a number of manufacturing applications. In addition, the natural presence of lauric acid, palmitic acid, erucic acid, etc. makes green lubricants a good candidate for better lubricity applications. The tribological performance studies on various applications of green lubricants suggest that they are appropriate for the majority of machining conditions providing better performance in terms of reduction in cutting force, tool wear, surface roughness, and cutting temperature and improving surface quality and tool life. The suggested remedies such as chemical alteration, gene modification and fractionalization are but a few methods found to be successful in overcoming the implementation issues of green lubricants. This paper has given a wide scope on standards of testing green lubricants in terms of bio-degradability, toxicity, and tribological performance. The results are found to be comparable to that of conventional lubricants. Further, research works in future can be extended to synthetic esters, water-based vegetable cutting oils and green nano lubricants.

REFERENCES

Abdalla, H. S., & Patel, S. (2006). The performance and oxidation stability of sustainable metalworking fluid derived from vegetable extracts. *Proc Inst Mech Eng, Part B: Eng Manuf*, 220, 2027–2040.

Abdalla, H. S., Baines, W., McIntyre, G., & Slade, C. (2007). Development of novel sustainable neat-oil metalworking fluids for stainless steel and titanium alloy machining. Part 1. Formulation development. *J Adv Manuf Technol*, 34, 21–33.

Aggarwal, A., Singh, H., Kumar, P., & Manmohan, S. (2008). Optimizing power consumption for CNC turned parts using response surface methodology and Taguchi's technique-A comparative analysis. *J Mater Process Technol*, 200(1), 373–384.

Alves, S. M., & Oliveira, J. F. G. (2006). Development of new cutting fluid for grinding process adjusting mechanical performance and environmental impact, *J Mater Process Technol*, 197, 185–189.

Andres, F. C. (2004). Experimental comparison of vegetable and petroleum base oils in Metalworking fluids using the tapping torque test. In *Proceedings of the Japan-USA Symposium on Flexible Automation*, Denver, CO.

Belluco, W., & De Chiffre, L. (2004). Performance evaluation of vegetable-based oils in drilling austenitic stainless steel. *J Mater Process Technol*, 148(2), 171–176.

Belluco, W. Walter, & De Chiffre, Leonardo (2000). Tribology in Metal Cutting. *IPL*, 55.

Bennett, E. O. (1983). Water based cutting fluids and human health. *Tribol Int*, 16(3), 45–60.

Bhatia, V. K., Chaudhry, A. B., Sivasankaran, G., Bisht, R. S., & Kashyap, M. M. (1990). Modification of jojoba oil for lubricant formulations. *J Am Oil Chem Soc*, 67, 1–7.

Bhuyan, S., Sundararajan, L., Yao, E., Hammond, G., & Wang, T. (2006). Boundary lubrication properties of lipid-based compounds evaluated using microtribological methods. *Tribol Lett*, Paper No: IJTC2006-12228, pp. 151–157; 7 pages.

Burton, G., Goo, C., Zhang, Y., & Jun, M. B. (2014). Use of vegetable oil in water emulsion achieved through ultrasonic atomization as cutting fluids in micro-milling. *J Manuf Process*, 16, 405–413.

Cetin, H., Ozcelik, B., Kuram, B., & Demirbas, E. (2011). Evaluation of vegetable based cutting fluids with extreme pressure and cutting parameters in turning of AISI 304L by Taguchi method. *J Clean Prod*, 19, 2049–2056.

Debnath, S., Reddy, M. M., & Yi, Q. S. (2014). Environmental friendly cutting fluids and cooling techniques in machining: A review, *J Clean Prod*, 83, 33–47.

Dian, N., Hamid, R. A., Kanagaratnam, S., Isa, W. R. A., Hassim, N. A. M., Ismail, N. H., & Sahri, M. M. (2017). Palm oil and palm kernel oil: Versatile ingredients for food applications. *J Oil Palm Res*, 29, 487–511.

Eziwhuo, S. J., Ossia, C. V., & Alibi, S. I. (2019). Evaluation of apricot kernel, avocado and African pear seed oils as vegetable based cutting fluids in turning AISI 1020 steel. *IOSRJ Eng*, 9, 10–19.

Fox, N. J., & Stachowiak, G. W. (2007). Vegetable oil-based lubricants – A review of oxidation. *Tribol Int*, 40(7), 1035–1046..

Gajrani, K. K., & Sankar, M. R. (2017). Past and current status of eco-friendly vegetable oil-basedmetal cutting fluids. *Mater Today Proceed*, 4, 3786–3795.

Goyan, R. L., Melley, R. E., Wissner, P. A., & Ong, W. C. (1998). Biodegradable lubricants. *Lubr Eng*, 54(7), 10–17.

Grushcow, J. (2005). High oleic plant oils with hydroxy fatty acids for emission reduction. In *Proc. World Tribology Congress III*, Washington, D.C., 485–486.

Herrmann, Christoph, Hesselbach, Jürgen, Bock, Ralf, & Dettmer, Tina. (2007). Coolants made of native ester – Technical, ecological and cost assessment from a life cycle perspective. Advances in Life Cycle Engineering for Sustainable Manufacturing Businesses, *Proceedings of the 14th CIRP Conference on Life Cycle Engineering*, Waseda University, Tokyo, Japan, Springer London, pp. 299–303.

Heshmat, H. (1991). High-temperature solid-lubricated bearing development-dry powder lubricated traction testing. *J Propuls Power*, 7(5), 814–820.

Heshmat, H. (1995). Quasi-hydrodynamic mechanism of powder lubrication: Part III: On theory and rheology of triboparticulates. *Tribol Trans*, 38(2), 269–276.

Heshmat, H., & Dill, J. F. (1992). Traction characteristics of high-temperature, powder-lubricated ceramics (Si_3N_4/SiC). *Tribol Trans*, 35(2), 360–366.

Hwang, Y. K., Lee, C. M. (2010). Surface roughness and cutting force prediction in MQL and wet turning process of AISI 1045 using design of experiments. *J Mech Sci Technol*, 24, 1669–1677.

Ismail, I., Oseh, J. O., Norddin, M. N. A. M., Ismail, A. R., Gbadamosi, A. O., & Agi, A., et al. (2019). Investigating almond seed oil as bio-diesel based drilling mud. *J Pet Sci Eng*, 181, 106201.

Itoigawa, F., Childs, T. H. C., Nakamura, T., & Belluco, W. (2006). Effects and mechanisms inminimal quantity lubrication machining of an aluminium alloy. *Wear*, 2006(260), 339–344.

Jagadeesh, K. Mannekote, Kailas, Satish V. (2012). The effect of oxidation on the tribological performance of few vegetable oils. *J Mater Res Technol*, 91–95, 2238–7854.

Jagadeesh, K., Mannekote Pradeep, L., Menezes Satish V. Kailas, & Sathwik Chatra, K. R., (2013). Tribology of green lubricants. 495–521. *Tribology for Scientists and Engineers*, Chapter: 10, Editors Pradeep L Menezes, Sudeep Ingole, Michael Nosonovsky, New York: Springer.

Joseph, P. V., Deepak, S., & Sharma, D. K. (2007). Study of some non-edible vegetable oils of Indian origin for lubricant application. *J Synth Lubr*, 24, 181–197.

Jost, H. P. (1992). Tribology the first 25 years and byond-achievements, shortcomings and future tasks. *Ind Lubr Tribol*, 44(2), 22–27.

Julieb, Z., Andres, F., Kimf, H., & Steven, J. (2003). Design of hard water stable emulsifier systems for petroleum and bio-based semi-synthetic metalworking fluids. *Environ Sci Technol*, 37, 5278–5288.

Katna, R., Singh, K., Agrawal, N., & Jain, S. (2017). Green manufacturing performance of a biodegradable cutting fluid. *Mater Manuf Process*, 32, 1522–1527.

Katna, R., Suhaib, M., & Agrawal, N. (2019). Nonedible vegetable oil-based cutting fluids for achining processes – A review. *Mater Manuf Process*, 35, 1–32.

Khan, M. M. A., & Dhar, N. R. (2006). Performance evaluation of minimum quantity lubrication by vegetable oil in terms of cutting force, cutting zone temperature, tool wear, job dimension and surface finish in turning AISI-1060 steel. *J Zhejiang UnivSci A*, 7(11), 1790–1799.

Koji, Kato, & Kosuke, Ito. (2005). Modern tribology in life cyele assessment. *Tribol Int Eng Ser*, 48(3), 495–506.

Kolawole, K. S., & Odusote, J. K. (2013). Performance evaluation of vegetable oil-based cutting fluids in mild steel machining. *Chem Mater Res*, 3(9), 35–45.

Kupongsak, S., & Lucharit, P. (2013). Process development for lipase extraction and the effect of extracted lipase on triglyceride base system. *Res J Pharm Biol Chem Sci*, 4, 1247–1254.

Lancaster, J. K. (1966). Anisotropy in the mechanical properties of lamellar solids and its effect on wear and transfer. *Wear*, 9, 169–188.

Lea, C. W. (2002). European development of lubricants from renewable sources. *Ind Lubr Tribol*, 54(6), 268–274.

Lee, S.W., & Nam, J. (2018). Machinability of titanium alloy (Ti-6Al-4V) in environmentally-friendly micro-drilling process with nanofluid minimum quantity lubrication using nanodiamond particles. *Int J Precis Eng Manuf-Green Tech* 5, 29–35. https://doi.org/10.1007/s40684-018-0003-z

Lee, T. S., & Choong, H. B. (2011). An investigation on green machining: cutting process characteristics of organic metalworking fluid, *Adv Mat Res*, 230–232, 809–813.

Liu, Z., Sharma, B. K., Erhan, S. Z., Biswas, A., Wang, R., & Schuman, T.P. (2015). Oxidation and low temperature stability of polymerized soybean oil-based lubricants. *Thermochim Acta*, 601, 9–16.

Loh, Soh Kheang, & May, Choo. (2012). Influence of a lubricant auxiliary from palm oil methyl esters on the performance of palm olein-based fluid. *J Oil Palm Res*, 24, 1388–1396.

Market Research Report. Metalworking fluids market size, share & trends analysis report by product (synthetic, bio-based), by application (near cutting, water cutting), by end use, by industrial end use, and segment forecasts, 2020–2027. 2020 (accessed 5 May 2020).

Matthew, T. S., Nader, S., Bigyan, A., & Lambert, A. D. (2007). Influence of fatty acid composition on the tribological performance of two vegetable-based lubricants. *J Synth Lubr*, 24, 101–110.

Neff, W. E., Mounts, T. L., Rinsch, W. M., Konishi, H., & El-Agaimy, M. A. (1994). Oxidative stability of purified canola oil triacylglycerols with altered fatty acid compositions as affected by triacylglycerol composition and structure. *J Am Oil Chem Soc*, 71(10), 1101–1109.

Michael, Nosonovsky, & Bharat, Bhushan (2010, 28 October). Green tribology: Principles, research areas and challenges. *Phil Trans R Soc A*, 368(1929), 3684677.

Ojolo, S. J., Amuda, M. O. H., Ogunmola, O. Y., & Ononiwu, C. U. (2008). Experimental determination of the effect of some straight biological oils on cutting force during cylindrical turning, *RevistaMateria*, 13(4), 650–663.

Okullo, J. B. L., Omujal, F., Agea, J. G., Vuzi, P. C., Namutebi, A., & Okello, J. B. A., et al. (2010). Physico chemical characteristics of shear butter oil. *African J Food Agric Nutr Dev*, 10, 2071–2084.

Ozcelik, B., Kuram, E., Cetin, M. H., & Demirbas, E. (2011). Experimental investigations of vegetable based cutting fluids with extreme pressure during of AISI 304L. *Tribol Int*, 44, 1864–1871.

Paswan, B. K., Jain, R., Sharma, S. K., Mahto, V., & Sharma, V. P. (2016). Development of Jatropha oil-in-water emulsion drilling mud system, *J Petrol Sci Eng*, 144, 10–18.

Perera, G. I. P., Herath, M., Perera Sanka, G. H. M., Muditha, P., & Medagoda, M. (2015). Investigation on white coconut oil to use as a metalworking fluid during turning. *Proc Inst Mech Eng B*, 229, 38–44.

Petlyuk, A. M., & Adams, R. J. (2004). Oxidation stability and tribological behavior of vegetable oil hydraulic fluids. *Tribol Trans*, 47(2), 182–187.

Pettersson, A. (2007). High-performance base fluids for environmentally adapted lubricants. *Tribol Int*, 40, 638–645.

Puttaswamy, Jeevan, Ramachandra, Jayaram. (2018). Experimental investigation on the performance of vegetable oil-based cutting fluids in drilling AISI 304L using taguchi technique. *Tribology Online*, 13(2), 60–66.

Rani, S, Joy, M. L., & Nair, K. P. (2015). Evaluation of physiochemical and tribological properties of rice bran oil – Biodegradable and potential base stoke for industrial lubricants. *Ind Crop Prod*, 65, 328–333.

Ruggiero, A., D'Amato, R., Merola, M., Valašek, P., & Müller, M. (2017). Tribological characterization of vegetal lubricants: Comparative experimental investigation on *Jatropha curcas L.* oil, rapeseed methyl ester oil, hydrotreated rapeseed oil, *Tribol Int*, 109, 529–540.

Ruggiero, A., D'Amato, R., Merola, M., Valášek, P., & Müller, M. (2016). On the tribological performance of vegetal lubricants: Experimental investigation on *Jatropha Curcas L.* oil. *Procedia Eng*, 49, 431–437.

Sharif, S., Sani, A. S. A., Rahim, E. A., & Sasahara, H. (2019). Machining performance of vegetable oil with phosphonium-and ammonium-based ionic liquids via MQL technique. *J Clean Prod*, 209, 947–964.

Sasaki, Shinya. (2010). Environmentally friendly tribology (Eco-tribology). *J Mech Sci Technol*, 24, 67–71.

Thangarasu, S. K., Shankar, S., Mohanraj, T., & Devendran, K. (2020). Tool wear prediction in hard turning of EN8 steel using cutting force and surface roughness with artificial neural network. *Proc Inst Mech Eng, Part C*, 234(1), 329–342.

Shao, X., Liu, W., & Xue, Q. (2004). The tribological behavior of micrometer and nanometer TiO$_2$ particle-filled poly (phthalazine ether sulfone ketone) composites. *J Appl Polym Sci*, 92(2), 906–914.

Erhan, S. Z., Sharma, B. K., Adhvaryu, A. T. A. N. U., & (2009). Friction and wear behavior of thioetherhydroxy vegetable oil, *Tribol Int*, 42, 2, 353–358.

Sharma, B. K., Adhvaryu, A., & Liu, Z. et al. (2006). Chemical modification of vegetable oils for lubricant applications. *J Amer Oil Chem Soc*, 83, 129–136.

Shashidhara, Y. M., & Jayaram, S. R. (2010). Vegetable oils as a potential cutting fluid—an evolution, *Tribol Int*, 43, 5–6, 1073–1081.

Shirani, A., Joy, T., Lager, I., Yilmaz, J. L., Wang, H. L., Jeppson, S., Cahoon, E. B., Chapman, K., Stymne, S., & Berman, D. (2020). Lubrication characteristics of wax esters from oils produced by a genetically-enhanced oilseed crop. *Tribol Int*, 146, 106234.

Shreeshail, M. L., Amol, C., Desai, I. G., Siddhalingeshwar, Krishnaraja G., & Kodancha, (2021). A study on influence of vegetable oils in milling operation and it's role as lubricant, *Mater Today: Proc*, 46, 2699–2713.

Sikdar S., Rahman M. H., & Menezes P. L. (2022). Synergistic study of solid lubricant nano-additives incorporated in canola oil for enhancing energy efficiency and sustainability. *Sustainability*, 14(1), 290.

Singh, Y., Sharma, A., & Singla, A. (2019). Non-edible vegetable oil–based feedstocks capable of bio-lubricant production for automotive sector applications—A review. *Environ Sci Pollut Res*, 26, 14867–14882.

Siniawski, M. T., Saniei, N., Adhikari, B., & Doezema, L. A. (2007). Influence of fatty acid composition on the tribological performance of two vegetable-based lubricants, *J Synth Lubr*, 24(2), 101–110.

Soni, Sunny, & Agarwal, Madhu. (2014). Lubricants from renewable energy sources – A review. *Green Chem Lett Rev*, 7, 359–382.

Sreejith, P., & Ngoi, B. (2000). Dry machining: Machining of the future. *J Mater Process Technol*, 101, 287–291.

Srikant, R. R., & Rao, P. N. (2017). *Use of Vegetable-Based Cutting Fluids for Sustainable Machining*. Springer, Cham, 31–46.

Srikant, R. R., Rao, D. N., & Rao, P. N. (2009). Influence of emulsifier content in cutting fluids on cutting forces, cutting temperatures, tool wear, and surface roughness. *Proc Inst Mech Eng Part J: Eng Tribol*, 223, 203–209.

Syahrullail, S., Kamitani, S., & Shakirin, A. (2014). Tribological evaluation of mineral oil andvegetable oil as a lubricant. *J Teknol*, 66, 37–44.

Tanilgan, K., Ozcan, M. M., & Ünver, A. (2007). Physical and chemical characteristics of five Turkish olive varieties and their oils. *Grasas Aceites*, 58,142–147.

Tzani, A., Karadendrou, M. A., Kalafateli, S., Kakokefalou, V., & Detsi, A.(2022). Current trends in green solvents: Biocompatible ionic liquids. *Crystals*, 12, 1776.

Verma, Dakeshwar, Dewangan, Yeestdev, Singh, Ajaya, Mishra, Raghvendra, Susan, Md, Salim, Rajae, Taleb, A., El-Hajjaji, Fadoua & Berdimurodov, Elyor. (2022). Ionic liquids as green and smart lubricant application: An overview. *Ionics*, 28, 1–10.

Vignesh, S., Mohammed Iqbal, U. (2022). Preparation and characterization of bio-based nano cutting fluids for tribological applications, *J Dispers Sci Technol*. 44:9, 1725–1737, DOI: 10.1080/01932691.2022.2038191

Winter, W., Öhlschläger, G., Dettmer, T., Ibbotson, S., Kara, S., & Herrmann, C. (2012). Using jatropha oil-based metalworking fluids in machining processes: A functional and ecological life cycle evaluation. In *Leveraging Technology for a Sustainable World*, Editors Dornfeld, D., Linke, B., Berlin, Heidelberg: Springer Berlin Heidelberg, 311–316.

Zhang, X., Wu, X., Yang, S., Chen, H., & Wang, D. (2000). Study of epoxidised rapeseed oil used as a potential biodegradable lubricant. *J Am Oil Chem Soc*, 77, 561–563.

Wu, H., Zhao, J., Xia, W., Cheng, X., He, A., Yun, J. H., Wang, L., Huang, H., Jiao, S., Huang, L., Zhang, S., and Jiang, Z. (2017). A Study of The Tribological Behaviour of TiO2 Nano-Additive Water-Based Lubricants, *Tribology International*, 109, 398–408.

Zeman, A., Sprengel, A., Niedermeier, D., & Späth, M. (1995). Biodegradable lubricants-studies on thermo-oxidation of metal-working and hydraulic fluids by differential scanning calorimetry (DSC), *Thermochimica Acta C*, 268, 9–15.

Zhang, Y., Li, C., Jia, D., Zhang, D., & Zhang, X. (2015). Experimental evaluation of MoS2 nanoparticles in jet MQL grinding with different types of vegetable oil as base oil. *J Clean Prod*, 87, 930–940.

Zimmerman, J. B., Clarens, A. F., Hayes, K. F., & Skerlos, S. J. (2003). Design of hard water stable emulsifier systems for petroleum and bio-based semi-synthetic metal-working fluids, *Environ Sci Tech*, 23, 5278–5288.

6 Nano-Additives-Assisted MQL Machining

Jaharah A. Ghani
Faculty of Engineering and Built Environment, Universiti
Kebangsaan Malaysia, Selangor, Malaysia

Nurul Nadia Nor Hamran
Faculty of Manufacturing and Mechatronics, Universiti
Malaysia Pahang, Pahang, Malaysia

6.1 INTRODUCTION

Due to the governmental pollution-preventing strategies that have set rigorous regulations to deal with environmental issues, researchers are making every effort to accomplish eco-friendly manufacturing (Sayuti et al., 2014). For example, to move towards sustainable development, lawful strategies and regulations in the USA are being tackled at county as well as organisational level, to support policies at the international level, such as the European Union (EU), the Organization for Economic Cooperation and Development (OECD) and the United Nations (UN) (Pervaiz & Abdel Samad, 2018; Stachurski et al., 2017). Malaysia itself is also firmly committed towards environmental sustainability such that as early as 1974, the Environmental Quality Act 1974, Act 127, was introduced by the government to outline the prohibition and control of pollution by owners of any industrial plant or process from releasing environmentally hazardous substances, pollutants, or wastes in its operations (Department of Environment, Ministry of Energy, Science, Technology Environment & Climate Change, 2019). It was reported that a reduction of 33% in GHG emissions had been achieved by the end of 2013. In continuation of the effort, the Malaysian government aims to pursue green growth for sustainability and resilience by reinforcing policy via a regulative and institutional framework, as well as launching the blueprint on *Sustainable Consumption and Production* (SCP) to stimulate industrial sectors to shift towards implementing green technology for their products and services (Economic Planning Unit, Prime Minister's Department, 2015).

Machining is one of the most critical processes in manufacturing. Thus, previous researchers have conducted rigorous research on sustainability machining

DOI: 10.1201/9781003363576-6

(Al-Zubaidi et al., 2013; Goindi et al., 2015, 2017; Najiha et al., 2015). Cutting fluids are commonly applied during machining due to their lubricating and cooling functions, which help increase the production rate of metalwork machining. Cutting fluid can be used in many different ways; however, flooding is the most common application method. This conventional method is achieved by directing the nozzle towards the work-tool area and utilizing the overflow of cutting fluid. Applying the flooding technique commonly results in a longer tool life but also has several adverse economic, environmental and health effects. Flooding is an undesirable cooling strategy due to the high consumption and cost of cutting fluid, cutting fluid waste disposal and complex treatment due to its toxicity and non-biodegradability, as well as its hazardous impact on the operator's health (Eltaggaz et al., 2018; Gupta et al., 2018; Stachurski et al., 2017; Yuan et al., 2018). Hence, utilizing a minimum quantity of cutting fluid and adequately selecting environmental and non-hazardous types of cutting fluid is in wide demand to replace the disadvantages of conventional flood cooling (Nor Hamran et al. 2020).

6.2 MACHINING WITH MINIMUM QUANTITY LUBRICANT (MQL)

Alternative machining conditions which are being introduced, such as dry machining, near dry machining – also known as machining with minimum quantity lubrication (MQL) – and cryogenic machining, have been explored by researchers, as seen in Figure 6.1 (Goindi et al., 2015, 2017; Hegab et al., 2019; Marques et al., 2016).

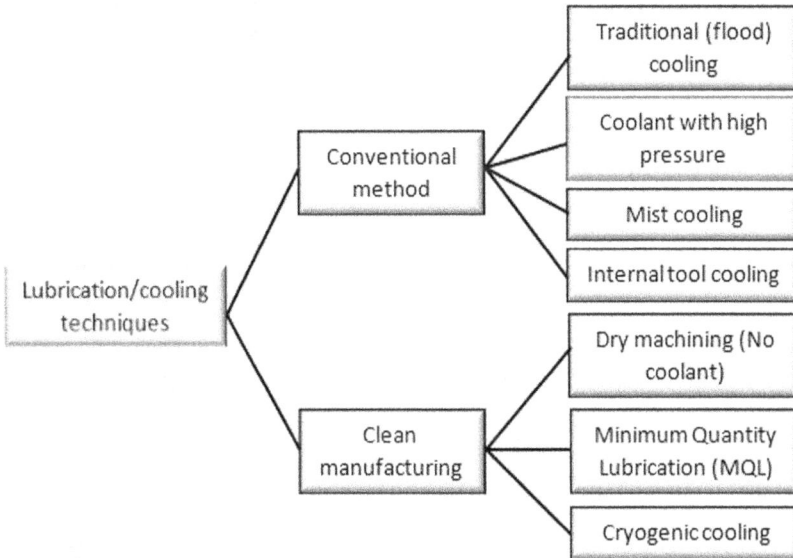

FIGURE 6.1 Method of cutting fluid application in machining operations (Debnath et al., 2014).

Additionally, machining using newly developed cutting fluids such as ionic liquid and solid lubricant (microfluid and nanofluid) is also emerging as viable and producing more excellent machining results than the conventional flooding technique (Gunda et al., 2016; Gupta & Laubscher, 2016).

MQL machining is a sustainable technique that sprays cutting fluid in minimal quantities directly into the cutting zone (Thakur et al., 2019). A review reveals that, compared with dry conditions, most studies demonstrate remarkable improvement in machining performance brought about by MQL machining during various processes. Figure 6.2 shows two categories of MQL delivery systems, namely, external application and internal application. Figure 6.3 shows the detail of the external and internal MQL systems.

Advancements in MQL can be divided into three categories: a combination of MQL with additives, MQL with cooled air/gas, and restructuring the MQL system (Nor Hamran et al., 2020). Furthermore, some research has examined a hybrid cryogenic MQL system which consists of a solid lubricant (SL) and cold compressed air (CCA) MQL system with solid lubricants. The types of solid lubricants used in the advancement of MQL machining are microfluid and nanofluid, while the types of cryogenic liquids used are CO_2 and LN2. Figure 6.4 shows the 266 research articles according to the categories of advancements in MQL systems.

Figure 6.4 shows that 49.6% of the publications are focused on MQL machining with nanofluid. Nanofluid is a heat transfer fluid introduced by S.U.S. Choi in 1995 that can provide efficient heat transfer compared to conventional cutting fluid (Che Sidik et al., 2017). Due to this beneficial thermal property, many researchers believe that by combining the MQL system with nanofluid, the new technique can enhance its cooling and lubrication performance.

Nano MQL machining also provides better lubrication as the particles can prevent cutting oil from being immediately released from the cutting zone because the nano-additives can retain the oil particles. A study conducted by Rahman et al. (2018) shows how the addition of nanoparticles (Al_2O_3, MSo_2 and rutile TiO_2) in vegetable oils (canola and extra virgin olive oil) during turning of Ti6Al4V EL1

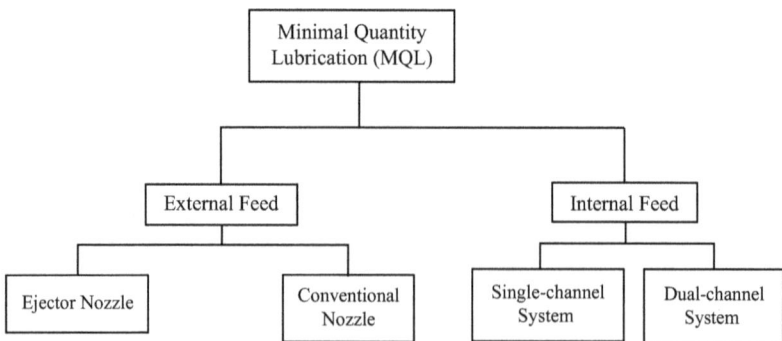

FIGURE 6.2   Categories of MQL delivery system (Katiyar et al., 2022).

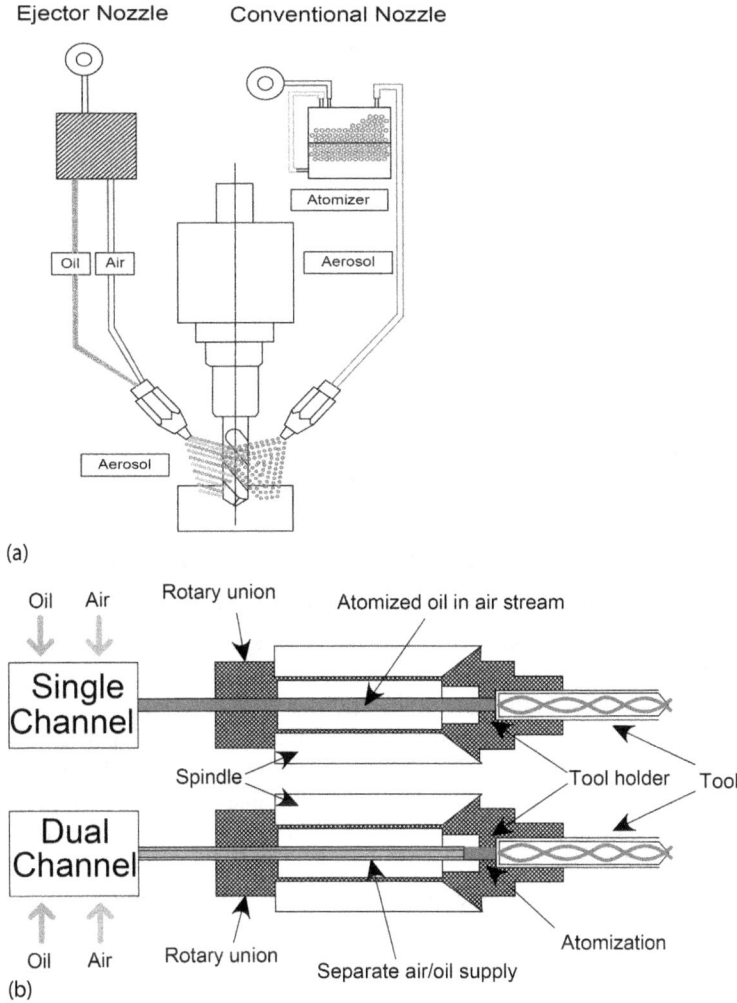

FIGURE 6.3 Schematic representation of (a) an external MQL system using ejector nozzle and conventional nozzle; and (b) an internal MQL system using single channel and dual channel (Katiyar et al., 2022).

provides a good surface finish – specifically, MQL machining with Al_2O_3 nanoparticles in canola oil provides 57.95% lesser Ra value than that of MQL machining – longer tool life and lower cutting temperature. Dambatta et al. (2019) also experimented with evaluating the tribological performance of grinding Si_3N_4 ceramic with different concentrations of SiO_2-based nanofluids and their study shows that the higher the concentrations, the better the surface quality and the more significant the reduction in grinding force.

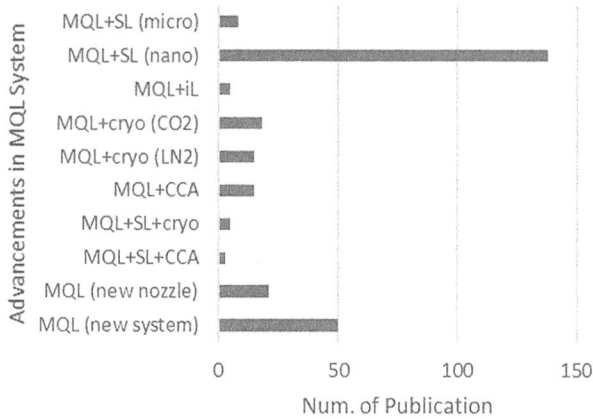

FIGURE 6.4 Distribution of research articles according to the categories of advancements from 2014 until 2019 (Nor Hamran et al. 2020).

6.3 TYPES OF NANOFLUID USED IN MQL SYSTEMS

Nanofluids are solid-liquid composite materials consisting of solid nanoparticles or nanofibers with sizes typically of 1 to 100 nm suspended in liquid with the criterion of agglomerate-free stable suspension for long durations without causing any chemical changes in the base fluid (Sridhara & Satapathy, 2011). Nanofluids are prepared by either a single-step direct evaporation or a two-step method. In the first method, the direct evaporation and condensation of the nanoparticulate materials in the base liquid are obtained to produce stable nanofluids. In the second method, the nanoparticles are obtained by different methods and then dispersed into the base liquid.

The many types of nanofluid currently used in MQL systems are as follows (Binayak Sen et al., 2019):

1. Al_2O_3 nanofluid
2. MoS_2 nanofluid
3. Diamond nanofluid
4. SiO_2 nanofluid
5. Carbon nanotube (CNT) nanofluid
6. Nanoplatelet

6.3.1 ALUMINA (AL_2O_3) NANOFLUID

Alumina (Al_2O_3) is the most common nanoparticle used by researchers in their experiments. Sridhara and Satapathy (2011) revealed that the thermal conductivity of nanofluids increases with an increasing volume fraction of nanoparticles; with a decreasing particle size, the shape of particles can also influence thermal conductivity of nanofluids, temperature, Brownian motion of the particle, interfacial layer, and with the additives. Table 6.1 shows some previous studies (Sridhara & Satapathy, 2011).

The application of Al_2O_3 nanofluid in machining has been studied by researchers (Vasu et al., 2011), and in a machining study by Vasu et al. (2011) which evaluated the performance of vegetable oil, Al_2O_3 nanofluid and dry condition were found to reduce the surface roughness, cutting force and cutting temperature compared with other lubricating mediums.

Mandal et al. (2012) examined the effect of volume fraction (%) of nanoparticles Al_2O_3 in MQL-assisted grinding of AISI 52100 with different volume percentages of 1, 3 and 5% with 40 and 80 nm dia. The result revealed that a higher nanoparticle volume fraction generated better machining performances. The effects of Al_2O_3-based nanofluids on the grinding of AISI 52100 also revealed a reduction in the grinding forces and coefficient of friction. However, a bigger diameter of nanoparticles resulted in a coarser surface finish (Mao et al., 2013; 2014). The influences of Al_2O_3 and CuO nanoparticles in grinding of Ti-6Al-4 V was conducted by Setti et al. (2012), and revealed that Al_2O_3 nanoparticles caused a lower coefficient of friction compared to dry, wet and CuO nanofluid environments.

TABLE 6.1
Selective Summary of the Thermal Conductivity Enhancement in Al_2O_3-Based Nanofluids (Sridhara & Satapathy, 2011)

Author (Year)	Base Fluid	Concentration	Particle Size (nm)	Enhancement Ratio	Method/ Parameters
Lee et al. 2009	Water	1.0 to 4.30	38.4	1.03 to 1.10	Two-step method
	Ethylene	1.0 to 5.0		1.03 to 1.18	
Wang et al. (1999)	Water	3.0 to 5.50	28	1.11 to 1.16	Two-step method
	Ethylene	5.0 to 8.0		1.25 to 1.41	
	glycol	2.25 to 7.40		1.05 to 1.30	
	Engine oil	5.00 to 7.10		1.13 to 1.20	
	Pump oil				
Eastman et al. (2001)	Ethylene glycol	1.00 to 5.00	35		Two-step method
Xie et al. (2002)	Water	1.80 to 5.00	60.4	1.07 to 1.21	Two-step method
	Ethylene	1.80 to 5.00	15	1.06 to 1.17	Solid
	glycolr	1.80 to 5.00	26	1.06 to 1.18	crystalline
	Ethylene	1.80 to 5.00	60.4	1.10 to 1.30	Phase effect
	glycolr	1.80 to 5.00	302	1.08 to 1.25	Morphology
	Ethylene	5.00	60.4	1.39	effect
	glycolr				pH effect
	Ethylene				
	glycolr				
	Pump oil				

A study conducted on hard milling under MQL conditions with nanofluid Al_2O_3 of 0.5% volume shows superior results (Minh et al. 2017). The improvement in tool life by almost 177–230% (depending on the type of nanofluid) and a reduction in surface roughness and cutting forces by almost 35–60% have been observed under MQL with Al_2O_3 nanofluids, due to their better tribological behaviour as well as cooling and lubricating effects.

6.3.2   MoS₂ NANOFLUID

Molybdenum disulfide (MoS_2) has a layered structure and has been known and used as a lubricant and catalyst for several years because of its unique properties such as anisotropy, chemical inertness and photo corrosion resistance, and exhibits good lubrication properties (Zeng et al., 2013). MoS_2 water-based and oil-based nanofluids have higher thermal conductivity and lower surface tension than base fluids (Su et al., 2015). Furthermore, the thermal conductivity and surface tension will increase and decrease, respectively, with mass fraction.

The nanofluids of nano and microparticles of boric and MoS_2 in sesame and coconut oil as lubricant in machining reduced the cutting temperature, cutting forces and surface roughness (Padmini et al., 2014). Furthermore, the performance of MoS_2/sesame, MoS_2/coconut, and MoS_2/canola in the machining of AISI 1040, in which 0.5% MoS_2 with coconut oil performed better, resulted in reducing cutting forces, temperature, tool wear and surface roughness by 37%, 21%, 44% and 39%, respectively, compared with dry machining (Padmini et al., 2016). Mixed MoS_2 nanoparticles in either palm, soybean or rapeseed oil were used in an MQL-assisted grinding operation that resulted in MoS_2 nanoparticles in soybeans performing the best in the grinding process (Zhang et al., 2015). Table 6.2 summarizes the literature on MoS_2-based MQL-assisted machining processes (Binayak Sen et al., 2019).

6.3.3   DIAMOND NANOFLUID

Due to its exceptional thermal, mechanical and electrical properties, the diamond has attracted attention for use as a nanoparticle in fluids. A number of researchers have investigated its application in machining (Nam & Lee, 2018; Nam et al., 2011; Shen et al., 2008).

A comprehensive study by Nam and Lee (2018) revealed that MQL machining Ti alloy with diamond nanoparticles reduced cutting force and chip adhesion. An investigation into the use of diamond particles (dia. 30 nm) in vegetable and paraffin oils in drilling Al 6061 showing a reduction in the cutting force (Nam et al., 2011). However, the paraffin oil-based nanofluid showed much better lubrication properties than vegetable oil-based nanofluids. Later, Nam et al. (2013) experimented with a larger diameter of diamond nanoparticles and found that it needed to be more efficient in drilling operations.

Shen et al. (2008) investigated the wheel wear and tribological characteristics in wet, dry and minimum quantity lubrication (MQL) grinding of cast iron with

TABLE 6.2
Summarized MoS$_2$-Based MQL-Assisted Machining
Processes (Binayak Sen et al., 2019)

				Critical Parameters	
Authors	W/P Materials	Nanoparticles Used	Operation	Machining Parameters	MQL Parameters
Padmini et al. (2014)	AISI 1040	H$_3$BO$_3$ and MoS$_2$	Turning	Cutting speed = 60 m/min feed = 0.14 mm/rev depth of cut = 0.5 mm	Not provided
Zhang et al. (2015)	45 Steel	MoS$_2$	Grinding	Wheel speed = 60 m/sec feed = 3000 mm/min depth of cut = 10 μm	Flow rate = 50 ml/h nozzle distance = 12 mm nozzle angle = 15^0 gas pressure = 6 bar
Padmini et al. (2016)	AISI 1040	MoS$_2$	Turning	Cutting speed = 40, 60, 100 m/secr XXfeed = 0.14, 0.17, 0.2 mm/rev XXdepth of cut = 0.5 mm	Flow rate = 10 ml/min

water-based Al$_2$O$_3$ and diamond nanofluids (200 nm carbon-coated diamonds and 100 nm non-coated mono-crystalline diamonds) in MQL grinding. The results revealed that during the nanofluid MQL grinding, a dense and complex slurry layer was formed on the wheel surface which could benefit the grinding performance by reducing grinding forces, improving surface roughness and preventing workpiece burning. Table 6.3 shows the diamond nanofluid used in machining (Binayak Sen et al., 2019).

6.3.4 SiO$_2$ NANOFLUID

Previous researchers have applied SiO$_2$ nanofluid in machining (Sarhan et al., 2012; Sayuti et al., 2014; Ooi et al., 2015; Aminullah et al., 2020). A milling operation using SiO$_2$ nanofluid as a lubricant was carried out by Sarhan et al. (2012), and the results revealed that SiO$_2$ nanofluid could significantly reduce cutting forces as well as specific energy consumption when compared to mineral oil. The effects of SiO$_2$ nanofluids in machining AISI4140 steel revealed that surface roughness could be minimized at 30 ° of MQL nozzle angle, 0.5% wt SiO$_2$ concentration and

TABLE 6.3

The Diamond Nanofluid Used in Machining (Binayak Sen et al., 2019)

| | | | | Critical Parameters | |
Authors	W/P Materials	Nanoparticle Used	Operation	Machining Parameters	MQL Parameters
Nam et al. (2011)	Aluminum 6061	Nano diamond	Micro-drilling	Cutting speed = 60,000 rpm feed = 50 mm/min XXdepth of cut = 0.4 mm	Not provided
Nam et al. (2013)	Aluminum 6061	Nano diamond	Micro-drilling	Cutting speed = 30,000, 60,000 rpm XXfeed = 10, 15 mm/min XXdepth of cut = 0.1, 0.5 mm	Not provided

low air pressure; minimum tool wear could be achieved at 60 ° MQL nozzle angle, 0.5% SiO_2 concentration and 2 bar air pressure (Sayuti et al., 2014).

SiO_2 nano cutting fluid was used for milling aluminum alloy Al6061-T6 in studies by Ooi et al. (2015) and Aminullah et al. (2020). Aminullah et al. (2020) studied three-volume concentrations (0.5, 1.0 and 1.5%) of SiO_2 using one-step and dilution methods. The results show that a 1.5% volume concentration produces the lowest surface roughness of 0.679 μm and the lowest cutting temperature of 29.3 °C.

6.3.5 CNT NANOFLUID

Carbon nano tube nanofluid has a better heat transfer rate due to its high thermal conductivity property, but its hydrophobic nature makes preparing stable nanofluid a significant challenge (Singh et al., 2020). Singh et al. (2020) varied the CNT concentration from 0.1 to 0.3 wt%, the surfactant/CNT ratio from 1 to 3, and the ultrasonication time from 0 to 180 min. The results show ultrasonication time and surfactant/CNT ratio are significant on the dispersion of CNT in a surfactant water solution with prepared CNT nanofluid was found to be stable for a longer duration. Carbon nano tube (CNT) nano machining particles were used in grinding wheels for cutting grains, indicating the best results (You and Gao, 2009).

An experimental study to determine the effect of CNT nanoparticles on turning operations conducted with varied concentrations (%) of CNT in MQL base fluid found that at 2% CNT inclusion, minimum tool wear and cutting temperature can be achieved (Rao et al., 2011). CNT nanofluids can effectively minimize the micro-cracks and surface roughness compared with dry machining conditions in grinding AISI D3 steel (Prabhu & Vinayagam, 2012).

6.3.6 NANOPLATELET

Graphene nanoplatelets (GnP) have excellent lubrication properties and high thermal conductivity, encouraging researchers to associate their application in MQL. The mix of GnP in the cutting fluid and an examination of its influence during a micro-milling operation revealed that it can significantly reduce cutting temperature and tool wear (Marcon et al., 2010). The effectiveness of GnP in surface-grinding operations indicates GnP-based cutting fluids performed flawlessly in terms of wheel life and surface finish in the grinding operation of D-2 steel (Alberts et al., 2009) and Inconel 718 (Ravuri et al., 2016). The MQL-assisted grinding process with 0.3% GnP in the base fluid effectively improves the surface quality of Inconel 718 (Pavan et al., 2017). In addition, GnP-based cutting fluids effectively minimized chipping and maximized tool life (Kwon & Drzal, 2015).

6.4 LUBRICATION MECHANISMS USING NANOFLUIDS IN MACHINING

Due to vacant bonds, nanoparticles are unstable, unsaturated and can attach easily to other atoms (Binayak Sen et al., 2019). In addition, nanoparticles can combine with the polar atoms of vegetable oil and produce high surface energy. The surface effect of the nanoparticle is shown in Figure 6.5.

Furthermore, the following are essential characteristics of nanoparticles when dispersed in the cutting zone (Lee et al., 2009):

- Incorporating different nanoparticles in the MQL fluid considerably diminished friction and wear rate, thus improving the lubricants' tribological aspect
- Nanoparticles have the propensity to build up a surface-protecting film
- The circular-shaped particles revolved between tool–workpiece interfaces and converted the sliding into rolling-sliding friction
- The nanoparticles demonstrated a mending effect in machining
- The high contact pressure reduced the compressive stress concentrations. Thus, nanoparticles uniformly endure a compressive force. The lubrication mechanism of nanofluids is displayed in Figure 6.6(a–d)

6.5 SUMMARY

Machining is one of the most critical processes in manufacturing, and due to the global demand to tackle sustainability issues, MQL is one of the best options. Implementing MQL machining, where the cutting fluid is sprayed in minimal quantity directly into the cutting zone, demonstrates a remarkable improvement in machining performance brought by MQL machining compared with dry conditions during various machining processes. Nevertheless, flood machining performed better when compared with MQL machining at certain cutting conditions. Therefore, the application of nanofluid-assisted MQL is another option

Polar Molecules

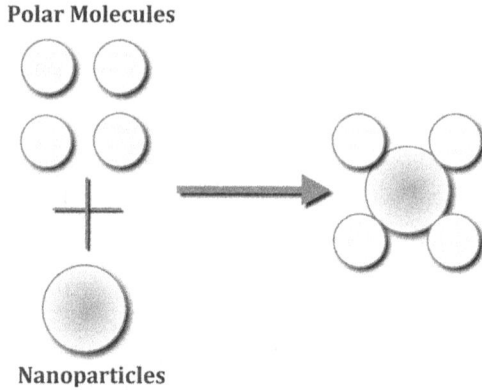

Nanoparticles

FIGURE 6.5   Schematic of nanoparticle surface effect (Binayak Sen et al., 2019).

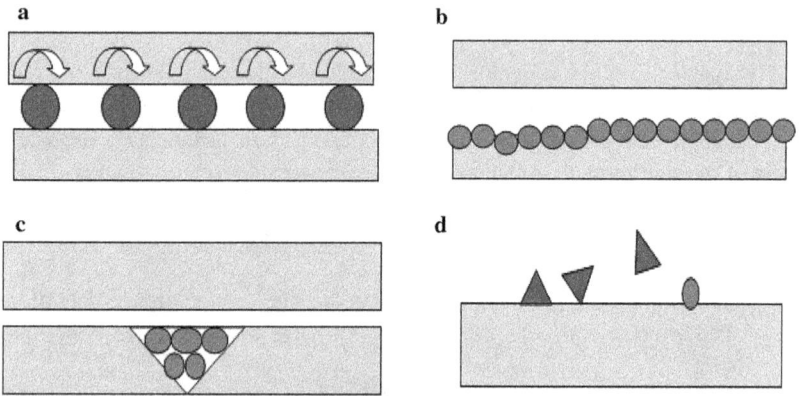

FIGURE 6.6   Lubrication mechanisms of nanofluid: (a) rolling effect; (b) surface protection film; (c) mending effect; (d) polishing effect (Kumar et al., 2023).

suggested since nanoparticles used in nanofluids have common properties such as high thermal conductivity, and the shape of the nanoparticles will also improve the tribological aspects when applied during the machining process. Hence, the machinability of the material being machined is improved. The common nanofluids used in MQL systems are Al_2O_3, MoS_2, diamond, SiO_2, CNT, and nanoplatelet.

REFERENCES

Aminullah, A.R.M., Hazim, A.A., Fikri, A.M.S., & Redhwan, A.A.M., 2020. An experimental evaluation of SiO_2 nano cutting fluids in CNC milling of aluminium alloy AL6061-T6. *International Journal of Synergy in Engineering and Technology*, 1(2), 124–130.
Alberts, M., Kalaitzidou, K., & Melkote, S., 2009. An investigation of graphite nanoplatelets as lubricant in grinding. *International Journal of Machine Tools and Manufacture*, 49(12–13), 966–970.

Al-Zubaidi, S., Ghani, J.A., & Che Haron, C.H., 2013. Prediction of tool life in end milling of Ti-6Al-4V alloy using artificial neural network and multiple regression models. *Sains Malaysiana*, 42, 1735–1741.

Shen, Bin, Shih, Albert J., & Tung, Simon C., 2008. Application of nanofluids in minimum quantity lubrication grinding. *Tribology Transactions*, 51, 730–737.

Sen, Binayak, Mia, Mozammel, Krolczyk, G.M., Mandal, Uttam Kumar, & Mondal, Sankar Prasad, 2019. Eco-friendly cutting fluids in minimum quantity lubrication assisted machining: A review on the perception of sustainable manufacturing. *International Journal of Precision Engineering and Manufacturing-Green Technology*, 8, 249–280. https://doi.org/10.1007/s40684-019-00158-6

Che Sidik, N.A., Samion, S., Ghaderian, J., & Muhammad Yazid, M.N.A.W., 2017. Recent progress on the application of nanofluids in minimum quantity lubrication machining: A review. *International Journal of Heat and Mass Transfer*, 108, 79–89. https://doi.org/10.1016/j.ijheatmasstransfer.2016.11.105

Dambatta, Y.S., Sayuti, M., Sarhan, A.A.D., Hamdi, M., Manladan, S.M., & Reddy, M., 2019. Tribological performance of SiO_2-based nanofluids in minimum quantity lubrication grinding of Si_3N_4 ceramic. *Journal of Manufacturing Processes*, 41, 135–147. https://doi.org/10.1016/j.jmapro.2019.03.024

Debnath, S., Reddy, M.M., & Yi, Q.S., 2014. Environmental friendly cutting fluids and cooling techniques in machining: A review. *Journal of Cleaner Production*, 83, 33–47. https://doi.org/10.1016/j.jclepro.2014.07.071

Department of Environment Ministry of Energy Science Technology Environment & Climate Change, 2019. Official Portal of Department of Environment Other Associated Legislation (Non Downloadable Files) Jabatan Alam Sekitar Rakan Alam Sekitar [WWW Document]. https://www.doe.gov.my/portalv1/en/tentang-jas/perundangan/akta-kaedah-peraturan-arahan-2 (accessed 12.20.19).

Minh, Duc Tran, Tran, Long, & Bao, Ngoc Tran, 2017. Performance of Al_2O_3 nanofluids in minimum quantity lubrication in hard milling of $60Si_2Mn$ steel using cemented carbide tools. *Advances in Mechanical Engineering*, 9(7), 1–9.

Eastman, J.A., Choi, S.U.S., Li, S., Yu, W., & Thomson, L.J., 2001. Anomalously increased effective thermal conductivities of ethylene glycol-based nanofluids containing copper nanoparticles. *Applied Physics Letters*, 78, 718–720. https://doi.org/10.1063/1.1341218

Economic Planning Unit Prime Minister's Deparment, 2015. Eleventh Malaysia Plan 2016-2020 Anchoring Growth On People [WWW Document]. Percetakan Nas. Malaysia Berhad. https://policy.asiapacificenergy.org/sites/default/files/11thMalaysiaplan.pdf

Eltaggaz, A., Hegab, H., Deiab, I., & Kishawy, H.A., 2018. Hybrid nano-fluid-minimum quantity lubrication strategy for machining austempered ductile iron (ADI). *International Journal on Interactive Design and Manufacturing*. https://doi.org/10.1007/s12008-018-0491-7

Goindi, G.S., Chavan, S.N., Mandal, D., Sarkar, P., & Jayal, A.D., 2015. Investigation of ionic liquids as novel metalworking fluids during minimum quantity lubrication machining of a plain carbon steel. *Procedia CIRP*, 26, 341–345. https://doi.org/10.1016/j.procir.2014.09.002

Goindi, G.S., Sarkar, P., Jayal, A.D., Chavan, S.N., & Mandal, D., 2017. Investigation of ionic liquids as additives to canola oil in minimum quantity lubrication milling of plain medium carbon steel. *International Journal of Advanced Manufacturing Technology*. https://doi.org/10.1007/s00170-017-0970-1

Gunda, R.K., Reddy, N.S.K., & Kishawy, H.A., 2016. A novel technique to achieve sustainable machining system. *Procedia CIRP*, 40, 30–34. https://doi.org/10.1016/j.procir.2016.01.045

Gupta, K., & Laubscher, R.F., 2016. Sustainable machining of titanium alloys: A critical review. *Proceedings of the Institution of Mechanical Engineers, Part B: Journal of Engineering Manufacture.* https://doi.org/10.1177/0954405416634278

Gupta, M.K., Mia, M., Singh, G.R., Pimenov, D.Y., Sarikaya, M., & Sharma, V.S., 2018. Hybrid cooling-lubrication strategies to improve surface topography and tool wear in sustainable turning of Al 7075-T6 alloy. *International Journal of Advanced Manufacturing Technology.* https://doi.org/10.1007/s00170-018-2870-4

Hegab, H., Kishawy, H.A., Umer, U., & Mohany, A., 2019. A model for machining with nano-additives based minimum quantity lubrication. *International Journal of Advanced Manufacturing Technology.* https://doi.org/10.1007/s00170-019-03294-0

Katiyar, J.K., Sahu, R.K., & Gupta, T.C.S.M. 2022. *Sustainable Lubrication* (1st ed.). CRC Press. https://doi.org/10.1201/9781003201199

Singh, Kriti, Sharma, S.K., & Gupta, Shipra Mital, 2020. Preparation of long duration stable CNT nanofluid using SDS. *Integrated Ferroelectrics*, 204(1), 11–22. https://doi.org/10.1080/10584587.2019.1674981

Kumar, A., Sharma, A.K., & Katiyar, J.K., 2023. State-of-the-art in sustainable machining of different materials using nano minimum quality lubrication (NMQL). *Lubricants*, 11, 64. https://doi.org/10.3390/lubricants11020064

Kwon, P., & Drzal, L.T., 2015. Nanoparticle graphite-based minimum quantity lubrication method and composition. Google Patents.

Lee, K., Hwang, Y., Cheong, S., Choi, Y., Kwon, L., & Lee, J., 2009. Understanding the role of nanoparticles in nano-oil lubrication. *Tribology Letters*, 35(2), 127–131.

Marcon, A., Melkote, S., Kalaitzidou, K., & DeBra, D., 2010. An experimental evaluation of graphite nanoplatelet based lubricant in micro-milling. *CIRP Annals*, 59(1), 141–144.

Marques, A., Guimarães, C., da Silva, R.B., da Penha Cindra Fonseca, M., Sales, W.F., & Machado, Á.R., 2016. Surface integrity analysis of inconel 718 after turning with different solid lubricants dispersed in neat oil delivered by MQL. *Procedia Manufacturing*, 5, 609–620. https://doi.org/10.1016/j.promfg.2016.08.050

Nor Hamran, N.N., Ghani, J.A., Ramli, R., & Che Haron, C.H., 2020. A review on recent development of minimum quantity lubrication for sustainable machining. *Journal of Cleaner Production*, 268, 122–165.

Najiha, M.S., Rahman, M.M., & Yusoff, A.R., 2015. Flank wear characterization in aluminum alloy (6061 T6) with nanofluid minimum quantity lubrication environment using an uncoated carbide tool. *Journal of Manufacturing Science and Engineering ASME..* https://doi.org/10.1115/1.4030060

Nam, J.S., Kim, D.H., & Lee, S.W., 2013. A parametric analysis on micro-drilling process with nanofluid minimum quantity lubrication. In *ASME 2013 International Manufacturing Science and Engineering Conference collocated with the 41st North American Manufacturing Research Conference, 2013* (pp. V002T004A016–V002T004A016): American Society of Mechanical Engineers.

Nam, J.S., Lee, P.-H., & Lee, S.W., 2011. Experimental characterization of micro-drilling process using nanofuid minimum quantity lubrication. *International Journal of Machine Tools and Manufacture*, 51(7–8), 649–652.

Nam, J., & Lee, S. W., 2018. Machinability of titanium alloy (Ti-6Al-4V) in environmentally-friendly micro-drilling process with nanofuid minimum quantity lubrication using nanodiamond particles. *International Journal of Precision Engineering and Manufacturing-Green Technology*, 5(1), 29–35. https://doi.org/10.1007/s40684-018-0003-z

Sarhan, A.A., Sayuti, M., & Hamdi, M. 2012. Reduction of power and lubricant oil consumption in milling process using a new SiO_2 nanolubrication system. *The International Journal of Advanced Manufacturing Technology*, 63(5–8), 505–512.

Sayuti, M., Sarhan, A.A.D., & Salem, F., 2014. Novel uses of SiO_2 nano-lubrication system in hard turning process of hardened steel AISI4140 for less tool wear, surface roughness and oil consumption. *Journal of Cleaner Production*, 67, 265–276. https://doi.org/10.1016/j.jclepro.2013.12.052

Stachurski, W., Sawicki, J., Wójcik, R., & Nadolny, K., 2017. Influence of application of hybrid MQL-CCA method of applying coolant during hob cutter sharpening on cutting blade surface condition. *Journal of Cleaner Production* https://doi.org/10.1016/j.jclepro.2017.10.059

Ooi, M.E., Sayuti, M., & Sarhan, A. A. 2015. Fuzzy logicbased approach to investigate the novel uses of nano suspended lubrication in precise machining of aerospace AL tempered grade 6061. *Journal of Cleaner Production*, 89, 286–295.

Padmini, R., Krishna, P.V., & Rao, G.K.M., 2014. Performance assessment of micro and nano solid lubricant suspensions in vegetable oils during machining. *Proceedings of the Institution of Mechanical Engineers, Part B: Journal of Engineering Manufacture*, 229(12), 2196–2204. https://doi.org/10.1177/0954405414548465

Pavan, R.B., Gopal, A.V., Amrita, M., & Bhanu, K.G., 2017. Experimental investigation of graphene nanoplatelets–based minimum quantity lubrication in grinding Inconel 718. *Proceedings of the Institution of Mechanical Engineers, Part B: Journal of Engineering Manufacture*. https://doi.org/10.1177/0954405417728311

Pervaiz, S., & Abdel Samad, W., 2018. Tool wear mechanisms of physical vapor deposition (PVD) TiAlN coated tools under vegetable oil based lubrication. In *Mechanics of Additive and Advanced Manufacturing* (Volume 9, pp. 101–107). https://doi.org/10.1007/978-3-319-62834-9_14

Prabhu, S., & Vinayagam, B.K., 2012. AFM investigation in grinding process with nanofluids using Taguchi analysis. *The International Journal of Advanced Manufacturing Technology*, 60(1–4), 149–160.

Rahman, S.S., Ashraf, M.Z.I., Amin, A.K.M.N., Bashar, M.S., Ashik, M.F.K., & Kamruzzaman, M., 2018. Tuning nanofluids for improved lubrication performance in turning biomedical grade titanium alloy. *Journal of Cleaner Production*. https://doi.org/10.1016/j.jclepro.2018.09.150

Rao, S. N., Satyanarayana, B., & Venkatasubbaiah, K., 2011. Experimental estimation of tool wear and cutting temperatures in MQL using cutting fuids with CNT inclusion. *International Journal of Engineering, Science and Technology*, 3(4), 2928–2932.

Ravuri, B.P., Goriparthi, B.K., Revuru, R.S., & Anne, V.G., 2016. Performance evaluation of grinding wheels impregnated with graphene nanoplatelets. *The International Journal of Advanced Manufacturing Technology*, 85(9–12), 2235–2245.

Thakur, A., Manna, A., & Samir, S., 2019. Multi-response optimization of turning parameters during machining of EN-24 steel with SiC nanofluids based minimum quantity lubrication. *Silicon*. https://doi.org/10.1007/s12633-019-00102-y

Vasu, V., & Pradeep Kumar Reddy, G., 2011. Effect of minimum quantity lubrication with Al_2O_3 nanoparticles on surface roughness, tool wear and temperature dissipation in machining Inconel 600 alloy. *Proceedings of the Institution of Mechanical Engineers, Part N: Journal of Nanoengineering and Nanosystems*, 225(1), 3–16.

Sridhara, Veeranna, & Lakshmi Narayan Satapathy, 2011. Al_2O_3-based nanofluids: A review. *Nanoscale Research Letters*, 6(1), 456. https://doi.org/10.1186/1556-276X-6-456

Wang, X., Xu, X, & Choi, S.U.S. (1999). Thermal conductivity of nanoparticle-fluid mixture. *Journal of Thermophysics and Heat Transfer*, 13, 474–480. https://doi.org/10.2514/2.6486

Xie, H, Wang, J, Xi, T, Liu, Y, & Ai, F., 2002. Thermal conductivity enhancement of suspensions containing nanosized alumna particles. *Journal of Applied Physics*, 91, 4568–4572. https://doi.org/10.1063/1.1454184

You, J., & Gao, Y., 2009. A study of carbon nanotubes as cutting grains for nano machining. *Advanced Materials Research*, 76, 502–507.

Yu, Su, Gong, Le, Li, Bi, & Chen, Dandan, 2015. An experimental investigation on thermal properties of molybdenum disulfide nanofluids. In *International Conference on Materials, Environmental and Biological Engineering (MEBE 2015)*. https://doi.org/10.2991/mebe-15.2015.197

Yuan, S., Hou, X., Wang, L., & Chen, B., 2018. Experimental investigation on the compatibility of nanoparticles with vegetable oils for nanofluid minimum quantity lubrication machining. *Tribology Letters*, 66. https://doi.org/10.1007/s11249-018-1059-1

Zeng Yuan-Xian, Zhong, Xiu-Wen, Liu, Zhao-Qing, Chen, Shuang, & Li, Nan, 2013. Preparation and enhancement of thermal conductivity of heat transfer oil-based MoS$_2$ nanofluids. *Journal of Nanomaterials*, 2013, 270490. https://doi.org/10.1155/2013/270490

Zhang, D., Li, C., Zhang, Y., Jia, D., & Zhang, X., 2015. Experimental research on the energy ratio coefcient and specifc grinding energy in nanoparticle jet MQL grinding. *The International Journal of Advanced Manufacturing Technology*, 78(5–8), 1275–1288.

Padmini, R., Krishna, P.V., & Rao, G.K.M., 2016. Effectiveness of vegetable oil based nanofluids as potential cutting fluids in turning AISI 1040 steel, *Tribology International*, 94, 490–501.

7 Tribological Aspects of Alternative Materials in the Manufacture of Green Lubricants for Industrial Applications

Ponnekanti Nagendramma, Ankit Pandey,
Atul Pratap Singh, and Anjan Ray
CSIR-Indian Institute of Petroleum, Dehradun,
Uttarakhand, India

7.1 INTRODUCTION

New demands placed on lubricants are changing rapidly. From the viewpoint of better performance throughout their whole life cycle, they should be eco-friendly and ultimately biodegradable [1].

The toughest issues for the lubricant sector are efficacy, eco-friendliness, and price. Concern for the environment has been demonstrated through constant research and development on alternative lubricants to minimize the impact. In any case, increasing lubricant biodegradability and decreasing eco-toxicity to lessen environmental effects has been a driving factor in production processes in recent years [2].

One of the most challenging tasks is the design of a universal biodegradable base stock that might replace mineral oils in the new generation of lubricants. In the past, chemists paid relatively little attention to the mineral oil base stock when trying to design a lubricant with specific performance parameters. Instead, they focused on the package; searching and testing combinations of additives that gave the base stock the desired properties [3]. Conversely, global demand for superior performance lubricants is being driven by performance warranties and environmental criteria for industrial uses and equipment. As a result, in order to fulfill current requirements for ecologically acceptable lubricants, a move to alternative base stock is necessary. Seed oil and polyol esters have been defined as the two basic groups of biodegradable lubricants.

Unlike mineral base oils, which are mostly governed by the supply of crude oil, synthetic lubrication oils are created by alchemical mixture of low molecular mass components as building blocks to construct high molecular weight molecules. The molecular structure is predictable and may be planned and controlled. The selection of non-conventional fluids is frequently driven by performance qualities that mineral oils do not provide [4].

1,3-Butylene glycol is an organic molecule with the formula $C_4H_{10}O_2$ that belongs to the secondary alcohols family and is one of four stable isomers of butylene glycol [5]. Butylene glycol is a translucent, almost colorless, thick, hygroscopic liquid. It has a unique, moderately sweet flavor and no odour [6–10].

This molecule is used as a coating solvent, a starting material for surfactants and different synthetic resins, anti-freeze and high-boiling-point solvent, animal food additives, humectants in tobacco composition, and an intermediary in the production of several other compounds. Because of its good moisture absorption, low volatility, low irritation, and low toxicity, high-quality, odourless butylene glycol has recently been employed as a solvent for toiletry items in the realm of cosmetics [11, 12]. It is also used in the production of anti-cancer colchicine derivatives and dual peroxisome proliferator-activated gamma and delta agonists that function as hypoglycemic medicines in the treatment of diabetes. Because of its antibacterial qualities, it is utilized in a variety of cosmetics and skincare products [13, 14].

The esterification of caprylic acid and oleic acid with butylene glycol to form corresponding esters was carried out with the help of an acid catalyst and an organic solvent. To increase the rate of reaction and finally develop high-quality esters, reactive distillation was necessary. The reaction temperature had an effect on the reaction rates, reactant molar ratio, and reaction duration. Progress was followed by determining the total acid number (TAN) value at regular intervals [15].

Alternative oil technology must keep pace with the rapid advancement of industrial oil lubricants as sustainable lubricant technologies are constantly improving. Creating unique formulations of alternative oils is crucial to enable industrial machinery to perform smoothly with little wear and maximum protection, hence increasing machinery life [16].

In developing novel alternative oil formulas, it is also essential to investigate formulations that minimize viscosity and boost friction qualities. Lubricant economy can be enhanced by lowering viscosity and improving friction characteristics in the boundary and blended lubrication regimes, machine performance, and equipment life [17].

Butylene glycol esters have two ester connections with two oxygen atoms each. Since oxygen increases molecular polarity, this raises lubricity and the viscosity index while lowering Noack evaporation. Nevertheless, due to the polarity of the oxygen atoms, compatibility with polymeric materials is challenging. Butylene glycol esters belong to a Group V type base oil with a high viscosity index, low viscosity and low friction which can increase fuel efficiency and minimize emissions of greenhouse gases [18]. These esters are biodegradable and easily blend with other oils and additives [19].

They are used in their native state or re-transformed into derivatives as additives for many industrial applications. BG-based esters (BGEs) have been cited only occasionally in lubricant applications.

According to the published and patented information, much work has been done on BGEs in food, personal care, pharmaceuticals, biotechnology, and agriculture. Several corporations are involved in the commercial production of BGEs (Table 7.1) [20–26].

They are used in their native state or re-transformed into derivatives as additives for many industrial applications. Butylene glycol-based esters have been cited only occasionally in lubricant applications. Figure 7.1 shows the flowchart for the synthesis of BGEs.

The interaction between lubricant molecules and surfaces strongly influences the lubricant's behavior under the boundary lubrication regime. In such cases, the fluids' chemical and physical reactivity with the surfaces characterizes the contact friction. Many synthetic fluids exhibit excellent chemical and physical

TABLE 7.1
Commercial Producers of BGEs

Country	Producer
India	Godavari Biorefineries
	Vizag Chemicals
	India Glycols Ltd
USA	Antares Chem Pvt Ltd
	Otto Chemie Pvt Ltd
	Rxsol Chemo Pharma International
Netherlands	The Peter Greven Group

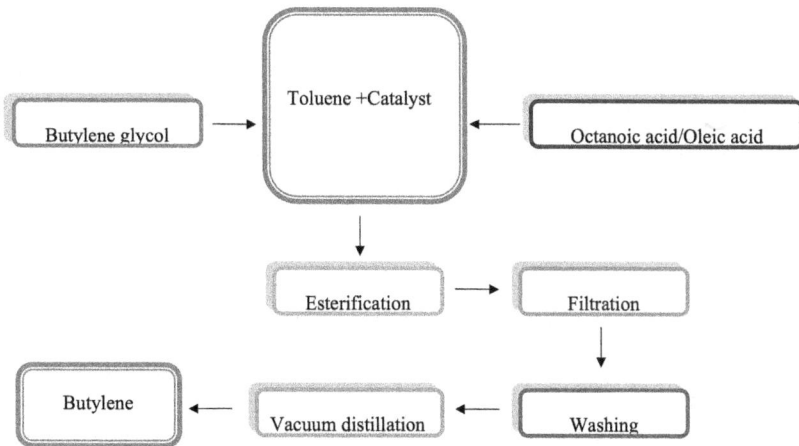

FIGURE 7.1 Flowchart for the synthesis of BGEs.

characteristics, including good oxidation and thermal stability. On the other hand, the same fluids may show moderate or poor lubricant performance under the boundary lubrication regime.

7.2 EXPERIMENTAL STUDIES OF BGES

A study on the esterification of butylene glycol was conducted in a round bottom flask equipped with a thermometer, a condenser for refluxing and a Dean and Stark receiver. A chemical reaction occurs between butylene glycol and fatty acids to produce an ester, eliminating water in a condensation reaction, referred to as Fischer esterification, as shown in Figure 7.2. All the reaction steps are reversible. Hence, according to Le Chatelier's principle, the forward reaction is favoured by taking an excess of butylene glycol or eliminating water [27]. A mixture of one mole of butylene glycol and 1.1 mole of fatty acids, and 25% (w/w %) of a heterogeneous catalyst by weight of the reaction mixture was refluxed with solvent toluene at 110–118 ± 1°C. The refluxing procedure continued till the theoretical amount of water was obtained in the receiver and the reaction took 8–9 hours to complete. The acid value of the product was noted every 60 minutes. Reduction in TAN values (Figure 7.3) indicated the reaction's progress and was taken to be concomitant with ester linkages formed. The product was chilled to room temperature before being filtered and rinsed with water until the pH reached neutral and toluene was collected through vacuum distillation. The product was percolated over basic alumina to remove traces of unreacted acid and improve its colour [28, 29].

7.2.1 SPECTROSCOPIC ANALYSIS OF BGES

The synthesized BGDC and BGDO esters were characterized by IR spectroscopy on a Perkin-Elmer IR spectrophotometer using KBr pellets. ^1H and ^{13}C NMR spectra were recorded using a Bruker Av III 500 MHz spectrometer.

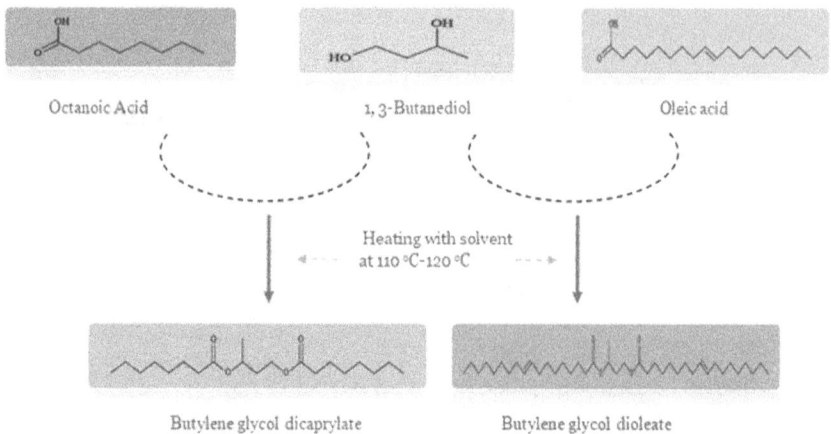

FIGURE 7.2 Synthesis of BGEs.

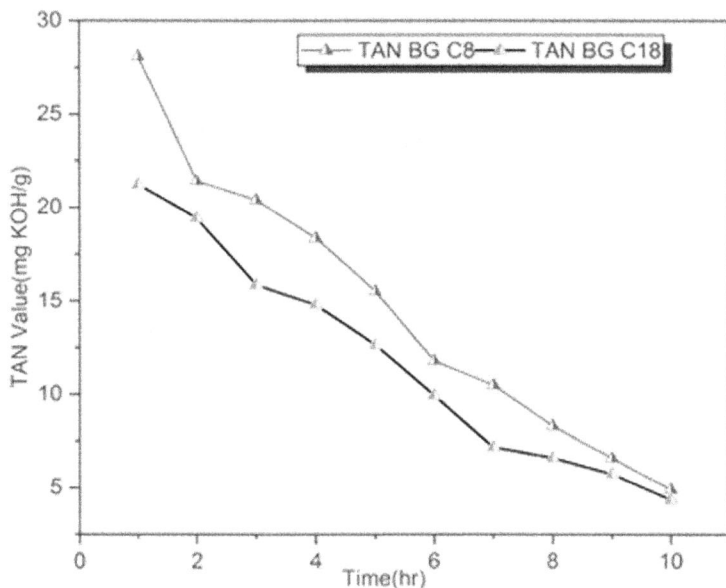

FIGURE 7.3   TAN of BGEs as a function of reaction time.

7.2.1.1   IR and NMR Studies

Both FTIR and FT-NMR analyses were used to characterize the developed BG esters. The peaks obtained from FTIR and FT-NMR spectroscopy of reactants and synthesized products are summarized in Table 7.2 and clearly show that the intended reactions have occurred.

The FTIR spectrum of BGDC and BGDO as shown in Figure 7.4 is consistent with the expected structure. The broad vibrational peaks of methylene and methyl units of the alkyl chain of BGEs are in the range of 2800–3000 cm^{-1}. The IR peaks which appeared at around 1741 cm^{-1} and 1153 cm^{-1} aredue to carbonyl and carbon-oxygen bond stretching of the ester group. The lower frequency strong band at1460 cm^{-1} is due to aliphatic esters. A peak at around 727 cm^{-1} is due to long alkyl chains present in the synthesized lube. The absence of a free $^-$OH stretch frequency at 3650–3590 cm^{-1} in the FTIR spectrum of all esters and low organic acidity in all fluids showed complete esterification. There were no peaks relating to –COOH and –OH groups, suggesting that all –CH$_2$OH groups of alcohols were esterified and –COOH groups were completely changed.

Proton shifts further confirmed the chemical structure of BGE esters in ^{1}H and ^{13}C NMR spectra (Figure 7.5–7.6). Functional groups and electronegative heteroatoms attract electrons, whereby the protons of the neighboring carbon atoms experience downfield shifts. ^{1}H NMR spectra of BG and fatty acid esters have a set of CH–, CH$_2$– and –OH PG protons. The –OCH$_2$ bond protons are confirmed from the signals of 4.11–4.32 ppm in BGEs. Protons located at the double bonds were assigned to the peaks at 5.17–5.3 ppm.

TABLE 7.2

IR and NMR Data of Butylene Glycol Esters

Analysis	Functional Group	Peaks (cm⁻¹)	Butylene Glycol	Oleic Acid	Octanoic Acid	BGC8	BGC18
FTIR	sp³C–H bend	2850–3000	Present	Present	Present	Present	Present
	C=O acid stretch (broad)	2400–3400	Absent	Present	Present	Present	Present
	C=O acyl ester	1735–1750	Absent	Absent	Absent	Present	Present
	C=O fatty acid	1755–1760	Absent	Absent	Present	Absent	Absent
	O–Hbend	1100–1450	Present	Present	Absent	Present	Present
	–O–H stretch	3600–3500	Absent	Absent	Present	Present	Present
¹H NMR (ppm)	t-CH₃	0.5–2	Present	Present	Present	Present	Present
	Methylene H	1.2–1.3	Present	Present	Present	Present	Present
	H adjacent C–O	4.11–4.32	Present	Absent	Present	Present	Present
	–O–H–	3.5–3.75	Present	Absent	Absent	Present	Absent
	–COOH	10–12	Absent	Present	Present	Absent	Absent
	–C–OO–CH	2–2.2	Absent	Absent	Absent	Present	Present
¹³C NMR (ppm)	–COOH–	160–173	Absent	Present	Present	Absent	Absent
	–C–OH	45–75	Present	Absent	Absent	Present	Present
	–COOR–	135–150	Absent	Absent	Absent	Absent	Absent
	sp³C–H	16–34	Present	Present	Absent	Present	Present
	sp³-CH₂	15–55	Present	Present	Present	Present	Present

7.2.2 Physico-Chemical Characterization of BGEs

Butylene glycol ester lubricants are made from C_8 and C_{18} acids, which have constant molecular structures and hence well-defined characteristics that may be adjusted to specific uses. The raw substances used to make ester-type base fluids can have an impact on a variety of lubricant qualities such as viscosity, flow properties, VI, lubricity, hydrolytic stability, thermal stability, biodegradability, and solvency.

The physico-chemical characteristics such as acid value, kinematic viscosity, VI, and density of synthesized esters were determined per the American Society of Testing Materials (ASTM) procedures D-974, D-445, D-2270 and D-4052, respectively. The dynamic/kinematic viscosity and density of BG ester were determined using the SVM3000 Anton Paar Stabinger viscometer at temperatures ranging from 30–100°C. The viscosity of an ester lubricant can be increased by increasing the molecule's molecular weight by prolonging the carbon chain of the acid or alcohol and the number of ester groups. Increased numbers of cyclic

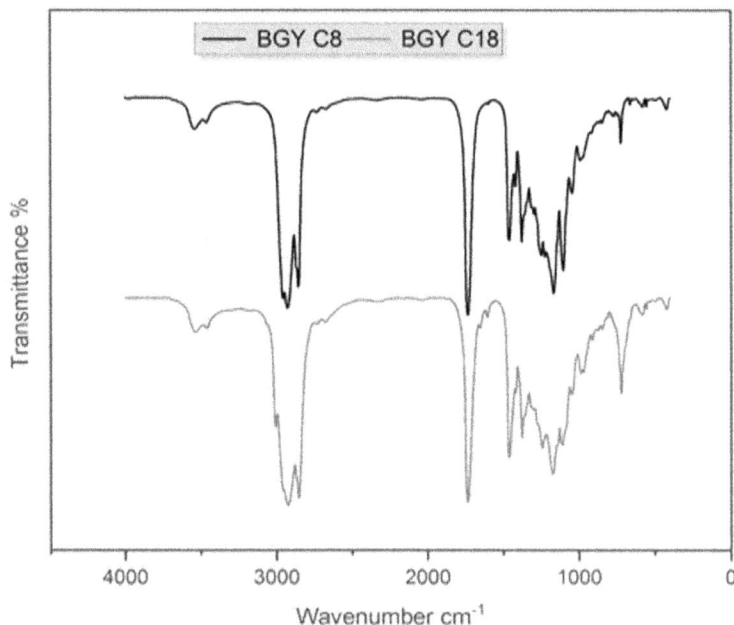

FIGURE 7.4 FTIR spectra of BGEs using Fourier transform spectrophotometer of 400–4000 cm⁻¹ range.

groups in the molecular backbone, as well as the quantity or degree of branching, can all have an effect on viscosity.

An ester lubricant's viscosity index (VI) can be increased by improving molecular linearity, acid and alcohol chain length and an absence of cyclic groups in the backbone significantly decreases the VI compared to aliphatic branches.

The lubricant's pour point can be decreased by increasing the number and location of the branches. The pour points of branches in the centre of the molecule are higher than those at the ends. Shortening the acid chain, lowering the molecule's internal symmetry, and increasing the linearity of the ester improve the VI but raise the pour point.

In general, increasing the molecular weight improves the overall lubricity. Normal acid- or alcohol-terminated esters have higher lubricities than branched acid/alcohol-terminated esters. Whereas mixed acid/alcohol-terminated esters have lubricities intermediate between normal acid/alcohol-terminated esters and branched acid/alcohol-terminated esters.

The ester linkage is very stable, and according to band energy estimates, it is more stable than the C–C bond. The thermal and oxidative stability of butylene glycol esters is determined by the amount and type of hydrogens present in decreasing stability order. Consequently, linear acid esters have a higher stability than branching acid esters and short chain acids have more stability than long chain acids.

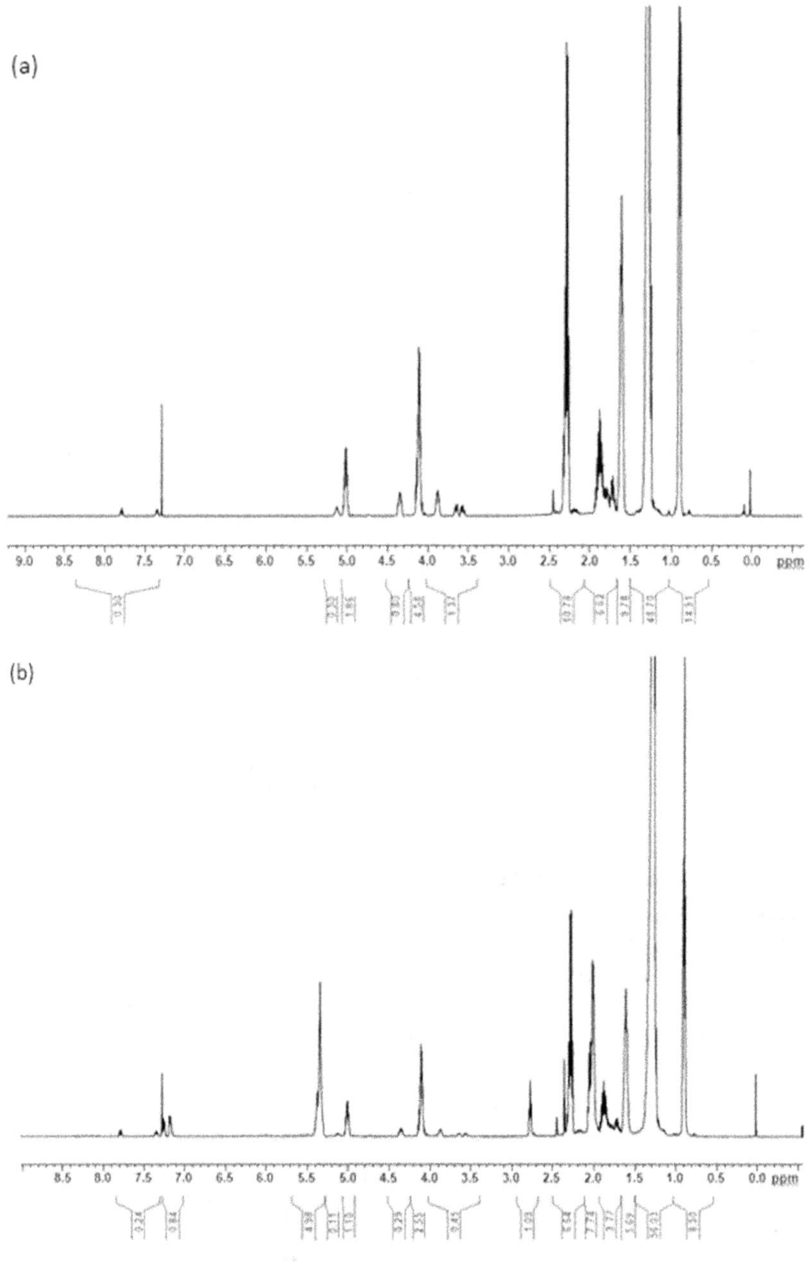

FIGURE 7.5 ¹H NMR of (a) BGDC and (b) BGDO esters.

(a)

(b)

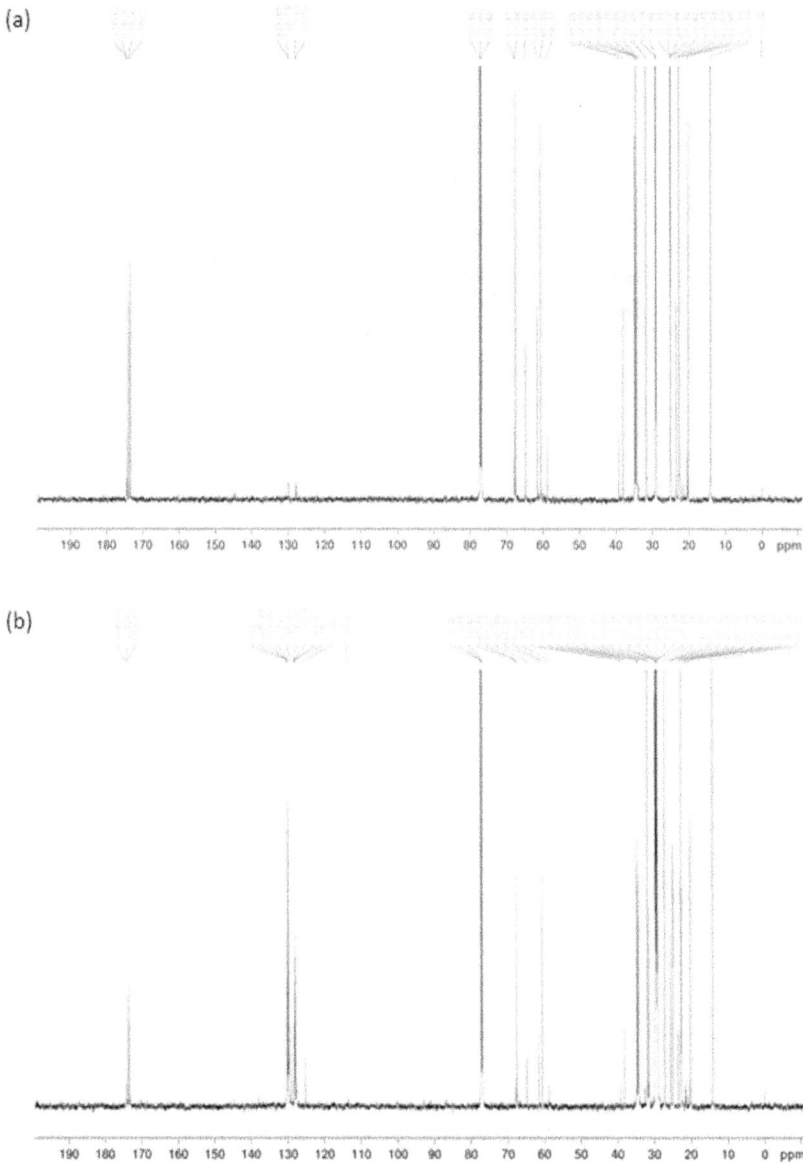

FIGURE 7.6 ^{13}C NMR of (a) BGDC and (b) BGDO esters.

The hydrolytic stability of esters is determined by two major factors: processing parameters and molecular geometry. All processing factors include the degree of esterification, acid value, catalyst engaged during esterification, and the level remaining in the ester after processing.

Molecular shape can affect hydrolytic stability in a variety of ways. By sterically limiting the acid part of the molecule, hydrolysis can be delayed. The length of the acid chain is also essential. Butylene glycol esters have higher hydrolytic stabilities.

Esters and mineral oils are often completely compatible. Esters can also be mixed with other synthetics like polyalphaolefins (PAOs). This allows esters a great deal of versatility and, combined with other oils, affords unrivalled opportunities to balance the expense of a lubricant mixture against its performance [30, 31]. Other factors such as density and concentration of additives in the base stock may also play a role [32, 33].

Esters' viscometric characteristics are determined by their molecular weight and acid branching. Butylene glycol C_8&C_{18} esters have viscosities of 18.96 and 12.31 cSt at 40°C, and 4.8 and 5.2 cSt at 100°C, respectively, as shown in Figure 7.7. Acidity is practically negligible, which, apart from indicating completion of the reaction, is also directionally preferred as the lubricant is less likely to induce corrosion on metal surfaces. Densities (0.869 and 0.796) smoothly decrease with increase in temperature and are also affected by the length of the alkyl chain [34]. The TAN value of the synthesized esters is very low. However, certain acid components tend to increase the TAN value, which may change kinematic viscosity and degree of oxidation of the compound, affecting its stability [35].

These unconventional base oils have been divided into four classes or groups by the American Petroleum Institute (API) along with Group I, representing the base stocks used in the bulk existing motor oils meeting the ILSAC (International

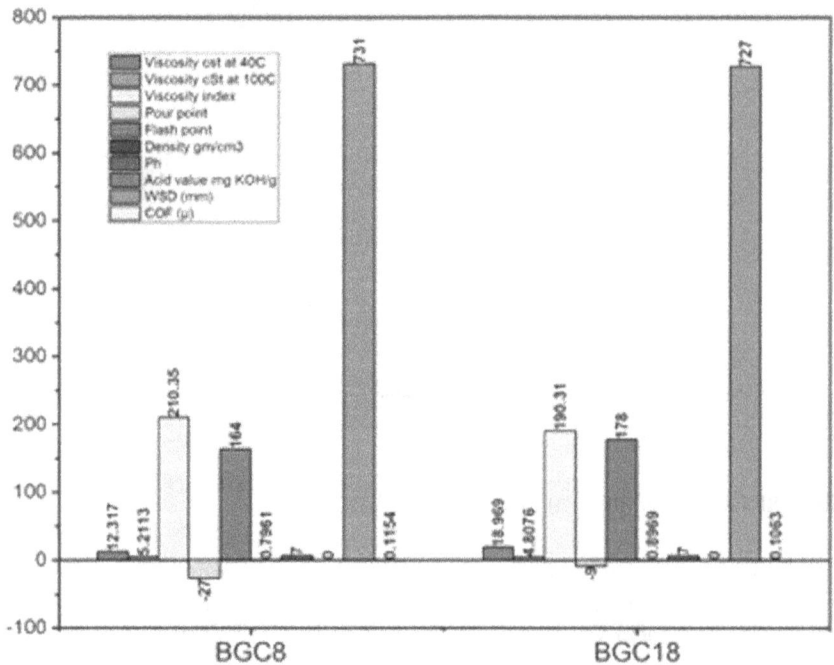

FIGURE 7.7 Physico-chemical characterization of BGEs.

Lubricant Standardization Advisory Committee) minimum performance standard for passenger car (petrol) engines: GFI and APISH qualities [36]. The synthesized butylene glycol C_8 and C_{18} esters are categorized as Group V.

7.3 TRIBOLOGICAL BEHAVIOUR OF BGES

Tribological behaviour represented by anti-wear and anti-friction properties of the synthesized esters was established based on wear scar diameter (WSD) and friction coefficient (COF) using a four-ball tribo tester (Figure 7.8).

For tribo-performance evaluations, the steel balls were cleaned in hexane using ultrasonic vibrations. The balls were then arranged in a tetrahedral geometric pattern, where the lowest three balls were put in a ball pot containing a sample of test lubricant. The upper fourth ball was attached to a rotating spindle and rotated over the bottom three stationary balls under a steady load for one hour. The standard test conditions of ASTM D-4172 B (Table 7.3) were maintained. The scars developed on the steel balls were studied using an optical microscope and their images captured [37–40].

7.3.1 TRIBO PERFORMANCE OF BGDC AND BGDO

The average values of the COF and WSD of the synthesized esters observed were 727 and 731(μm). As summarized in Figures 7.9–7.10, the friction and wear results reveal that the synthesized esters' yields improved tribological performance. The smoother surface of steel balls is observed in Figure 7.11. The

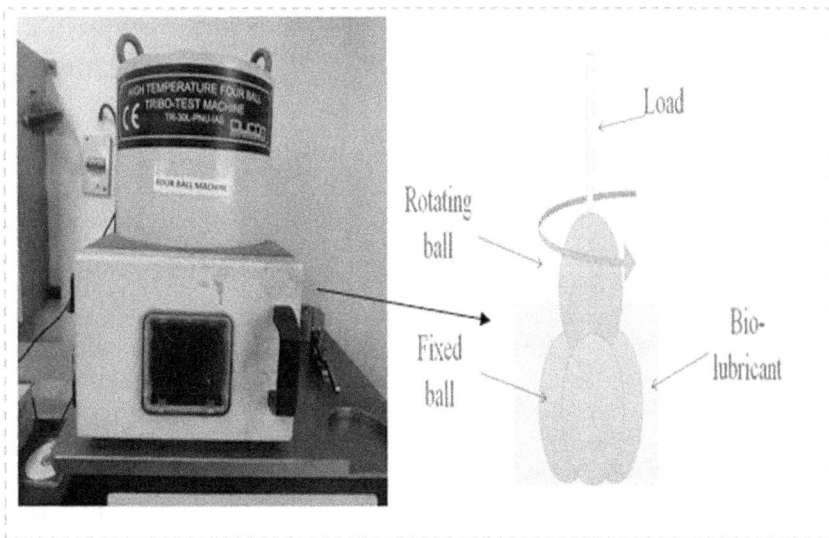

FIGURE 7.8   Four-ball tribo tester.

TABLE 7.3

Operating Parameters of Four-Ball Tester

S. No	Operating Parameter	
1	Load (N)	392 N
2	Speed (RPM)	1200
3	Lubricant temperature (°C)	75°C
4	Test duration	3600 s

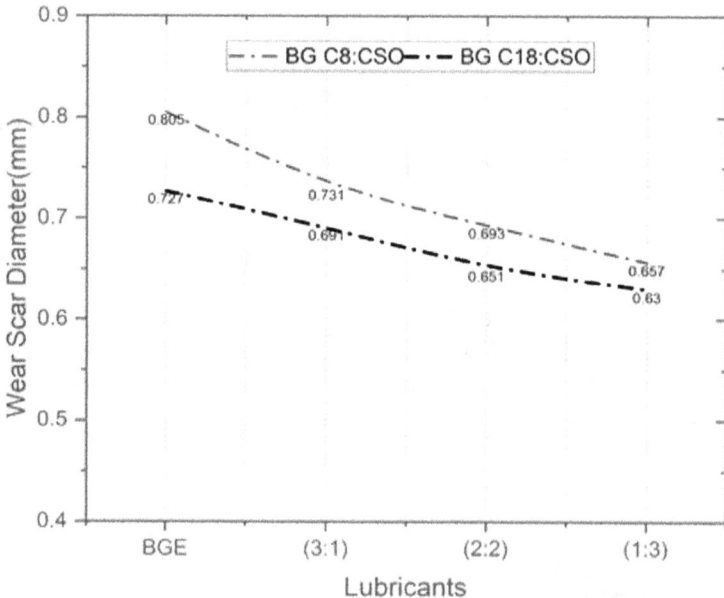

FIGURE 7.9   Wear behaviour of BGEs blend with CSO.

BGE adsorbs metal surfaces and forms boundary lubricating films. These adhered lubricant molecules protect the steel test balls' surfaces from further damage, resulting in smaller WSDs.

The presence of large molecular surface area in one plane enhances anti-wear properties. In BGE the molecule has a branched and staggered structure with a high molecular weight, resulting in improved lubricity. Therefore, the tribo-properties of BGE are better than conventional oils.

Meanwhile, in BGEs, the chemical modifications preserve their beneficial properties, such as high viscosity, superior lubricity, strong corrosion resistance, and minimum evaporation losses. [41]. The lower friction and wear of BGE ester is due to stronger boundary film formation. Hence, the studied esters can be considered as potential friction and wear reducers and can be used as industrial lubricants.

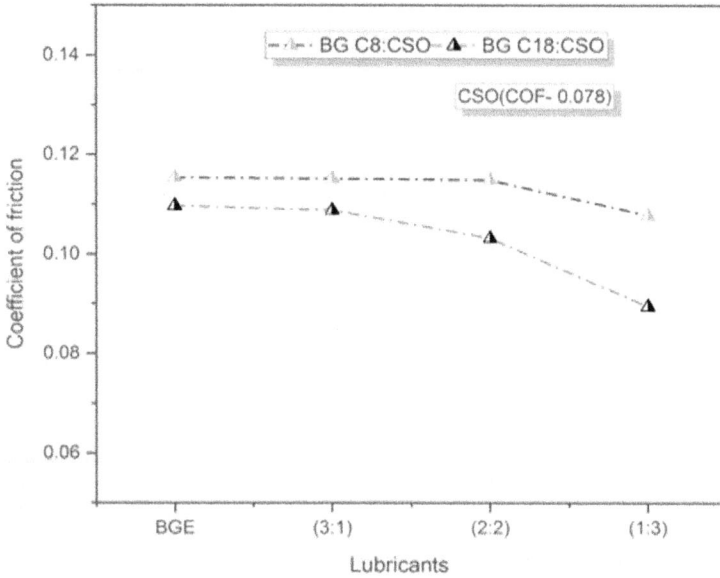

FIGURE 7.10   Frictional behavior of BGEs and BGEs blend with CSO.

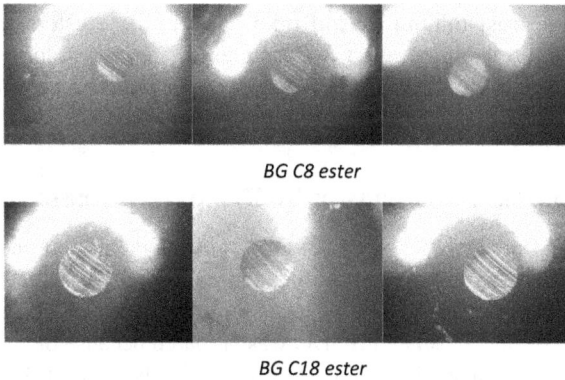

BG C8 ester

BG C18 ester

FIGURE 7.11   Wear scar images for BGEs.

Post-experimental studies were performed using SEM to examine the surface morphology of wear scars on the used ball test specimens lubricated with BGDO. Figure 7.12 shows the SEM images of the used ball specimens lubricated with BGDO.

The figure shows that the wear scars have a smooth finish with small abrasion lines along the sliding direction, which supports the results obtained for WSD and COF. In any test specimen, the wear scars do not correspond to any extreme wear damage. Surface smoothing and rubbing wear are measured [42, 43].

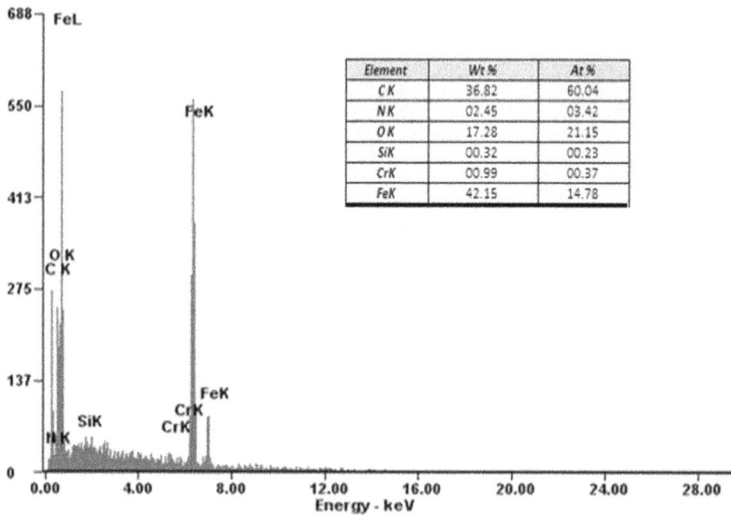

Element	Wt %	At %
C K	36.82	60.04
N K	02.45	03.42
O K	17.28	21.15
Si K	00.32	00.23
Cr K	00.99	00.37
Fe K	42.15	14.78

FIGURE 7.12   SEM images of BGDO ester.

Pure esters do not dominate aliphatic hydrocarbons or mineral oils as boundary lubricants, but the hydrolysis or oxidation products of many esters are useful, wear preventive, rust inhibiting, and mild. Hence, these esters develop fairly good boundary lubricating properties during use. Esters are soluble in vegetable oils, making them useful blending components for increasing the oil characteristics [44].

7.3.2   TRIBOLOGICAL PROPERTIES OF BGDO AND COTTONSEED OIL BLENDS

Blending based on vegetable oil and synthetic esters is growing in popularity in various applications. We report the initial studies carried out on cottonseed oil as biocompatible blends with BGDO esters as biolubricants.

A blend of cottonseed oil and BGDO was prepared because of the higher compatibility with vegetable oils. They were taken in the ratio 1:3, 2:2, and 3:1, respectively, in a conical flask and heated up to 80°C with constant stirring for about 60 minutes. The cottonseed oil was blended with synthetic BG ester to investigate the tribo-performance behaviour. The results indicate that the cottonseed oil and synthetic BGDO ester form a compatible blend and showed significant enhancement of performance properties, as can be seen in Figures 7.13–7.15, respectively.

There are various factors which contribute to choosing cottonseed oil for friction studies. Long and heavy fatty acid molecules with dipolar nature create a dense film which provides cottonseed oil with a greater ability to sustain under high pressure and create a lubricating film. By minimizing wear, friction, and heat dissipation, this film helps to enhance surface quality and overall productivity.

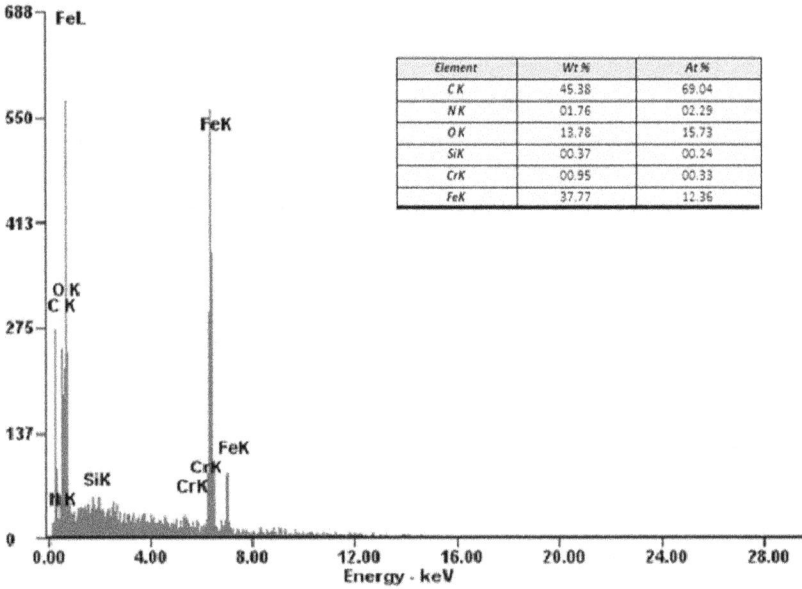

Element	Wt %	At %
C K	45.38	69.04
N K	01.76	02.29
O K	13.78	15.73
Si K	00.37	00.24
Cr K	00.95	00.33
Fe K	37.77	12.36

FIGURE 7.13 SEM images of blend of BGDO ester with cottonseed oil (1:3).

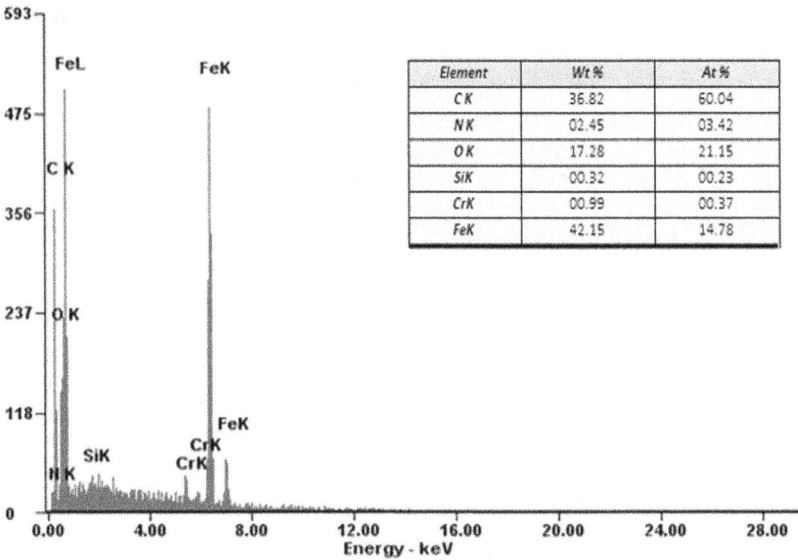

Element	Wt %	At %
C K	36.82	60.04
N K	02.45	03.42
O K	17.28	21.15
Si K	00.32	00.23
Cr K	00.99	00.37
Fe K	42.15	14.78

FIGURE 7.14 SEM images of blend of BGDO ester with cottonseed oil (2:2).

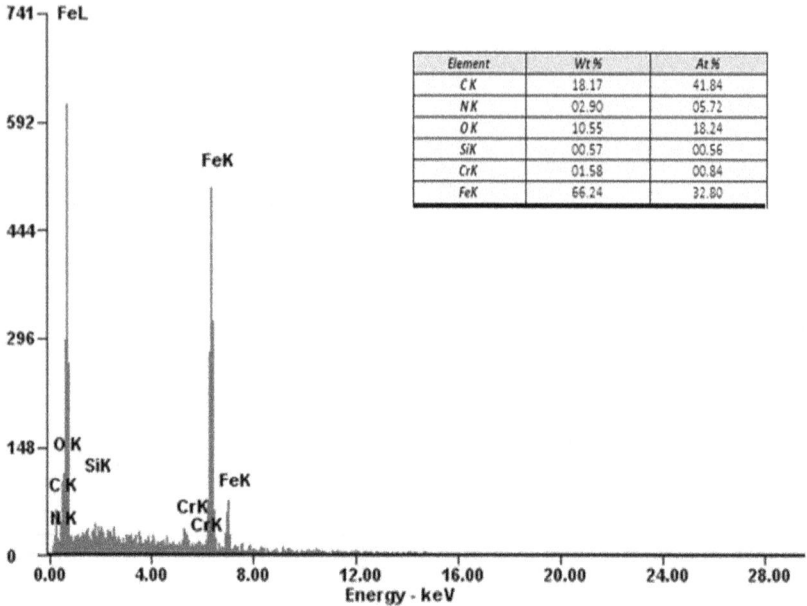

Element	Wt %	At %
C K	18.17	41.84
N K	02.90	05.72
O K	10.55	18.24
Si K	00.57	00.56
Cr K	01.58	00.84
Fe K	66.24	32.80

FIGURE 7.15   SEM images of blend of BGDO ester with cottonseed oil (3:1).

The high molecular weight and boiling point of cottonseed oil prevents loss due to evaporation. Lower smoke generation and fire dangers with high flash points result in lower metal removal rates [45].

The superior tribological characteristics of cottonseed oil blends are as a result of its chemical composition of triacylglycerol molecules composed of esters. The long chains of polar fatty acids, the presence of cyclopropane ring, and tocopherols as natural antioxidants within the oil provide a longer shelf life.

Structural factors that impact the viscosity such as chain length, unsaturation, and presence of certain polar groups like esters also increases the lubricity of oil. Lubricity is a property of lubricants that has to do with how easily the surfaces glide over each other. Cottonseed oil shows good lubricity because of lubricating film formation due to chemi-absorption on metal surfaces [46]. This, thus, indirectly impacts the values of WSD and coefficient of friction. Esters formed by linear acids or alcohols often have higher lubricity than those prepared by alcohol and branched acids.

Synthetic oil being extremely pricey and non-environment friendly, thus compels us to look into alternative options. Vegetable oils, being biodegradable choices, thus provide a good choice for developing blends. When compared with pure cottonseed oil, an esterified product and its blending show improved properties in respect of pour point and tribological performance [47, 48].

The use of biodegradable products in environmentally sensitive areas is increasing rapidly. To meet the demand for environmentally friendly processes and products, it is essential to embrace innovative functional technologies. More and more lubricants are required for their ecological and economic benefits and not just for their technical abilities

7.4 FUTURE SCOPES

In the future, there is potential for the development of new and improved butylene glycol esters with even better lubricating properties and a wider range of applications in machinery. Furthermore, study into the use of butylene glycol esters in combination with other additives, such as anti-wear agents and viscosity improvers, as well as its blend with cottonseed oil might lead to additional advances in lubricant performance. Its low viscosity gives butylene glycol the potential to be utilized as a lubricant, reducing friction between moving parts. Moreover, it has remarkable thermal stability and can withstand degradation at high temperatures. In addition, further research on the safety and environmental effect of butylene glycol as a lubricant would be beneficial. This could include studies on its biodegradability, toxicity, and potential for accumulation in the environment. Research could also investigate the effects of butylene glycol on the performance of different types of machinery and how it compares to other lubricants. Overall, the future of butylene glycol esters as a lubricant looks promising, and they are expected to play a growing role in the lubricant industry in years to come.

7.5 CONCLUSIONS

A preliminary study to assess the lubrication behavior of synthetic lubricant base stock as well as its blends with cottonseed oil as a prospective replacement of conventional oils has been presented. The characteristic anti-friction and anti-wear properties have been experimentally studied. Thus, we conclude:

- A potential eco-friendly and biodegradable base oil developed using BG and fatty acid in the presence of an acidic catalyst results in product yields exceeding 90%. The synthesized BGEs are biodegradable and have excellent physico-chemical characteristics.
- Tribological testing of the synthesized BGEs revealed that they performed well in terms of anti-friction and anti-wear properties.
- Butylene glycol esters are also used as lubricants due to their excellent lubricating properties and low toxicity. The usage of cottonseed oil as a biodegradable and environment-friendly blend is an important alternative.

Our initial studies indicate that BGEs possesses suitable properties as a biodegradable lubricant for industrial applications. Products are widely used, either alone or as part of formulations and as multifunctional lubricants. Further detailed studies of BGEs are needed for the effective development of formulations for lubrication in industrial applications.

ACKNOWLEDGEMENTS

The authors gratefully acknowledge the director of CSIR-IIP for granting permission to publish the research findings.

REFERENCES

1. http://www.lubrizol.com/referencelibrary/lubtheory/base.htm
2. John, A. Moore. (1997). Resistance slows synlube growth, *Hart's Lubricant and World*, 7(3), 33–35.
3. Woods, Mike. (1997). Think green: Biodegradable lubes glow with promise. *Lubes-n-Greases*, 3(7), 14.
4. Van der Waal, G., D. Kenbeek. (1993). Testing, application and future developments of environmentally friendly ester base fluids. *Lubrication Science*, 10, 67–83.
5. Karacaoglu, Sokrati. (1940). https://atamankimya.com/
6. Powers, Justin L. (1981). *Food Chemicals Codex*, 3rd ed., Washington, DC: National Academy Press, 771.
7. Weast, R.C. (1979). *CRC Handbook of Chemistry and Physics*, 59th ed., Boca Raton, FL: CRC Press, 2666.
8. Hawley, G.G. (1971). *The Condensed Chemical Dictionary*, New York: Van Nostrand Reinhold, 971.
9. Windholz, M. (ed.). (1976). *The Merck Index*, 9th ed., Rahway, NJ: Merck and Go.
10. CTFA. (1981). Submission of unpublished data. CTFA cosmetic ingredient chemical description. Butylene glycol, (CTFA Code 2-1, 7-175).
11. Galvão, A.C., Francesconi, A.Z. (2009). Experimental study of methane and carbon dioxide solubility in 1, 4 butylene glycol at pressures up to 11 MPa and temperatures ranging from 303 to 423 K. *The Journal of Supercritical Fluids*, 51, 123–127.
12. Tsuji, Yasuo et al. (2002).1,3 butylene glycol of high purity and method for producing the same, United State Patent, US6,376,725 B1.
13. Diegenant, C., Constandt, L., Goossens, A. (2000). Allergic contact dermatitis due to 1,3-butylene glycol. *Contact Dermatitis*, 43(4), 234–235.
14. Magerl, A., Pirker, C., Frosch, P.J.(2003). Allergic contact eczema from shellac and 1,3-butylene glycol in an eyeliner. *Journal der Deutschen Dermatologischen Gesellschaft*, 1(4), 300–302.
15. Nagendramma, Ponnekanti. (2011). Study of pentaerythritol tetraoleate ester as industrial gear oil. *Lubrication Science*, 23, 355–362.
16. Yang, Jung, Cho, Soon-Haeng, Park, Jongki, Lee, Kwan-Young. (2007). Esterification of acrylic acid with 1,4 butanediol in a batch distillation column reactor over amberlyst 15 catalyst. *The Canadian Journal of Chemical Engineering*, 85, 883–888.
17. Lee, David. (2018). *Base oil basics: Quality starts at the base*. Chevron: Lubricants.
18. Esche, C. et al. (2018). Esters for engine oils. *Tribology and Lubrication Technology*, 74(11), 80–82.
19. Fitch, Bennett. (2017). *Understanding the differences between base oil formulations*. Machinery: Lubrication.
20. https://godavaribiorefineries.com (2006).
21. http://www.vizagchemical.com/ (1985).
22. https://www.indiaglycols.com (1983).
23. https://www.antareschem.com (2011).

24. https://www.ottokemi.com (2006).
25. http://rx-sol.com/ (1995).
26. https://www.peter-greven.de/en/company (1923).
27. Miner, C.S., Dalton, N.N. (1953). *Chemical Properties and Derivatives of Glycerine.* New York: Reinhold Publication Corp, 4–21.
28. Nagendramma, P. (2011). Study of pentaerythritol tetraoleate ester as industrial gear oil. *Lubrication Science*, 23, 355–362.
29. Sharma, N., Bari, S.K., Nagendramma, P., Thakre, G.D., Ray, A.(2021). Tribological investigations of sustainable bio-based lubricants for industrial applications. *Green Tribology: Emerging Technologies and Applications*, CRC Press, New York, USA, 71–97.
30. Nowicki, J., Mosio-Mosiewski, J. (2013). Esterification of fatty acids with C8-C9 alcohols over selected sulfonic heterogeneous catalysts. *Polish Journal of Chemical Technology*, 15, 42–47.
31. McNutt, J., Quan, S.H.(2016). Development of biolubricants from vegetable oils via chemical modification. *Journal of Industrial and Engineering Chemistry*, 36, 1–34.
32. Yan, F., He, W., Jia, Q., Wang, Q., Xia, S., Ma, P. (2018). Prediction of ionic liquids viscosity at variable temperatures and pressures. *Chemical Engineering Science*, 184, 134–140.
33. Housel, T. (2014). Synthetic esters: Engineered to perform. *Machinery Lubrication*, 4, 1–10. www.machinerylubrication.com/Articles/Print/29703
34. Ghatee Md, H., Zare, M., Moosavi, F., Zolghadr, A.R. (2010). Temperature-dependent density and viscosity of ionic liquids 1-alkyl-3-methylimidazolium iodides: Experiment and molecular dynamics simulation. *Journal of Chemical & Engineering Data*, 55(9), 3084–3088.
35. Wolak, A. (2018). Changes in lubricant properties of used synthetic oils based on the total acid number. *Measurement and Control*, 51(3-4), 65–72.
36. Nagendramma, P. (2004). Development of eco-friendly/biodegradable synthetic (polyol&complex) ester lube base stocks. Thesis, Srinagar, India: HNB Garhwal Central University.
37. Nagendramma, P., Shukla, B.M., Adhikari, D.K. (2016). Synthesis, characterization and tribological evaluation of new generation materials for aluminum cold rolling oils. *Lubricants*, 4, 23.
38. Nutiu, R., Maties, M., Nutiu, M. (1990). Correlation between structure and physical and rheological properties in the class of neopentanepolyol esters used as lubricating oils. *Journal of Synthetic Lubrication*, 7(2), 145–154.
39. Flider, F.J. (1993). Polyglycerol Esters as Functional Fluids and Functional Fluid Modifiers. US5380469A United States.
40. Rothfuss, N. E., Petters, M. D. (2016). Influence of functional groups on the viscosity of organic aerosol. *Environmental Science & Technology*, 51(1), 271–279.
41. Bahlakeh, G., Ramezanzadeh, B. (2017). A detailed molecular dynamics simulation and experimental investigation on the interfacial bonding mechanism of an epoxy adhesive on carbon steel sheets decorated with a novel cerium-lanthanum nanofilm. *ACS Applied Materials &Interfaces*, 9, 17536–17551.
42. Osama, M., Singh, A., Walvekar, R., Khalid, M., Gupta, T.C.S.M., Gupta, W.W. (2017). Recent developments and performance review of metalworking fluids. *Tribology International*, 114, 389–401.
43. Anna, M., Gradkowski, M. (2007). Antiwear action of mineral lubricants modified by conventional and unconventional additives. *Tribology*, 27, 177–180.

44. Gairing Max, F., Mike, Frend, A. Reglitzky, Purmer Piet, P. (1994). Environmental Needs and New Automotive Technologies Drive Lubricants Quality. *Proc. 14th World Petroleum Congress* 3, 99.
45. Agarwal, S.M., Lahane, S., Patil, N.G., Brahmankar, P.K. (2014). Experimental Investigations into wear characteristics of M2 steel using cottonseed oil. *Procedia Engineering*, 97, 4–14.
46. Siraskar, G., Jahagirdar, R.S. (2018). Cottonseed oil and esterifies cottonseed oil as lubricant in IC. *IJARIIE*, 4, 433–440.
47. Siraskar, G., Wakchaure, V.D., Jahagirdar, R.S., Tiwari, H.U. (2020). Cottonseed tri-methylolpropane (TMP) ester as lubricant and performance characteristics for diesel engine. *International Journal of Engineering and Advanced Technology*, 3, 761–768.
48. Wagner, H., Luther, R., Mang, T. (2001). Lubricant base fluids based on renewable raw materials. Their catalytic manufacture and modification. *Applied Catalysis A: General*, 221, 429–442.

8 Potential Use of Vegetable Oil-Based Nano-Fluids for Machining Processes

Ananthan D. Thampi, P. Pranav, Edla Sneha, and S. Rani
College of Engineering Trivandrum, Trivandrum, India

8.1 INTRODUCTION

Cutting fluids (CFs) are lubricants utilized at the workpiece–tool contact. It removes the heat produced in the cutting zone and also flushes out the chips developed. This keeps the machining area clean. The use of CFs has increased production and raised manufacturing standards, which in turn has led to more efficient working conditions in the industry (Zimmerman et al., 2003).

Nowadays, green strategy plays an important role in determining the product's market. Thus, the demand for non-toxic and environment-friendly CFs is increasing by the day. Presently, mineral-based CFs are used by the vast majority of manufacturing companies globally due to their low cost, availability, and high shelf life. These mineral oil-based lubricants can cause various types of environmental contamination and can affect the health of skilled operators who come into direct contact with them. Several disease-causing toxins formed in these water-soluble CFs by different micro-organisms are reported to be very harmful to operators and workers (de Groot and Flyvholm, 2020; Lima and Elsner, 2020). Agricultural products are indirectly polluted as a result of contamination and incorrect use of petroleum-based lubricants (Pranav et al., 2021).

DOI: 10.1201/9781003363576-8

Different machining techniques, such as turning, drilling, grinding, and milling are commonly used in the industry. The use of CFs contributes to the reduction of friction during machining. It was reported that petroleum-based products account for 85% of all lubricants used globally (Pop et al., 2008; Shashidhara and Jayaram, 2010). Several occupational safety agencies have suggested a reduction in the usage of petroleum products, and these authorities are also supporting greener technology to help society live more sustainably (Singh and Gupta, 2006). Many investigations into vegetable oil (VO)-based CFs are already underway due to the demand for eco-friendly cutting fluids. Green CFs are generally made by combining vegetable oils with water and adding suitable emulsifiers. Usually, this results in a stable emulsion of oil in water. Methods like acoustic/ultrasonic emulsification are also used to diffuse these immiscible oils in water (Burton et al., 2014).

In the lubricating mechanism of CFs, the base fluid used is quite important. Vegetable oil-based CFs must have qualities that are comparable to or superior to those of mineral oil-based CFs. Vegetable oils are one of the most promising mineral oil-based CF replacements. They must be chosen for their capacity to produce a protective tribo-film on mating surfaces, as well as their ability to withstand high temperatures, corrosion, and oxidation. Chemical modification or the addition of chemical components known as additives might boost these capacities even more (Prasannakumar et al., 2022; Sneha et al., 2020; Thampi et al., 2020; Trajano et al., 2014). Blending nano-additives with base oil is thought to be a promising technique to enhance lubricant characteristics using nanotechnology (A. Singh et al., 2020; Wu et al., 2018). The tribological properties of these nano-fluids are said to be dependent on the size and shape of the nanoparticles (Wu et al., 2007). This chapter addresses a nano-fluid made of vegetable oil that can be employed in the machining process.

8.2 CLASSIFICATION OF CUTTING FLUIDS

The most prevalent lubricant used in machining processes is CFs. The cooling and lubricating actions of CFs, as well as the elimination of chips formed at the cutting area, are their most notable features. The cooling action reduces the temperature developed in the machining zone. However, the lubricating action helps the chips to flow easily across the rake face and prevents build-up edge formation. The workpiece, tool material, and machining process implemented influences the selection of CFs. Based on the type of machining, CFs are mainly classified as:

- Emulsifiable and neat CFs
- Synthetic and semi-synthetic CFs
- Gaseous CFs

Emulsifiable CFs are the most widely used types of CF. The emulsion formed by mixing it with water is used for general purposes and high-speed machining processes. The constituents of emulsifiable CFs are base oil, emulsifiers, and additives. Neat CFs, also known as straight oils, are used as is, without mixing with water. The constituents of neat CFs are base oil (mainly mineral oils) and additives such as sulfur, chlorinated paraffin, etc. Synthetic CFs are known to provide a better cooling effect. They are prepared from alkaline organic/non-organic compounds chemically. Additives are added to enhance the desired properties. These types of CF have commonly been used to machine both ferrous and non-ferrous metals. A mixture of emulsifiable and synthetic fluid creates a semi-synthetic type of fluid. They are mainly used for heavy-duty machining processes. The most common gases used for gaseous fluids are air, nitrogen, argon, and helium. The negative aspect of these gaseous CFs, which makes it less attractive for machining operation, is its poor cooling capacity.

8.3 TEST METHODS FOR DEVELOPING GREEN CUTTING FLUID

The test methods implemented for developing a green cutting fluid are explained in detail in this section. The common tests conducted for developing a green CF are shown in Table 8.1.

TABLE 8.1
Common Tests Conducted for Green Cutting Fluids

Sl. No.	Name of the Test Conducted	Standards Used	Description
1.	Tribological properties	ASTM D4172, ASTM G99, ASTM D2783	To measure COF, wear, and extreme pressure
2.	Viscosity	ASTM D446	To measure kinematic viscosity
3.	Density	ASTM D4052	To evaluate the density of the fluid
4.	Chemical properties	IS: 548 (Part 1) – 1964	To measure acid, peroxide, and iodine values
5.	Pour and cloud points	ASTM D97, ASTM D2500	To evaluate low-temperature properties
6.	Thermal stability	ASTM E1131	To evaluate the maximum temperature levels
7.	Thermal conductivity	ASTM D2717	To evaluate the amount of heat transfer occurring in the fluid medium
8.	Flash and fire points	ASTM D1310-14/ ASTM D92-05	To evaluate the volatility of a lubricant
9.	Corrosion Stability	ASTM D130	To evaluate the corrosive nature
10.	Emulsion Stability	ASTM D3707	To evaluate the ability to form a better emulsion
11.	Oxidative stability	AOCS Cd-12-57 with TGA curve (oxygen environment), ASTM D2272	To evaluate the oxidative stability of the lubricant

8.3.1 TRIBOLOGICAL PROPERTIES

Tribological characteristics, such as wear scar diameter (WSD), extreme pressure, COF, and weight loss due to wearing, are commonly examined using four-ball testers or pin-on-disc tribometers as per ASTM D4172, ASTM D2783, and ASTM G99 standards, respectively. These characteristics are critical for determining the performance of a suitable lubricant. The worn surface image of the pin/balls needs to be obtained using an optical microscope or scanning electron microscope (SEM) which gives a detailed idea about the wear mechanism happening. The roughness of the interacting surface also needs to be determined.

8.3.2 PHYSICO-CHEMICAL PROPERTIES

The physical characteristics are density, color, and viscosity. The density of the sample can be evaluated using a density meter or specific gravity bottle. The dynamic/kinematic viscosity can be measured using a rheometer or cannon-Fenske viscometer.

Chemical properties such as acid, peroxide, and iodine value can be evaluated using a manual titration method as per IS: 548 (Part 1) – 1964. The amount of free fatty acids can be determined by evaluating the acid value. The oxidation of oil results in the formation of peroxides. The amount of peroxides present in the oil can be determined by evaluating the peroxide value. Both the acid and peroxide values help to determine the oxidation stability of oil. The critical oxidation sites in the oils are unsaturated bonds. The amount of unsaturation present in the oil can be determined by evaluating the iodine value.

8.3.3 THERMAL PROPERTIES

The oil's cold flow properties like pour and cloud points are assessed using ASTM D97, and ASTM D2500, respectively. These properties are especially important while utilizing the lubricants during the winter months and in colder countries. Lubricity will be negatively affected if the cloud and pour points of the lubricants are poor. Vegetable oils tend to produce macro-crystalline formations at low temperatures for an extended time period due to a bend in the triglyceride. Even worse, these structures self-stack and lose kinetic energy (Erhan et al., 2006).

Thermogravimetric analysis (TGA) determines the change in weight of the sample as a function of temperature. This method is used to evaluate the breakdown of oil samples at temperatures ranging from ambient temperature to 800 °C or higher. A simultaneous thermal analyzer is used to conduct a test in an oxygen environment. As an outcome, the temperature at which mass deterioration begins and when total degradation occurs can also be determined. High-temperature properties like flash and fire points of a lubricant help to determine the volatility of the oil sample. The ASTM D1310-14/ASTM D92-05 standards are employed to evaluate the flash and fire points. The lubricants should have high flash and fire points to guarantee safe operation.

For a lubricant, better thermal conductivity is required for proper cooling. If the thermal conductivity is too low, the heat transfer rate will be low which will cause heating of the workpieces while machining and eventually result in poor surface finish and burns.

8.3.4   EMULSION STABILITY

The emulsion stability test was carried out in accordance with ASTM D3707 guidelines. Emulsion samples were created by varying the amount of emulsifier used. After proper mixing, samples were collected into vials and securely kept in a hot air oven for 48 hours at a temperature of 85 °C. The sample was extensively evaluated by visual examination for any oil-water layer separation. If a distinct layer forms, it is assumed that the corresponding layer is less stable. The absence or reduction of layer formation suggests improved emulsion stability.

8.3.5   CORROSION STABILITY

The corrosion stability of the oil samples can be assessed using an ASTM D130 copper strip corrosion tester. The copper strip is polished with emery paper and silica powder, then washed and dried before being placed in a test tube, and 30 mL of test samples are poured into the test tube to submerge the copper strip. After that, the test tube is put in a steel tube that is maintained at a constant temperature of 100 °C for a period of three hours. The color of the copper strip after three hours is graded by comparing it to the color chart provided for corrosion stability.

The corrosive stability of the emulsion needs to be evaluated as per the ASTM D4627. For this test, a petri dish with filter paper and 4 g of cast iron chips is filled with 25 ml of the prepared emulsion and covered with a lid. The dish is kept in a dark place for 24 hours, after which the filter paper needs to be cleaned in water to eliminate all cast iron chips. The weight loss of chip and stain markings on the filter paper are then observed to understand the corrosion stability of the sample. If the weight loss of the chip is minor, as are the stain marks on the filter paper, the emulsion is considered to be corrosive resistant.

8.3.6   OXIDATIVE STABILITY

Vegetable oils have low oxidative stability due to polyunsaturated acids like linoleic and linolenic acids (Wu et al., 2000). The stability of the samples against oxidation can be assessed using the hot oil oxidation test (HOOT) as recommended by AOCS Cd-12-57 standards. To speed up oxidation, the experiment can be carried out for 120 hours using a hot air oven (100 °C). The acid value, peroxide value, and viscosity of the samples will increase after HOOT as oxidative byproducts are formed. The sample with the lowest % increment after 120 hours is considered to have greater oxidative stability.

8.4   NEGATIVE ASPECTS OF MINERAL OIL-BASED CUTTING FLUIDS

The improper disposal of these mineral oil-based CFs have a negative impact on soil, water, and air. These create hazardous effects on agricultural products, aquatic life, etc. (Adhvaryu et al., 2005; Wilson, 1998). Enzymes and toxins produced in water-soluble CFs by microbial action are reported to be carcinogenic in nature, and their use could expose employees to cancer (Zeman et al., 1995). The alkaline nature of cutting fluids encourages the growth of harmful bacteria and after a time period, these emit a bad odor due to oxidization (Rebeccal et al., 1998). In CFs, Pseudomonas bacteria grow at an exponential rate, and turns aggressive in the event of an injury. These bacteria can break down the oil into inorganic chemicals (Rao et al., 2007).

The additives widely used in most of the commercial CFs are reported to be toxic and non-renewable in nature, and this leads to an ecological imbalance in nature (De Chiffre and Belluco, 2002). With an increase in temperature, chlorinated paraffin-like additives are converted to dioxin, which is categorized as hazardous (Klocke and Eisenblätter, 1997; Randegger-Vollrath, 1998). Due to environmental issues, the usage of biocides in CFs is restricted (Shokrani et al., 2012). The biocides found in mineral oils also induce occupational diseases such as oil acne, dermatitis, allergies, and respiratory infections in workers (Jabbar et al., 2017; Lima and Elsner, 2020; Park, 2019; Vijay et al., 2007; Warshaw et al., 2017). Recent studies related to CFs have reported cases of bladder cancer and deaths from lung cancer (Colin et al., 2018; Lima and Elsner, 2020).

8.5   VEGETABLE OILS AS BASE STOCK FOR CUTTING FLUIDS

These vegetable oils are a mixture of triglyceride structures, which is a glycerol molecule with three long-chain fatty acids along with ester linkages (Fox and Stachowiak, 2007). Free fatty acids are also present in it. The glycerol part in the triglycerides and polar head of the fatty acid has the ability to adhere to the contact surface to form an effective protective film between them. The fatty acid chains will enhance the strength of film formed (Prasannakumar et al., 2022). The polarity in fatty acid chains creates a dipole in vegetable oils which will provide an oiliness that improves the anti-wear properties (Maleque et al., 2003; Wagner et al., 2001). The fatty acid profile (chain length, saturation level) will differ among vegetable oils. The commonly present fatty acids in vegetable oils are stearic acid, which has zero double bonds; oleic acid, which has one double bond; linoleic acid, which has two double bonds; and linolenic acid, which has three double bonds (Castro et al., 2005). The structures of these fatty acids are shown in Figure 8.1.

The significant problem of using vegetable oils is their poor low-temperature property and oxidation stability (Adhvaryu et al., 2004; Frewing, 1942). The reason for these negative aspects is also because of the presence of saturated and unsaturated fatty acids in vegetable oils. Many industries are trying

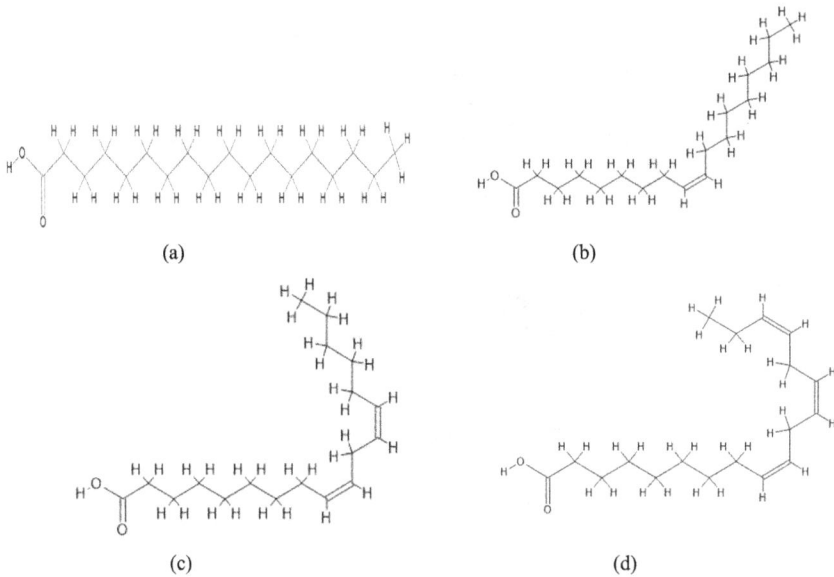

FIGURE 8.1 Examples of saturated and unsaturated fatty acids (a) stearic acid, (b) oleic acid, (c) linoleic acid, (d) linolenic acid.

to commercialize vegetable oil-based cutting fluid, but economic feasibility is one main challenge faced by the industries when compared with the commercially available products. The biodegradable nature of these vegetable oils is the main reason for continuing research works in this area (Peralta Álvarez et al., 2016; Shashidhara and Jayaram, 2010).

Leading firms such as Blazer in Switzerland, Renewable Lubricants in the United States, and others are attempting to commercialize vegetable oil-based working fluids (Shashidhara and Jayaram, 2006, 2010). The better performance of Inconel-like superalloys is increasing the demand for potential alternative cutting fluids. Furthermore, an alternative cutting fluid must be developed to obtain high-dimensional accuracy with improved surface properties (Şirin and Kıvak, 2019). The performance of the vegetable oil-based CFs in most of the research works has been experimentally evaluated based on the cutting forces, the workpiece's surface roughness, tool wear and change in its morphology, and cutting temperature generated. To improve machining performance, suitable additives need to be used. Nanoparticles are considered a desirable additive for this purpose (de Souza et al., 2019).

8.6 LUBRICATION MECHANISM OF NANOPARTICLES

When various nanoparticles were incorporated into the lubricating oil, the tribological properties were reported to be improved by four different lubricating mechanisms (Majeed et al., 2020; Sneha et al., 2021a; Thampi et al., 2021).

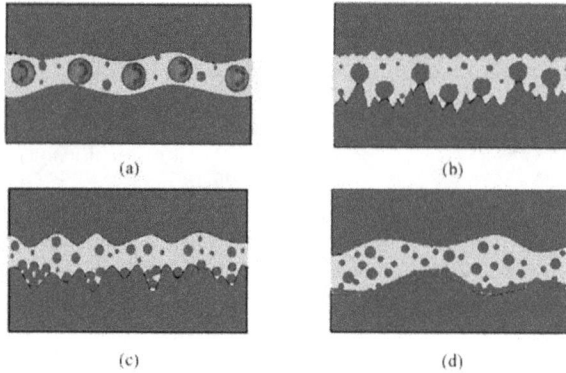

FIGURE 8.2   Lubrication mechanism of nanoparticles (Shafi et al., 2018; Singh et al., 2020).

These mechanisms are mentioned below and shown in Figure 8.2. The nanoparticles will:

a. roll between contact surfaces and convert the sliding friction to rolling friction (rolling effect);
b. polish the contact surface and reduce surface roughness (polishing effect);
c. form a physical film and compensate for the mass loss (mending effect); and
d. form a protective film between the friction pairs.

The usage of 0.5 weight % Al_2O_3/SiO_2 nanoparticles caused a rolling effect on the contact surface, which caused a significant drop of 50% in friction (Jiao et al., 2011). It was reported that when copper nanoparticles were used as an additive, they formed a protective layer that reduced shear force (Pan and Zhang, 2010). An LaF_3 nanoparticle covered with sulphur and a phosphate-based organic compound was reported to have formed an effective tribo-film by depositing LaF_3 nanoparticles on the surfaces in contact causing a chemical reaction between the sulphur and phosphate (Zhou et al., 2001). Surface-modified Y_2O_3 nanoparticles were reported to have entered the micro-cracks on worn surfaces and thereby enhanced the properties (Yang et al., 2012). The variation in WSD in regard to surface roughness was found to be the least for diamond and SiO_2 nanoparticles. The presence of nanoparticles polished the rubbed surface, resulting in a reduction of the surface roughness, but the wear scar remained constant (Peng et al., 2009). Generally, the shape, crystal **structure**, size, and dispersal of nanoparticles **showcase** a crucial role in the tribological properties of nano-fluids.

8.7   EFFECT OF NANOPARTICLES ON THE TRIBOLOGICAL PROPERTIES OF VEGETABLE OILS

Kumar et al. (2020) mixed CuO nanoparticles with canola oil and the least COF and specific wear rate were obtained at 0.1% weight. According to Xu et al, the tribological characteristics of rapeseed oil with hollow spherical MoS_2 precipitated in nano TiO_2 were superior to the samples with MoS_2 only (Xu et al., 2015). Both TiO_2 nanoparticles and microparticles were introduced to chemically modified rapeseed by Arumugam and Sriram. When compared to micro-sized TiO_2, it was discovered that TiO_2 nanoparticles improved lubricating qualities. With nanoscale TiO_2 and 6.9% with micro TiO_2, the COF of epoxidation followed by hydroxylation of rapeseed oil was lowered by 15.2% and 6.9%, respectively (Arumugam and Sriram, 2013).

When Thottackkad et al. introduced CuO nanoparticles on coconut oil, the findings showed an optimal range of 0.34% weight of CuO nanoparticles produced the lowest friction coefficient and wear rate. The viscosity and fire point rose after the addition of nano-CuO, but the surface roughness decreased (Thottackkad et al., 2012). Suthar et al. employed Al_2O_3 nanoparticles as a jojoba oil additive, achieving the lowest COF and wear at a range of 0.1% weight. Copper nanoparticles were added to pongamia oil by Rajubhai et al. Based on the minimal COF and wear induced, the optimal range of copper nanoparticles was found to be 0.075% weight (Rajubhai et al., 2020).

Boshui et al. found that stearic acid-modified cerium borate nanoparticles improved the anti-wear and anti-friction ability of rapeseed oil (Boshui et al., 2015). Copper and hBN nanoparticles were added to epoxidized olive oil by Kerni et al. After adding 0.5% weight nanoparticles, some lubricant parameters such as wear volume and COF increased (Kerni et al., 2019). According to Mahara et al., adding 0.3% weight SiO_2 nanoparticles to neem oil resulted in minimal friction and wear (Mahara and Singh, 2020).

Rajaganapathy et al. created nano-lubricants using CuO and TiO_2 nano-additives in palm oil and brassica oil. When compared to other nano-lubricants, palm oil containing 0.5% wt. of CuO nano-additive had the lowest COF and specific wear rate (Rajaganapathy et al., 2021). The least COF and wear rate for castor oil combined with TiO_2 nanoparticles occurs at 0.2% concentration, according to Singh et al. (2020a). To improve the lubricating properties of jojoba oil, Zaid et al. employed TiO_2 nanoparticles. When 0.3% wt. TiO_2 nanoparticles were introduced to jojoba oil, the COF and wear obtained from the test findings were reported to be the minimum (Zaid et al., 2021).

Gulzar et al. used CuO and MoS_2 nano-additives to improve the anti-wear and extreme pressure characteristics of chemically treated palm oil. According to the findings, these nanoparticles increased the measured characteristics by 1.5 times. MoS_2 outperformed CuO in terms of characteristics and dispersion stability (Gulzar et al., 2015). In rice bran oil, Rani incorporated TiO_2, CeO_2, and ZrO_2 nano-additives. The findings showed that 0.5% wt. CeO_2 and 0.3% wt. TiO_2 had the greatest decrease in COF and wear, respectively (Rani, 2016). Singh et al. added SiC nanoparticles in varying concentrations after adjusting moringa oil through two steps of transesterification. At 0.5% weight SiC nanoparticle, the COF and wear were found to be the lowest (Singh et al., 2020c).

Cortes et al. investigated the tribological and rheological properties of sunflower oil using SiO_2 and TiO_2 as nano-lubricant additives. The findings revealed that the rheological characteristics of nanoparticles vary depending on their type and concentration. The friction and wear volume loss were decreased by 93.7 and 70.1%, respectively, with TiO_2 and SiO_2 nanoparticles (Cortes et al., 2020). Rastogi et al. infused jatropha oil with SiO_2 nanoparticles. After the addition of 0.6% nanoparticles, optimal COF and wear loss were obtained (Rastogi et al., 2021). Singh et al. studied the tribological properties of copper nano-additive-induced desert date oil. Adding copper nanoparticles to a limit of 0.9% improved COF and wear of the prepared bio-lubricant. ZnO nanoparticles were added to epoxidized *Euphorbia lathyris* oil by Singh et al. Due to good anti-wear characteristics and dispersion stability, the optimal ZnO nanoparticles concentration was found to be 0.5% (Singh et al., 2020b).

Sneha et al. investigated the anti-wear properties of halloysite nanoclay in rice bran oil with turmeric oil. When compared, rice bran oil with 0.1% wt. halloysite nanoclay had the lowest COF and wear scar diameter. The WSD decreased to 0.491 mm when 1.5% wt. turmeric oil was added as an anti-oxidant ingredient to rice bran oil with halloysite nanoclay (Sneha et al., 2021b).

8.8 VEGETABLE OIL-BASED NANO-FLUIDS FOR VARIOUS MACHINING PROCESSES

Machining is a critical activity in manufacturing industry, and it requires the use of different CFs. Turning, milling, drilling, and grinding are some of the major machining processes implemented in the industry. The usage of vegetable oil-based CFs in machining are mentioned below:

8.8.1 Drilling Operation

Minimal quantity lubrication (MQL) micro-drilling operation in aluminium 6061 against DIXI 1138 with nano-diamond (2% vol.) added vegetable oil-based CF resulted in a greater decrease in torque and thrust forces (Nam et al., 2011). The micro-drilling procedure of Ti-6Al-4 V with 0.4% wt. of nano-diamond particles (35 nm), and a reduced feed rate (10 mm/min) improved the machinability of palm oil-based cutting fluid (Nam and Lee, 2018). Drilling AISI 321 stainless steel with an HSS drill tool was performed with nano-graphene sunflower oil-based CF. The nano-fluid showed increased cooling capabilities. The drilling forces obtained were observed to be higher for ordinary MQL conditions when compared to nano-fluid MQL conditions. This is mainly because of the cutting zone's weak heat dissipation capabilities and inadequate lubricating impact (Pal et al., 2020). Drilling AISI 4140 steel with carbide tool was performed using coconut oil-based copper nano-fluid. The high temperature in the machining was prevented by the addition of copper due to its high thermal conductivity and effective heat transfer coefficient. As a result, MQL with copper nano-fluid as a coolant reduces tool wear and therefore surface roughness (Muthuvel et al., 2020).

8.8.2 MILLING OPERATION

The impact of dry/MQL milling operation on flank wear was studied with commercial vegetable oil-based CF, and nano-graphene dispersed in vegetable oil-based CF. The study indicated that the flank wear reduced significantly while implementing MQL and due to the cooling/lubricant effects of nano-graphene, the CF with nano-graphene resulted in a less damaged rake face (Uysal, 2016). The addition of graphene to vegetable-based CF- LB2000 improved the oil film's anti-friction and load-bearing ability in the machining zone. Furthermore, the milling process' vibration was greatly reduced (Li et al., 2018). TC4 alloy is widely used in various engineering fields. Experiments were conducted on this titanium alloy under various lubrication conditions with several MQL conditions. The surface micro-hardness was mainly affected by both the CF ability and its flow rate. The addition of graphene particles in vegetable oil-based CF enhanced the cooling and lubrication effect, thus, improved the machining performance (Li et al., 2019). The milling of Inconel superalloy with hBN, graphite, and MoS_2 nanoparticle in vegetable oil-based CF reported that 0.5% hBN nanoparticles improved machining performances and tool life (Şirin and Kıvak, 2019).

8.8.3 TURNING OPERATION

Modified jatropha oil (MJO) with hBN additives were used to conduct performance tests on MQL-based turning operations. It was observed that MJO has good stable lubricity and can replace traditional CFs. The inclusion of hBN particles improved lubrication performance, according to the findings. Combined with 0.05% wt. of hBN particles, MJO enhanced kinematic viscosity and the viscosity index, and showed excellent tribological and machining properties (Talib et al., 2017; Talib and Rahim, 2018). Coconut oil-based working fluid with 0.8% wt. Al_2O_3 nanoparticles and cocamidopropyl betaine minimized tool wear and increased tool life by 40.17% while machining Inconel alloy (Ali et al., 2020).

8.8.4 GRINDING OPERATION

The lubrication capabilities of CuO in canola oil were examined in the surface grinding of AISI 1045 hardened steel. While grinding using CuO nano-fluids in MQL condition, the specific tangential force and force ratio was found to be reduced. The existence of a lubricating film reduced metal-to-metal contact in MQL grinding with CuO nano-fluids, lowering heat retention on the workpiece. Furthermore, even at a high depth of cut and linear velocity of the workpiece, the rolling action of nanoparticles produced a smooth surface (Shabgard et al., 2017). The grinding of Inconel 718 with palm and groundnut oil-based CFs using the MQL method was tested. In comparison to pure MQL/flood grinding, palm oil with 0.5% Al_2O_3 has the lowest grinding energy and COF values, whereas MQL with 1% Al_2O_3 in palm oil has a reduced surface roughness value, which is attributed to greater lubrication that allows the chips to slide away from the surface more easily (Virdi et al., 2019). CuO nanoparticles at different concentrations were dispersed in the vegetable oils to grind Inconel alloy. The sunflower oil-based nano-fluids have shown superior cooling lubrication compared to pure-flood lubrication due to the sunflower oil's ability to form a stable film in the machining zone, which improves the wetting property and heat-carrying capacity (Virdi et al., 2020).

8.9   CONCLUSIONS AND FUTURE SCOPE

Vegetable oils are a promising alternative for mineral-based oils due to their superior lubricating properties and environmentally beneficial characteristics. In machining, cooling and lubrication are critical for decreasing friction at the tool–workpiece interface. Hence nano-fluids can provide superior coolant and lubricant properties.

- Vegetable oil-based CFs have shown promising results compared with commercial lubricants. However, the tribological and thermal conductivity properties were found to be improved with the addition of nanoparticles.
- Research on non-edible vegetable oils such as jatropha, karanja, tamanu, and polanga oils is increasing nowadays due to less competition and usage of these oils in daily activities. Therefore, extensive research needs to be conducted on non-edible oils.
- Vegetable oils such as mahua, rapeseed, palm, and neem oils reduced thrust force and increased tool life when drilling. Nanoparticles such as CuO and nano-graphene were added to the workpiece to increase cooling and surface roughness. When machining high-strength materials like titanium alloy, palm oil-based nano-fluid may be the best option. The natural anti-oxidants present in vegetable oils increases the shelf life of the oil.
- While performing a milling operation vegetable oils are mixed with hBN, graphite, and MoS_2 nanoparticle and hBN blends were found to enhance machining characteristics and increase tool life. Non-edible oils, such as jatropha oil-based nano-fluids have shown promising results as a metalworking fluid for milling operations.
- While performing turning operations, denser oils such as castor oil provide the most cooling and lubrication effect. Palm kernel oil was found to be ideal for cutting aluminum. Non-edible oils have reduced chip thickness and improved surface smoothness, whilst neem oil improved cooling.
- The blending of various vegetable oils has given better performance while performing grinding. The use of nanoparticles enhanced the G-ratio significantly due to the increased material removal. CuO nanoparticles were used in grinding to reduce force ratio and specific tangential force.

The economic, technical and environmental aspects of nano-fluids need to be evaluated for the use of nano-fluids in the industrial sector. The results obtained after evaluating the performance of nano-lubricants show that they have great potential to be used as a metalworking fluid in industrial applications. However, research groups should concentrate on improving the oxidation and thermal stability of these innovative lubricants. Nanoparticles suspension stability must be examined and reported in the case of nano-fluids. The environmental impact of nanoparticles' use needs to be evaluated as industries are moving towards greener cutting fluids. The economical aspect of the nano-fluids needs to be identified since CFs need to be produced on a large scale. Also, the long-term stability of the nano-fluids needs to be evaluated so that functionality is not lost. The application of nano-fluids has to be tested on various workpiece materials, such as ferrous and non-ferrous materials, since the nanoparticles will have a different reaction based on the workpiece. The solubility of nanoparticles, as well as the optimum percentage of nanoparticles in the vegetable oils, needs to be evaluated. Reducing the percentage of additives used in CF's multifunctional nanoparticles will have a great future in the field of bio-lubricants.

REFERENCES

Adhvaryu, A., Erhan, S.Z., Perez, J.M., 2004. Tribological studies of thermally and chemically modified vegetable oils for use as environmentally friendly lubricants. *Wear* 257, 359–367.

Adhvaryu, A., Liu, Z., Erhan, S.Z., 2005. Synthesis of novel alkoxylated triacylglycerols and their lubricant base oil properties. *Ind. Crop Prod.* 21, 113–119.

Ali, M.A.M., Azmi, A.I., Murad, M.N., Zain, M.Z.M., Khalil, A.N.M., Shuaib, N.A., 2020. Roles of new bio-based nanolubricants towards eco-friendly and improved machinability of Inconel 718 alloys. *Tribol. Int.* 144, 106106.

Arumugam, S., Sriram, G., 2013. Preliminary study of nano-and microscale TiO$_2$ additives on tribological behavior of chemically modified rapeseed oil. *Tribol. Trans.* 56, 797–805.

Boshui, C., Kecheng, G., Jianhua, F., Jiang, W., Jiu, W., Nan, Z., 2015. Tribological characteristics of monodispersed cerium borate nanospheres in biodegradable rapeseed oil lubricant. *Appl. Surf. Sci.* 353, 326–332.

Burton, G., Goo, C.-S., Zhang, Y., Jun, M.B.G., 2014. Use of vegetable oil in water emulsion achieved through ultrasonic atomization as cutting fluids in micro-milling. *J. Manuf. Process.* 16, 405–413.

Castro, W., Weller, D.E., Cheenkachorn, K., Perez, J.M., 2005. The effect of chemical structure of basefluids on antiwear effectiveness of additives. *Tribol. Int.* 38, 321–326.

Colin, R., Grzebyk, M., Wild, P., Hédelin, G., Bourgkard, È., 2018. Bladder cancer and occupational exposure to metalworking fluid mist: a counter-matched case–control study in French steel-producing factories. *Occup. Environ. Med.* 75, 328–336.

Cortes, V., Sanchez, K., Gonzalez, R., Alcoutlabi, M., Ortega, J.A., 2020. The performance of SiO$_2$ and TiO$_2$ nanoparticles as lubricant additives in sunflower oil. *Lubricants* 8, 10.

De Chiffre, L., Belluco, W., 2002. Investigations of cutting fluid performance using different machining operations. *Tribol. Lubr. Technol.* 58, 22.

deGroot, A.C., Flyvholm, M.-A., 2020. Formaldehyde and formaldehyde-releasers. *Kanerva's Occup. Dermatology* 521–542. https://doi.org/10.1007/978-3-319-68617-2_37

de Souza, M.C., de Souza Gonçalves, J.F., Gonçalves, P.C., Lutif, S.Y.S., de Oliveira Gomes, J., 2019. Use of Jatropha and Moringa oils for lubricants: Metalworking fluids more environmental-friendly. *Ind. Crop Prod.* 129, 594–603.

Erhan, S.Z., Sharma, B.K., Perez, J.M., 2006. Oxidation and low temperature stability of vegetable oil-based lubricants. *Industrial Crops and Products* 24(3), 292–299.

Fox, N.J., Stachowiak, G.W., 2007. Vegetable oil-based lubricants—A review of oxidation. *Tribol. Int.* 40, 1035–1046.

Frewing, J.J., 1942. The influence of temperature on boundary lubrication. *Proc. R. Soc. London. Ser. A. Math. Phys. Sci.* 181, 23–42.

Gulzar, M., Masjuki, H.H., Varman, M., Kalam, M.A., Mufti, R.A., Zulkifli, N.W.M., Yunus, R., Zahid, R., 2015. Improving the AW/EP ability of chemically modified palm oil by adding CuO and MoS_2 nanoparticles. *Tribol. Int.* 88, 271–279.

Jabbar, M.A., Hashim, Z., Zainuddin, H., Munn-Sann, L., 2017. Respiratory health effects of metalworking fluid among metal machining workers. *Asia Pacific Environ. Occup. Heal. J.* 3, 15–19.

Jiao, D., Zheng, S., Wang, Y., Guan, R., Cao, B., 2011. The tribology properties of alumina/silica composite nanoparticles as lubricant additives. *Appl. Surf. Sci.* 257, 5720–5725.

Kerni, L., Raina, A., Haq, M.I.U., 2019. Friction and wear performance of olive oil containing nanoparticles in boundary and mixed lubrication regimes. *Wear* 426, 819–827.

Klocke, F., Eisenblätter, G., 1997. Dry cutting. *Cirp Ann.* 46, 519–526.

Kumar, V., Dhanola, A., Garg, H.C., Kumar, G., 2020. Improving the tribological performance of canola oil by adding CuO nanoadditives for steel/steel contact. *Mater. Today Proc.* 28, 1392–1396.

Li, M., Yu, T., Yang, L., Li, H., Zhang, R., Wang, W., 2019. Parameter optimization during minimum quantity lubrication milling of TC4 alloy with graphene-dispersed vegetable-oil-based cutting fluid. *J. Clean. Prod.* 209, 1508–1522.

Li, M., Yu, T., Zhang, R., Yang, L., Li, H., Wang, W., 2018. MQL milling of TC4 alloy by dispersing graphene into vegetable oil-based cutting fluid. *Int. J. Adv. Manuf. Technol.* 99, 1735–1753.

Lima, A.L., Elsner, P., 2020. Metal Industry. *Kanerva's Occup. Dermatology* pp. 2123–2125.

Mahara, M., Singh, Y., 2020. Tribological analysis of the neem oil during the addition of SiO_2 nanoparticles at different loads. *Mater. Today Proc.* 28, 1412–1415.

Majeed, F.S.A., Yusof, N.B.M., Suhaimi, M.A., Elsiti, N.M., 2020. Effect of paraffin oil with XGnP and Fe_2O_3 nanoparticles on tribological properties. *Mater. Today Proc.* 27, 1685–1688.

Maleque, M.A., Masjuki, H.H., Sapuan, S.M., 2003. Vegetable-based biodegradable lubricating oil additives. *Ind. Lubr. Tribol.* 55(3), 137–143.

Muthuvel, S., Naresh Babu, M., Muthukrishnan, N., 2020. Copper nano-fluids under minimum quantity lubrication during drilling of AISI 4140 steel. *Aust. J. Mech. Eng.* 18, S151–S164.

Nam, J., Lee, S.W., 2018. Machinability of titanium alloy (Ti-6Al-4V) in environmentally-friendly micro-drilling process with nano-fluid minimum quantity lubrication using nanodiamond particles. *Int. J. Precis. Eng. Manuf. Technol.* 5, 29–35.

Nam, J.S., Lee, P.-H., Lee, S.W., 2011. Experimental characterization of micro-drilling process using nano-fluid minimum quantity lubrication. *Int. J. Mach. Tool Manuf.* 51, 649–652.

Pal, A., Chatha, S.S., Sidhu, H.S., 2020. Experimental investigation on the performance of MQL drilling of AISI 321 stainless steel using nano-graphene enhanced vegetable-oil-based cutting fluid. *Tribol. Int.* 151, 106508.

Pan, Q., Zhang, X., 2010. Synthesis and tribological behavior of oil-soluble Cu nanoparticles as additive in SF15W/40 lubricating oil. *Rare Met. Mater. Eng.* 39, 1711–1714.

Park, R.M., 2019. Risk assessment for metalworking fluids and respiratory outcomes. *Saf. Health Work* 10, 428–436.

Peng, D.X., Kang, Y., Hwang, R.M., Shyr, S.S., Chang, Y.P., 2009. Tribological properties of diamond and SiO₂ nanoparticles added in paraffin. *Tribol. Int.* 42, 911–917.

Peralta Álvarez, M.E., Marcos Bárcena, M., Aguayo González, F., 2016. A review of sustainable machining engineering: optimization process through triple bottom line. *J. Manuf. Sci. Eng.* 138. https://doi.org/10.1115/1.4034277

Pop, L., Puşcaş, C., Bandur, G., Vlase, G., Nuţiu, R., 2008. Basestock oils for lubricants from mixtures of corn oil and synthetic diesters. *J. Am. Oil Chem. Soc.* 85, 71–76.

Pranav, P., Sneha, E., Rani, S., 2021. Vegetable oil-based cutting fluids and its behavioral characteristics in machining processes: A review. *Ind. Lubr. Tribol.* 73(9), 1159–1175.

Prasannakumar, P., Edla, S., Thampi, A.D., Arif, M., Santhakumari, R., 2022. A comparative study on the lubricant properties of chemically modified Calophyllum inophyllum oils for bio-lubricant applications. *J. Clean. Prod.* 339, 130733.

Rajaganapathy, C., Vasudevan, D., Murugapoopathi, S., 2021. Tribological and rheological properties of palm and brassica oil with inclusion of CuO and TiO₂ additives. *Mater. Today Proc.* 37, 207–213.

Rajubhai, V.H., Singh, Y., Suthar, K., Surana, A.R., 2020. Friction and wear behavior of Al-7% Si alloy pin under pongamia oil with copper nanoparticles as additives. *Mater. Today Proc.* 25, 695–698.

Randegger-Vollrath, A., 1998. Determination of chlorinated paraffins in cutting fluids and lubricants. *Fresenius J. Anal. Chem.* 360, 62–68.

Rani, S., 2016. The tribological behavior of TiO₂, CeO₂ and ZrO₂ nano particles as a lubricant additive in rice bran oil. *Int. J. Sci. Eng. Res.* 7, 708–712.

Rao, D.N., Srikant, R.R., Rao, P.N., 2007. Effect of emulsifier content on microbial contamination of cutting fluids. *Int. J. Mach. Mach. Mater.* 2, 469–477.

Rastogi, P.M., Kumar, R., Kumar, N., 2021. Effect of SiO₂ nanoparticles on the tribological characteristics of jatropha oil. *Mater. Today Proc.* 46, 10109–10112.

Rebeccal, G., Rogere, M., Peter, A., 1998. Biodegradable lubricants. *Lubr. Eng.* 7, 10–16.

Shabgard, M., Seyedzavvar, M., Mohammadpourfard, M., 2017. Experimental investigation into lubrication properties and mechanism of vegetable-based CuO nano-fluid in MQL grinding. *Int. J. Adv. Manuf. Technol.* 92, 3807–3823.

Shafi, W.K., Raina, A., Ul Haq, M.I., 2018. Friction and wear characteristics of vegetable oils using nanoparticles for sustainable lubrication. *Tribol. - Mater. Surf. Interfaces*, 12(1), 27–43.

Shashidhara, Y.M., Jayaram, S.R., 2006. Vegetable Oil Based Lubricants for Industrial Applications—A Review, in: *Proceedings of International Conference on Industrial Tribology-06*, 30th Nov.–2nd Dec.

Shashidhara, Y.M.Ã., Jayaram, S.R., 2010. Vegetable oils as a potential cutting fluid—An evolution. *Tribol. Int.* 43, 1073–1081. https://doi.org/10.1016/j.triboint.2009.12.065

Shokrani, A., Dhokia, V., Newman, S.T., 2012. Environmentally conscious machining of difficult-to-machine materials with regard to cutting fluids. *Int. J. Mach. Tool Manuf.* 57, 83–101.

Singh, A., Chauhan, P., Mamatha, T.G., 2020. A review on tribological performance of lubricants with nanoparticles additives. *Mater. Today Proc.* 25, 586–591.

Singh, A.K., Gupta, A.K., 2006. Metalworking fluids from vegetable oils. *J. Synth. Lubr.* 23, 167–176.

Singh, Y., Chaudhary, V., Pal, V., 2020a. Friction and wear characteristics of the castor oil with TiO2 as an additives. *Mater. Today Proc.* 26, 2972–2976.

Singh, Y., Sharma, A., Singh, N.K., Chen, W.-H., 2020b. Development of bio-based lubricant from modified desert date oil (balanites aegyptiaca) with copper nanoparticles addition and their tribological analysis. *Fuel* 259, 116259.

Singh, Y., Sharma, A., Singh, N.K., Noor, M.M., 2020c. Effect of SiC nanoparticles concentration on novel feedstock Moringa Oleifera chemically treated with neopentylglycol and their trobological behavior. *Fuel* 280, 118630.

Şirin, Ş., Kıvak, T., 2019. Performances of different eco-friendly nano-fluid lubricants in the milling of Inconel X-750 superalloy. *Tribol. Int.* 137, 180–192.

Sneha, E., Akhil, R.B., Krishna, A., Rani, S., Kumar, S.A., 2020. Formulation of bio-lubricant based on modified rice bran oil with stearic acid as an anti-wear additive. *Proc. Inst. Mech. Eng. Part J J. Eng. Tribol.* 235 (9), 1350650120977381.

Sneha, E, Sarath, V. S., Rani, S., Bindu Kumar, K., 2021b. Effect of turmeric oil and halloysite nano clay as anti-oxidant and anti-wear additives in rice bran oil. *Proc. Inst. Mech. Eng. Part J J. Eng. Tribol.* 235, 1085–1092.

Sneha, Edla, Akhil, R.B., Krishna, A., Rani, S., Kumar, S.A., 2021a. Formulation of bio-lubricant based on modified rice bran oil with stearic acid as an anti-wear additive. *Proc. Inst. Mech. Eng. Part J J. Eng. Tribol.* 235, 1950–1957.

Talib, N., Nasir, R.M., Rahim, E.A., 2017. Tribological behaviour of modified jatropha oil by mixing hexagonal boron nitride nanoparticles as a bio-based lubricant for machining processes. *J. Clean. Prod.* 147, 360–378.

Talib, N., Rahim, E.A., 2018. Experimental evaluation of physicochemical properties and tapping torque of hexagonal boron nitride in modified jatropha oils-based as sustainable metalworking fluids. *J. Clean. Prod.* 171, 743–755.

Thampi, A.D., John, A.R., Rani, S., Arif, M.M., 2020. Chemical modification and tribological evaluation of pure rice bran oil as base stocks for biodegradable lubricants. *J. Inst. Eng. Ser. E* 102, 1–6.

Thampi, A.D., Prasanth, M.A., Anandu, A.P., Sneha, E., Sasidharan, B., Rani, S., 2021. The effect of nanoparticle additives on the tribological properties of various lubricating oils–Review. *Mater. Today Proc.* 47, 4919–4924.

Thottackkad, M.V., Perikinalil, R.K., Kumarapillai, P.N., 2012. Experimental evaluation on the tribological properties of coconut oil by the addition of CuO nanoparticles. *Int. J. Precis. Eng. Manuf.* 13, 111–116.

Trajano, M.F., Moura, E.I.F., Ribeiro, K.S.B., Alves, S.M., 2014. Study of oxide nanoparticles as additives for vegetable lubricants. *Mater. Res.* 17, 1124–1128.

Uysal, A., 2016. Investigation of flank wear in MQL milling of ferritic stainless steel by using nano graphene reinforced vegetable cutting fluid. *Ind. Lubr. Tribol.* 68(4), 446–451.

Vijay, V., Yeatts Jr, J.L., Riviere, J.E., Baynes, R.E., 2007. Predicting dermal permeability of biocides in commercial cutting fluids using a LSER approach. *Toxicol. Lett.* 175, 34–43.

Virdi, R.L., Chatha, S.S., Singh, H., 2019. Experiment evaluation of grinding properties under Al2O3 nano-fluids in minimum quantity lubrication. *Mater. Res. Express* 6, 96574.

Virdi, R.L., Chatha, S.S., Singh, H., 2020. Processing characteristics of different vegetable oil-based nano-fluid MQL for grinding of Ni-Cr alloy. *Adv. Mater. Process. Technol.* 8(1), 1–14.

Wagner, H., Luther, R., Mang, T., 2001. Lubricant base fluids based on renewable raw materials: their catalytic manufacture and modification. *Appl. Catal. A. Gen.* 221, 429–442.

Warshaw, E.M., Hagen, S.L., DeKoven, J.G., Zug, K.A., Sasseville, D., Belsito, D. V., Zirwas, M.J., Fowler Jr, J.F., Taylor, J.S., Fransway, A.F., 2017. Occupational contact dermatitis in North American production workers referred for patch testing: retrospective analysis of cross-sectional data from the North American Contact Dermatitis Group 1998 to 2014. *Dermatitis* 28, 183–194.

Wilson, B., 1998. Lubricants and functional fluids from renewable sources. *Ind. Lubr. Tribol.* 50(1), 6–15.

Wu, H., Zhao, J., Cheng, X., Xia, W., He, A., Yun, J.-H., Huang, S., Wang, L., Huang, H., Jiao, S., 2018. Friction and wear characteristics of TiO_2 nano-additive water-based lubricant on ferritic stainless steel. *Tribol. Int.* 117, 24–38.

Wu, X., Zhang, X., Yang, S., Chen, H., Wang, D., 2000. The study of epoxidized rapeseed oil used as a potential biodegradable lubricant. *J. Am. Oil Chem. Soc.* 77, 561–563.

Wu, Y.Y., Tsui, W.C., Liu, T.C., 2007. Experimental analysis of tribological properties of lubricating oils with nanoparticle additives. *Wear* 262, 819–825.

Xu, Z.Y., Xu, Y., Hu, K.H., Xu, Y.F., Hu, X.G., 2015. Formation and tribological properties of hollow sphere-like nano-MoS_2 precipitated in TiO_2 particles. *Tribol. Int.* 81, 139–148.

Yang, G.-B., Chai, S.-T., Xiong, X.-J., Zhang, S.-M., Yu, L.-G., Zhang, P.-Y., 2012. Preparation and tribological properties of surface modified Cu nanoparticles. *Trans. Nonferrous Met. Soc. Chin.* 22, 366–372.

Zaid, M., Kumar, A., Singh, Y., 2021. Lubricity improvement of the raw jojoba oil with TiO2 nanoparticles as an additives at different loads applied. *Mater. Today Proc.* 46, 3165–3168.

Zeman, A., Sprengel, A., Niedermeier, D., Späth, M., 1995. Biodegradable lubricants—studies on thermo-oxidation of metal-working and hydraulic fluids by differential scanning calorimetry (DSC). *Thermochim. Acta* 268, 9–15.

Zhou, J., Wu, Z., Zhang, Z., Liu, W., Dang, H., 2001. Study on an antiwear and extreme pressure additive of surface coated LaF3 nanoparticles in liquid paraffin. *Wear* 249, 333–337.

Zimmerman, J.B., Clarens, A.F., Hayes, K.F., Skerlos, S.J., 2003. Design of hard water stable emulsifier systems for petroleum-and bio-based semi-synthetic metalworking fluids. *Environ. Sci. Technol.* 37, 5278–5288.

9 Roles of Tribology in Friction Stir Welding and Processing

Sanjeev Kumar
Subharti Institute of Technology and Engineering,
Meerut, India

Jitendra Kumar Katiyar
SRM Institute of Science and Technology, Kattankulathur,
India

9.1 INTRODUCTION

Tribology contributes to the formation of energy-efficient technologies in various categories; such as component manufacturing, offering components longer life (due to reduced friction during use), and energy-saving components applied in interdisciplinary fields (Spikes, 1998). It incorporates a range of fundamental engineering subject areas, for example, mechanics (solid and fluid), internal combustion engines, heat transfer, automobiles, and lubrication (Jin and Fisher, 2008). Tribology may be defined as the science, technology, and engineering of interacting surfaces in relative motion and includes the study and practical application of the principles of wear, lubrication, and friction. Tribology plays an interdisciplinary role in prospective areas linked with materials, chemistry, physics, and even biology (from orthopaedic and dental implants to ophthalmic devices), introducing energy-efficient technology and functioning of joints (Spikes 1998; Affatato and Grillini, 2013).

Presently, aluminium alloy is mostly used throughout sectors because of its properties. The third generation of Al-Li alloy shows remarkable properties over Al alloy. It is frequently used in aerospace industries for its excellent properties, such as higher corrosion resistance, improved fatigue crack growth resistance, low density, and higher mechanical strength, etc. The joining of alloys like Al, Mg, Cu, Polymer, etc. through conventional welding methods, confer many defects such as the formation of oxide layers, solidification, distortion, etc. These can be removed by using FSW/P, which is extremely productive, emits no fumes, and requires no filler material, making it an environment-friendly process. It is also a solid-state hot shear and forging process which was introduced by Wayne M. Thomas of The Welding Institute, Cambridge, U.K., in December 1991.

DOI: 10.1201/9781003363576-9

A non-consumable rotating tool is rotated along the longitudinal axis to join similar or dissimilar plates. The different types of process parameters, like tool rotational speed (TRS), traverse speed or welding speed (TTS), tool tilt angle (TTA), tool design, axis load, plunge depth, and material, are responsible for the quality of joints (Affatato and Grillini, 2013). A tool has different parts, such as a shank, shoulder, and tool pin. The shoulder is an extended part of the shank which connects with the tool pin in the abutting surface of the workpiece during welding. The tool pin is capped by the shoulder. It moves inside the workpiece to perform a proper weld at solidus temperature by friction and plastic deformation, material mixing, diffusing the oxide layer and preventing defects caused by high pressure and temperature. Heat is generated between the tool and workpiece by pure sliding adhesion and sticking deformation (Kumar et al., 2020a).

The process of friction stir welding can be separated mainly into four phases which are shown in Figure 9.1: (i) Plunging, (ii) dwelling, (iii) welding, and (iv) dwelling and pulling out or retracting (Kumar et al., 2020; Khairuddin et al., 2012). In the plunging stage (shown in Figure 9.1(a), the rotating tool penetrates the abutting edge of the workpiece and produces heat. The generating force during this stage is higher compared to the welding stages. Afterwards, the tool continues rotating in its position as shown in Figure 9.1(b), the dwelling stage. The temperature level is increased beneath the position of the workpiece. The traverse speed is constant during this stage. The tool subsequently travels along the joining line to form the joint by stirring the material in the welding stage shown in Figure 9.1(c). After joining the material, the force acting on the tool has reduced and pulling out the tool from the workpiece is shown in Figure 9.1(d). This is also called the retracting phase.

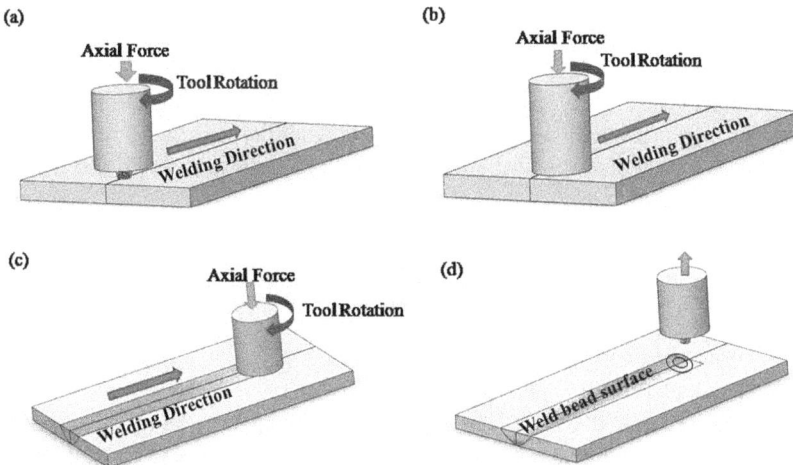

FIGURE 9.1 Different process of friction stir welding (a) plunging (b) dwelling (c) welding and (d) retracting.

The friction stir process (FSP) is a solid process used to produce composites for modifying the microstructure and improving material properties caused by longer heat exposure. The different types of mechanisms which affect the joint quality beyond the process parameters are discussed below:

9.2   COEFFICIENT OF FRICTION

The coefficient of friction (COF), often symbolized by the Greek letter μ, is a dimensionless scalar quantity defined as the ratio of the force acting between two bodies to the force causing friction between them (Halina and Motyka, 2019). Generally, it is assumed that the value of COF is less than one, but this is not true for all. The coefficient of friction depends on the materials; for example, for contact between rubber and other surfaces, the COF values lie between 1 and 2, while lower in the case of ice on steel (Schulson, 2001). A COF value of zero indicates that no friction is induced between the contact surfaces. The COF value increases with increased temperature and decreased load. The COF would decrease because the material becomes softer and weaker as the temperature rises. No straightforward method can be found for determining COF values (Nandan et al., 2005).

Friction stir welding/processing is used to join different types of material by the friction induced between the tool and the workpiece. It is challenging to construct a simple method for determining the COF in FSW due to the complicated kinematics of the FSW process. Previous studies have acknowledged the COF as a variable in FSW, but they have ignored the change and assumed a constant value for the duration of the entire FSW cycle. Different researchers have observed different COF joining of material in FSW. Kumar et al. (2009) observed that the value of COF generated between the FSW tool (die steel) and the workpiece (Al alloy) varies from 0.15 to 1.4, with different ranges of TRS from 200 rpm to 1,400 rpm with an increment of 400 rpm. The COF value was found to be 0.6 below the critical temperature; while conversely, the COF value was observed at 1.4 above the critical temperature at specific contact pressure. They applied the formulae to reach the COF given below, where L is the tangential load measured through the load cell, D is the distance of the load cell from the axis of the pin, and N is the normal force.

$$\mu = \frac{450L}{DN} \qquad (9.1)$$

Meyghani et al. (2019) investigated the effect of COF in thermal analysis during FSW at different TRS (800, 1,200, and 1,600 rpm) and TTS (40, 70, and 100 mm/min). The preeminent temperature prediction was observed at a COF of 0.3, TRS of 1,600 rpm, and TTS of 40 mm/min. The COF value is calculated using under given formula, where P_0 represents the axial pressure, τ_y is the shear stress, τ_0 is the contact shear stress at the pin and the bottom of the shoulder surface, τ_1 is the contact shear stress at the pin side area, and α is the thread angle in case of threaded cylindrical pin profile. In contrast, α is the cone angle in the conical pin profile.

$$\mu = \frac{\tau_0 - \tau_1}{1 - \dfrac{\tau_1 - \tau_0 \sin\alpha}{(1-\sin\alpha)\tau_y} P_0 (1 - \sin\alpha)} \tag{9.2}$$

Kumar et al. (2022) observed the connection between the COF and varying TRS from 600 rpm to 1,800 rpm. The relationship between COF and TRS was calculated from this formula, where ω is the tool angular velocity, R_s is the shoulder radius (m), and δ is the local variation in fractional slip. The COF values decrease with an increase in TRS from 600 rpm to 1,800 rpm, as shown in Figure 9.2.

$$\delta = 0.31 \exp\left(\frac{\omega R_s}{1.87}\right) - 0.026 \tag{9.3}$$

$$\mu = 0.5 \exp\left(-\delta \omega R_s\right) \tag{9.4}$$

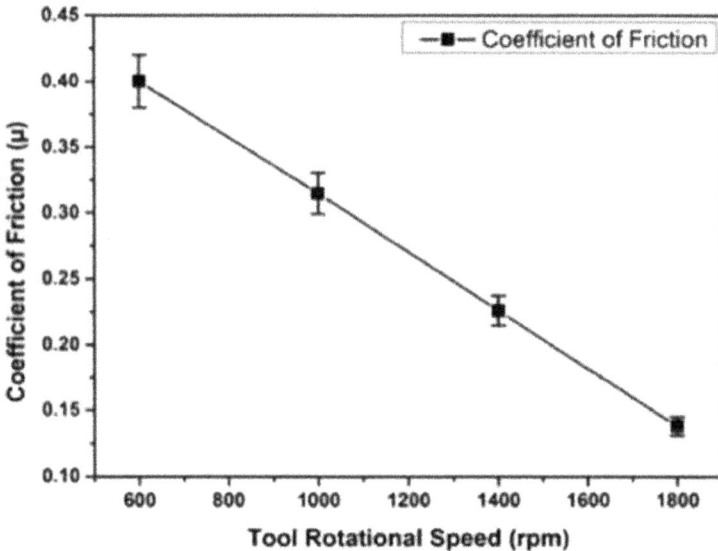

FIGURE 9.2 Coefficient of friction on varying TRS from 600 RPM to 1,800 RPM (Kumar et al. 2022).

9.3   MECHANISMS OF WEAR

Wear is the gradual loss of substance brought on by the mechanical interaction of two contacting surfaces. These surfaces typically move relative to one another under load, sliding, or rolling (Eyre, 1976). The local mechanical failure of highly strained interfacial zones causes wear, and this form of failure is frequently influenced by external conditions. Surface deterioration results in the formation of wear particles through a series of actions characterized by adhesion and particle transfer mechanisms, by a process of direct particle synthesis resembling machining or, in some circumstances, a surface fatigue form of failure. The different types of wear mechanism discussed and included below are also shown in Figure 9.3 (Majumdar and Manna, 2015).

9.3.1   ADHESIVE WEAR

Adhesive wear is the unintended transfer of material from one surface to another. The loss of material between contact surfaces is due to the adhesion properties. The atomic forces between the materials of contacting surfaces under relative pressure outweigh the inherent material characteristics of each surface. The frictional force induced between the surface and adhesive wear is directly proportional to each other. An increase in frictional forces typically results in a significant increase in wear (Rigney, 1997).

9.3.2   ABRASIVE WEAR

The type of wear mechanism known as abrasive wear causes the material on the surface to disintegrate under the influence of the hard surface sliding across the soft surface, resulting in the loss of material (Moore, 1974) The material is removed from the surface by the hard particles when the hard surface slides over the smooth surfaces. This type of wear may occur due to micro-cutting (sharp grit cut in the material); fracture (due to the crack initiated at the respective surface

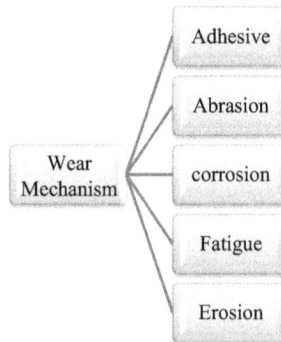

FIGURE 9.3   Classification of wear mechanism.

and forming a brittle fracture); fatigue (repetitive strain develop on the metal surface by the grit); grain pull-out (the grain boundary is weak and pulls out the grits); two body (the grits are attached on the surface during contact); and three body (grits are free to roll when abrasion occurs) (Kayaba, 1984; Stachowiak and Batchelor, 2013; Phillips, 1975; Vingsbo and Hogmark, 1981). This nature of wear happens when solid particles are put onto the surface of a material whose hardness is equivalent to or lower than the loaded particles (Yang et al., 1994).

9.3.3 CORROSION WEAR

This type of wear is formed as a result of the chemical or electrochemical response to a corrosive medium. The new surface is consequently subject to extra corrosive deterioration. It could be brought on by general oxidation, chemical vapours, dampness, combustion products, or unintentionally by lubricating additives (Rigney and Hirth, 1979). It is frequently referred to as corrosion-mechanical wear because it typically occurs concurrently with one of the types of mechanical wear.

9.3.4 FATIGUE WEAR

Surface fatigue failure occurs due to the strain induced on the surface by repeated stress and cyclic load from hard grit on the surface, which grows and forms cracks. It frequently occurs when strong localized forces are repeatedly applied to a tiny contact region, as in rolling or sliding contact (Jin and Fisher, 2008).

9.3.5 EROSION WEAR

Erosive wear is caused by the impact of particles against a solid surface. This is a function of particles' velocity, impact angle, and abrasive size. The wear rate is directly proportional to the velocity and particle size.

9.4 FACTORS AFFECTING WEAR

Sliding speed, load, temperature, hardness, elastic modulus, and material composition are the main determinants of wear. Contact temperature has an impact on wear resistance (Meshref et al., 2020). Abrasive wear would increase because the hardness and yield strength decrease with increasing temperature (Rymuza, 2007). For the majority of the materials, the yield strength and hardness decrease with increasing temperature. Metals' yield strength decreases at high temperatures due to the dislocation movement, which facilitates easier plastic deformation. The typical load has a significant effect on wear rate. The shear force and frictional thrust both increase as even more load is applied, accelerating the rate of wear (ASM International, 1992). In a study by Nuruzzaman et al. (2011) the rate of abrasive wear significantly increased between 0 and 2.5 m/s of sliding speed. The rise in wear may be due to frictional heating. Given that wear is also dependent on

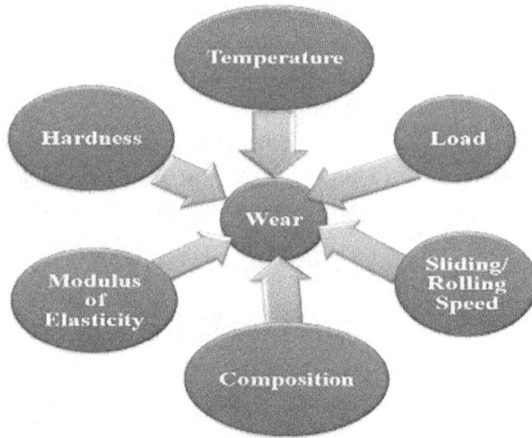

FIGURE 9.4   Factors affecting wear in FSW.

the force being applied to the surfaces, the presence of lubricant, and the surface roughness of the interacting surfaces, it cannot be generalized that an increase in sliding speed will necessarily increase wear. A higher number indicates more stiffness and the elastic modulus is a measure of a material's resistance to elastic deformation (Lu et al., 2006). The composition of the materials also affects the wear behaviour. For example, reinforcement is applied during the friction stir processing to modify the grain structure which is responsible for enhancement in strength. The factor affecting wear in FSW is shown in Figure 9.4.

9.5   SELECTION OF TOOL MATERIALS USED IN FSW

The joint quality and wear of the tool are two crucial factors in the selection of tool material which affects the joint quality. The interaction with degraded tool material may potentially have an impact on the weld microstructure and strength. In addition to the possible negative impacts on the weld microstructure, substantial tool wear raises the processing cost of FSW. The tool materials wear drastically on low yield strength at high temperatures due to the significant heating of the tool during FSW. At high temperature, the tool experiences the stresses which depend on the strength of the workpiece material in FSW. For a specific workpiece and processing conditions, temperatures of the workpiece depend on the tool's material qualities. The tool's thermal stresses affect the coefficient of thermal expansion. The tool material may be degraded due to improper interaction between the tool and the workpiece [28]. Different types of tool material are used to join the specific type of workpiece, which is enlisted in tabulated form as shown below in Table 9.1. The tool material is selected for the process of FSW/P on the basis of their characteristics and properties. The tool material should be high resistance to wear, good strength, good creep resistance, good

TABLE 9.1

**Particular Alloys are Used as Specific Tool Materials
for Joining (Mishra and Mahoney, 2007)**

Alloy	Tool Material
Aluminium alloys	H13 tool steel, tungsten carbide, WC-CO, MP159
Magnesium alloys	H13 tool steel, tungsten carbide, nickel alloys, PCBN
Copper and copper alloys	H13 tool steel, tungsten alloys, nickel alloys, PCBN
Titanium alloys	Tungsten alloy
Stainless steel	PCBN, tungsten alloy
Low alloy steel	Tungsten carbide, PCBN
Nickel alloys	PCBN (polycrystalline cubic boron nitride)

thermal fatigue resistance, good fracture toughness, low coefficient of thermal expansion, good machinability, and low cost (Zhang et al., 2012).

9.5.1 TRIBOLOGICAL PROPERTIES OF TOOLS IN FSW

The FSW tool's wear is a key aspect in achieving high-quality FSW joints since a tool that has been worn over its tolerance limits cannot effectively mix the joining material, leading to defects like lazy S or kissing bonds. Kumar et al. (2022) observed lazy S-type defects at higher TRS and TTS, which may be one of the reasons for the tool wear. As at higher TRS and TTS, the interaction of the tool and workpiece is low, high force is induced on the tool during joints of base materials. The production of wear particles, which mix with the joint material and weaken the joint strength, is another element of FSW tool wear. Studying FSW tool wear mechanisms is thus important, especially for high melting point materials (Thompson and Babu, 2011; Prater et al., 2014). However, it turns out that FSW of aluminium alloys is similarly sensitive to FSW tool wear. The literature related to tool wear is shown below as tabulated form in Table 9.2.

9.5.2 FACTORS AFFECTING TRIBOLOGICAL PROPERTIES DURING JOINING

Tool wear intensity depends upon the workpiece and tool's material, design, and welding process parameters. The proper selection of tool geometry is responsible for the joint quality. The causes of defect generation in FSW joints have already been examined, as well as the feasibility of defect detection using ultrasonic testing.

9.5.2.1 Tool Rotational Speed

The most valuable parameter among process parameters, that is, TRS, plays a significant role in frictional heat generation, the flow of plasticized material during FSW, and the tool's wear. The TRS is also responsible for the quality of joints. Proper selection of TRS will reduce the tool's wear and improve joint strength.

TABLE 9.2
Literature Survey of the Tool Wear

Material	Parameters	Remarks	Ref.
AA2024-T3	Different FSW tool shoulder sizes (6,12, and 16 mm)	With increasing shoulder dia. the wear rate and tensile strength increases; however, hardness decreased	Sharma et al. (2022)
AA6061 and AA 7075	Spindle speed – 600, 800 and 1,000 rpm. TTS – 40 mm/min. Tool pin – cylindrical Reinforcement – hexagonal boron nitride	The reinforced weld revealed improved wear resistance compared to unreinforced ones	Madhusudan et al. (2022)
AA5083 and AA6061	Reinforcement – aluminium matrix composite	The tensile properties and micro-hardness of reinforcement using welded alloy improved after multi-pass FSW	Rani et al. (2022)
AA6161 and AA5083	Reinforcement – nano-sized Al_2O_3 nanoparticles	The mechanical properties (tensile strength and hardness) increased when the number of multi-passes was increased, achieving higher mechanical properties (tensile strength – 272.7 MPa and hardness – 145.7HV) at the 3rd pass in FSW due to uniform dispersion of Al_2O_3	Rani et al. (2022a)
AA 7175	Reinforcement – marble dust	The micro-hardness (MMC samples) is enhanced but tensile strength slightly decreased	Chaudhary et al. (2022a)
Al-Zn-Mg-Si-Cu alloy	Reinforcement – SiC and B_4C, load applied – 3, 5, and 10 N	Using SiC and B_4C reinforced 20% increased hardness COF decreased with increasing load The wear rate decreased (30–55% when using SiC and B4C) compared to non-reinforcement in FSP	Ayvaz et al. (2022)
Carbon nanotube (CNTs) – reinforced poly-ether ether-ketone (PEEK) matrix composites	Deformation-driven processing	The COF and wear rate of 3.0 wt% CNTs/PEEK were, respectively, 7.32% and 6.71% lower compared to pure PEEK Due to its self-heating, low-cost, and high-performance properties, this method offers a highly effective method for creating high wear resistant polymeric composites	Cui et al. (2022)

TABLE 9.2 (Continued)

Material	Parameters	Remarks	Ref.
SiCp/Al	Tool pin, high-speed steel pin and high-speed steel + AlCrN coated pin	A novel method (based on hydrodynamic pressure) was proposed for calculating tool wear. The wear rate after coating is less but will increase when the coating is worn away	Zuo et al. (2022)
AA6026	Tool material – HSS, TRS – 400–600 rpm TTS – 15–25 mm/min. Frequency –10–14 KHz Process – FSW and ultrasonic-assisted friction stir welding	Tool wear is reduced 25% using ultrasonic-assisted friction stir welding which increases tool life	Chowdhary et al. (2022)
Ti_6Al_4V alloy	Tool – Ni-based super alloy (ZhS6U)	Abrasion and adhesion types of wear were observed on tool's pin area Anti-wear Ti_2Ni/Ni4Ti3 inter-metallics with fine MC carbide network layer generated	Zykova et al. (2022)
HSLA steel	Tool – WC TRS – 400, 600, 800, and 1,000 rpm	Weld toe radius increased with increasing TRS The nominal stress amplitude and hardness increased with increasing TRS	Yamamoto et al. (2022)
TA5 alloy	Tool – Co-based	Diffusion wear and abrasive wear was observed on the surface tool The stir zone (SZ) is divided into two zones: i) the contaminated zone (CZ) and ii) non-contaminated zone (NCZ) The CZ has higher micro-hardness values due to the presence of α and β phases The tensile strength of SZ is 9.1% higher than base material	Du et al. (2022)
AA2198-T8	Refill friction stir spot welding used to join	Adhesive wear was observed at the tool pin tip The variation behaviour of both the SZ area and the inverse of the hook height fit the LSS behaviour along the welding cycles	De Castro et al. (2022)
AISI 316 L	Tool – TWC tool, TRS – 600, 800, and 1,000 rpm TTS – 15, 20, and 25 mm/min. Pre-heat the specimen before joining using nichrome coil	The pre-heating of specimen reduced the tool wear by40% in comparison with normal FSW of specimen and increased the tensile strength (504 N/mm^2) and joint efficiency (97%)	Singh et al. (2021)

(Continued)

TABLE 9.2 (Continued)

Material	Parameters	Remarks	Ref.
Inconel 625	Tool-polycrystalline cubic boron nitride (PCBN) TRS –200,1,000, and 1,200 rpm TTS –1 and 1.5 mm/s	Thermal softening of W-Re binder phase and diffusion of W precede wear and failure of tool	Vicharapu et al. (2021)
Similar and dissimilar material of MS and SS	Joined both by using FSW and electrically assisted friction stir welding (EAFSW) process TTS – 11, 20, 24 mm/min. TRS – 550–1,050 rpm	The % of tool wear rate decreased with increased TRS and decreased TTS	Sengupta et al. (2021)
AA6061-T6 plates	Tool-square (SQ) and threaded with three intermittent flat faced (TIF) tool pins; both uncoated and AlTiN coated tools	TIF tool revealed lower profile damaged compared to square tool pin AlTiN coated tool reduced the damage of tool pin profile	Garg and Bhattacharya, (2021)
High carbon steel	Tool – tungsten carbide with nickel binder phase (WC-Ni)	Tool's life is increased at higher TRS (300 rpm) and lower TTS (150 mm/min.) because of minimum tool wear up to 1,000 mm	Vicharapu et al. (2021b)
Pure Mg sheet	FSP – Grooving (5×5 mm), Reinforcement – SiC (micro and nano-sized), groove depth increased from 0.5 mm to 0.8 mm TRS – 900–1,100 rpm TTS – 25–15 mm/min	At TRS of 1,100 rpm, TTS of 15 mm/min with three lead and return steps improved properties After used reinforcement the mechanical properties improved (hardness 26% higher compared to pure Mg) and tensile strength improved (2–4% on the micro-composite samples) of welded samples Wear rate decreased (0.05 mg/m for micro and 0.02 mg/m for nano-composite) to the base material	Sadooghi and Rahmani (2021)
AA 7075-T6 aluminium	Tool – 1.2344 steel with and without an anti-wear coating	After joining length after200m length, the steel tool gets worn while there is no effect on coated tool with AlCrN	Lacki et al. (2020)
AA7075	Reinforcement – SiC nano powder	ANOVA revealed that QPM are fitting for predicting tool wear The TRS and pass number of 52.9% and 13.1%, respectively, had more effect on tool wear At a TRS of 900 rpm, TTS of 50mm/min and three passes, a higher micro-hardness was achieved (127.24VHN)	MollaRamezani et al. (2019)

TABLE 9.2 (Continued)

Material	Parameters	Remarks	Ref.
DH36 and EH46 steel	Tool – PCBN TRS/TTS for DH36 – 200 rpm/100 mm/min and 550 rpm/400 mm/min. TRS/TTS for EH46 – TRS –120,150,200 rpm TTS – 50,100 mm/min	The FSW tool wear increases with increased TRS, plunged depth and TTS Tool wear may be reduced with a suitable combination of process parameters	Almoussawi et al. (2018)
AA6061 and SiCp reinforced 6092 Al matrix composite	Tool offset – to 3	The tool pin offset to Al alloy side considerably reduced tool wear As tool offset increased from 0to 3 mm, pin and shoulder wear length reduced from 0.17 mm to almost 0, and from 0.28 to 0.02 mm, respectively	Almoussawi et al. (2018)
AA6092/17.5 SiCp	TRS – 1,000 rpm, 1,500 rpm, and 2,000 rpm, const TTS – 2 mm/s, TTA-2°	The higher tool wear rate was observed at higher TRS and exhibited low joint efficiency than others; however, at 1,500 rpm revealed a higher joint efficiency and minimum wear rate	Acharya et al. (2018)

Prabhu and Kumar (2020) investigated the effect of on tool wear (with or without pin offset) of joining dissimilar materials, pure Cu and an AA6061-T6 Al alloy. Tool wear occurs due to the adhesion and weldment layer and it has been observed that the parameter TRS also affects the wear rate. The TRS values decrease when the galling wear is observed. However, with increasing TRS, notched wear is observed.

Fall et al. (2016) discussed the tool wear and wear rate of friction stir welded Ti-6Al-4 V alloy using conical tungsten carbide tool in FSW at varying TRS (500, 600, 700,1,000, 1,250, 1,500 rpm) and a constant TTS of 100 mm/min. The tool wear and wear rate was determined before and after FSW, using weight of tool (accuracy level-1mgr). The wear was measured after 10 cm weld length for all processes. The wear was observed 2 mm below the centre of the tool pin root and higher wear was achieved at lower TRS. The sound joint was observed between the TRS ranges of 1,000–1,500 rpm. Pin centre and edge of pin are the most wear affected area as shown in Figure 9.5.

Wu et al. (2014) studied the effect of varying TRS ranges from 400 to 1,200 rpm with a constant TTS of 100 mm/min on tool wear, microstructure, and mechanical properties of joining Ti-6Al-4 V plated using pcBN tool. The higher tool wear, observed in plunging stage of FSW at 800 and 1200 rpm because the higher force acts on the tool in this stage. However, greatest wear is observed at lower TRS (400 rpm) from the distance of 7 mm (plunging). The tools wear

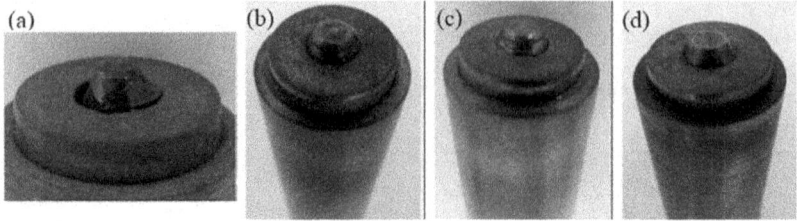

FIGURE 9.5   Tool wear after joining at different TRS (a) 500 rpm, (b) 1,000 rpm, (c) 1,250 rpm, and (d) 1,500 rpm. (Fall et al., 2016).

particles TIB and α-Ti (N) lead to an increment in hardness and tensile strength but a loss of ductility of the stir zone. An onion ring structure was observed at 1200 rpm.

9.5.2.2   Tool Traverse Speed

The tool traverse speed is the second most effective parameter of FSW which affects the tool wear during FSW. Prado et al. (2003) joined AA 6061 T6Al and 6061-T6+20% Al_2O_3 MMC plates using threaded tool geometry at varying traverse speed (1, 3, 6, and 9 mm/s), constant higher TRS (1000 rpm). It is particularly intriguing to notice that tool optimization seems to increase with increasing traverse speeds, which in this case achieved as much as 9 mm/s and were quite favourable in terms of commercial welding speed. The tool wear was tracked. it may be due to the loss of tool material from abrasive wear. The research of Prado et al. (2001) is expanded by Shindo et al. (2002) and Fernandez et al. (Fernandez and Murr, 2004) to account for welding speed. Due to a wear-optimized probe shape, which prevents material flow from removing the probe's surface through embedded hard material particles, tool wear can be reduced at welding speeds below 6 mm/s, resulting in a static area without a significant increase in tool wear. It should be emphasized that, in additional altered material flow circumstances, the energy input also corresponds with welding speed, and the reduced wear could thus be accounted for by the decreased frictional power. However, Thompson observed that the tool wear because too much shear deformation was removing too much material. Poelman (2011) found that the plastic deformation caused the tool mushroom in initial stage of FSW, which affected the tool wear rate. In this kind of arrangement, the tool height is lowered while the tool radius is raised. Michael (2012) observed excessive tool wear has a deleterious effect on surface characteristics and tensile strength; as a result, process improvement adjustments are required to avoid weld seam defects like inadequate penetration. Phulera et al. (2021) compared the effect of plasma nitriding and without nitriding on mechanical and tribological properties of friction stir welded AA2024 Al alloys at varying TRS (1000 rpm) and TTS (14 and 20 mm/min). The plasma nitriding of both plate at TRS of 1000 rpm and TTS of 14 mm/min exhibited higher tensile strength. The plasma nitriding joint shows the low wear rate due to the filled the micro-void by nitrogen and hardened the welded surface. Shindo et al. (2002) generated a self-optimized shape without threads, using threaded steel pin tools for joining the Al359+20%SiC metal matrix composite at varying TTS from 6mm/s to 11 mm/s. This shape continues to produce excellent, homogenous welds without extra tool wear or shape change at fixed welding speeds beyond 6 mm/s. At 6 mm/s and 9 mm/s, this self-optimized form is slightly different. Zero wear rates are predicted by extrapolating linear wear rate data above weld speeds of roughly 11 mm/s as shown in Figure 9.6. With reduction the tool diameter, the weld width zone decreased.

Kumar et al. (2020) investigated the effect of varying process parameters like TRS (750, 1,000, and 1,400 rpm), TTS (56, 80 and 112 mm/min) on AA6061-T6 plates with using different reinforcement μSiC and μSi_3N_4 micro-particles. A rectangle groove is made on workpiece to apply this reinforcement for processing through the square tool pin. The maximum tensile strength was observed at the combination of 1,400 rpm (TRS)/56mm/min (TTS) and 1,000 (TRS)/112mm/min (TTS) without reinforcement and with μSiC micro-particles. The reinforcement

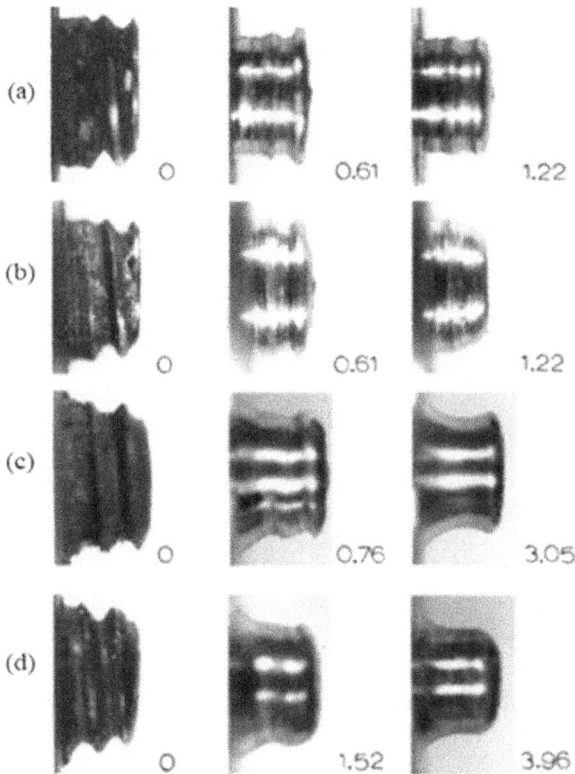

FIGURE 9.6   Tool wears at varying TTS (a) 1 Mm/S (b) 3 Mm/s (c) 6 Mm/s (d) 9 mm/s at constant TRS of 1,000 rpm (Shindo et al. 2002).

μSiC used in FSW have higher mechanical properties and wear resistance compared to μSi₃N₄ used reinforcement.

9.5.2.3 Tool-Pin Geometry

The tool pin geometry plays an important role in FSW. Different types of tool pins are used in FSW. Someone have good nature of mixing the plasticized material and someone adhere the more material during FSW. The shoulder generates 80–85% of the total heat however the tool pin generates only 5 to 15%. While the tool pin deform and wear due to the flow stress of plasticized material, temperature and strain rate of workpiece during stir to join the different types of material. Amirafshar and Pouraliakbar, (2015) studied the mechanical and tribological properties of the processed surface using different types of tool pin design (cylindrical, conical, square, and triangular) at a constant TRS of 400 rpm, tool traverse speed of 100 mm/min, and axial force 5KN using friction stir processing techniques. The tribological behaviour of processed steel is acquired through the pin-on-disc wear testing method. The weight loss of processed material is increased with increasing sliding distance and observed minimum in case of using a square tool pin. The lower wear rate is achieved using a square tool pin. However higher wear rate is found by conical tool pin, and a similar trend was observed in the graph between the friction coefficient and sliding distance. Lower grain size is observed using a square tool pin higher and achieves a higher hardness value. Kumar et al. (2021) used different type of tool pin profile (threaded taper, triangular and hybrid tool (couple of triangular and threaded)) to join third generation of Al-Li alloy at constant TRS 1400 rpm, TTS 180 mm/min and TTA of 2° and observed triangular tool pin is broken during experimentation due to the higher stress exert on tool pin. While hybrid tool pin profile exhibited improved mechanical properties.

9.5.2.4 Tool-Pin Coating

Adesina et al. (2018) investigated the influence of AlCrN coated W-25%Re-Hfc and uncoated W-25% Re-Hfc composite tool material on mechanical and tribological properties of friction stir welded specimen. The specific wear rate of coated sample is lower than the uncoated. The wear resistance of coated samples shows the improved mechanical properties (tensile strength 264.2±5), low COF and improves the tool life through this coating (AlCrN).

Wanga et al. (2014) discussed the characterization of the different types (W–1.1% La₂O₃) and grades (WC-Co based) of FSW tool pin joins in a Ti-6Al-4VV titanium alloy. Degradation of the tool was observed in (W–1.1% La₂O₃) due to the plastic deformation; this may be reduced by increasing the FSW tool pin dia. Fracture failure was observed in the CY16 FSW tool, while microcracking was observed in WC411. Lacki et al. (2020) joined a 7075-T6 aluminium alloy using using a tool with wear coating (AlCrN) and one without a coating and found the AlCrN coating tool had an increased life compared to the tool without. The application of the AlCrN coating significantly reduced the consumption of the base material and increased the tool life. The application of the anti-wear AlCrN coating led to a reduction in the COF compared to that observed during testing of the sample without the coating (μ = 0.40–0.45). Figure 9.7 shows the tool and tool's shoulder surface wear from application with and without an anti-wear coated tool.

9.5.2.5 Axial Load

Soundararajan et al. (2020) successfully joined base material AA8011, heat treated alone and heat treated and shot-peened with varying axial load and constant TRS and TTS. The heat treated and shot-peened specimen exhibited high tensile strength and hardness values compared to the others. An increase in the load and sliding distance, led to an increase in the wear rate (25%), however, the COF (16%) decreased compared to the base material AA8081 due to the reduction of internal stress and the presence of refining particles.

9.5.2.6 Medium of FSW

Abdollahzadeh et al. (2021) discussed the effect of vibration (65 HZ), which was generated by motor and cooling medium (water and lubricant oil) applied during

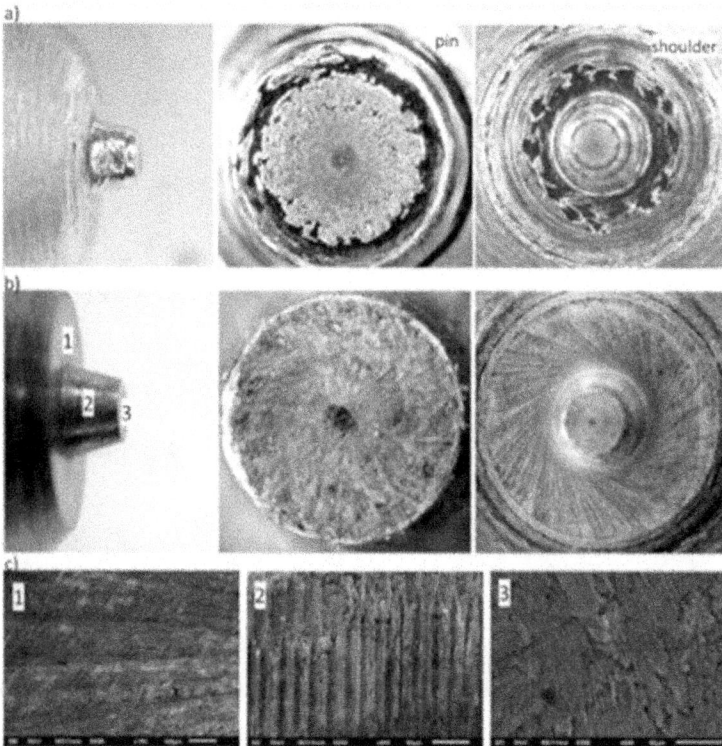

FIGURE 9.7 Status of the tool after the joining surface using (a) without coated tool (b) with coated tool (c) frictional surface of tool with respective surface (Lacki et al. 2020).

the FSW of 5083Al alloys at a constant TRS (1,250 rpm) and TTS (90 mm/min.). The welding performed under the medium of lubricant oil exhibited higher temperature. A refined grain structure was observed under the vibration and water cooling medium and showed higher joint efficiency (87%) and better wear resistance compared to FSW as shown in Figure 9.3. Phillips et al. (2021) joined AA5083–H131material at different TRS (circular symbol; 200 rpm, square; 300 rpm, and triangle; 400rpm) with a varying deposition ratio 0.55, 0.65, and 0.75, respectively, shown in Figure 9.8. Increasing the TRS produced higher heat into deposit at a constant TTS.

9.5.2.7 Reinforcement in FSP

Chaudhary et al. (2022) developed a friction stir powder deposition (FSPD) process and applied it on AA 6061-T6 material plates to investigate the effect of process parameters on the nature of deposition, microstructure, and mechanical properties. Compared to the base material, the superior deposition exhibits an increased micro-hardness (0.84 times) and wear rate (1.07 times). The maximum deposition efficiency (22%), was achieved at the maximum deposition height (0.45 mm). Prakash et al. (2015) discussed the influence of the different percentages of reinforcement Al_2O_3 (0, 2, 4, 6, 8, and 10 wt%) with 0.5 wt% Gr applied on received 6061 Al alloys at constant process parameters such as TRS of 700 rpm, TTS of 60 mm/min and axial load 7KN. By increasing the %age composition of

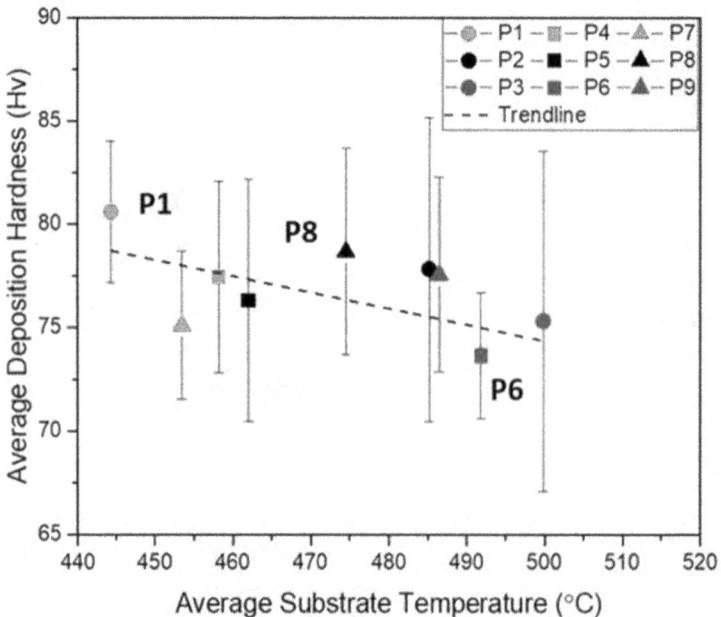

FIGURE 9.8 Average deposition hardness v/s average substrate temperature at varying TRS (Phillips et al. 2021).

Al_2O_3 with 0.5 % Gr, the wear rate decreases primarily and then increases. A low wear rate (0.81 g/s) was observed at 6% of Al_2O_3 with 0.5% Gr reinforce used on AA6061 alloy, which decreased by 1.65 times of the base material. Further, this composite showed the lower grain size and observed a higher hardness value (165 BHN). Sadooghi and Rahmani (2021) discussed the tribological and mechanical properties of friction stir processing of different sizes (micro- and nano-sized) of SiC particles as a reinforcement applied on a 5×5 mm groove size, 7 mm thick, pure Mg sheet at a constant TRS of 900 rpm and TTS of 25 mm/min. The nano-sized reinforcement particle exhibited higher tensile strength (225.18MPa) and higher average hardness (82.1HV) compared to the parent metal. The wear rate decreased with an increase in the distance and was lower compared to the parent metal. The nano-composite samples showed higher wear resistance than pure and micro-composite samples due to the variant pin surfaces.

Prasad et al. (2021) discussed the mechanical and tribological properties of friction stir welded 4mm thick stir cast AA7075 (0.4 and 8 wt%) /ZrB2 situ composite plates. The wear rate of AA7075 (8 wt%) is 33% less compared to parent metal. However, the 4 wt% of AA7075 exhibits higher joint efficiency (93%) and a minimum grain size (3.2μm) is observed using 8 wt% of AA7075. Wear resistance of the materials (only NZ) improved and COF decreased after friction stir welding. The average COF decreased with increasing wt% of reinforcements and the decreasing trend is sizeable. The reduction in COF of welded composites may be attributed to higher hardness as compared to un-welded composites. Tarasov et al. (2014) discussed applying a coated FSW tool (mixed layer of inhomogeneous composition of oxides and silicides) to join an Al alloy. The aluminium alloy stuck to the working surface of the tool during FSW under the influence of high mechanical stress and temperature, and the iron/aluminium reaction initiated diffusion. The tool material distorted and dissociated from the welding surface during FSW due to the development of the shear stress which affected the quality of the joints.

9.5.2.8 Weld Length During FSW/FSP

The weld length plays a significant role in tool wear. With increasing length, the tool wear increases due to the generation of friction between the tool and workpiece, as shown in Figure 9.9. The deposition of material on a threaded tool may affect the material mixing and flowability properties. Hasieber et al. (2020) suggested a novel method to detect weight, the shoulder, and probe deviation during the joining of 8 mm thick and 80 m long AA6060-T66 sheets. Shoulder and probe wear might be individually considered. According to the length of the weld seam, it was discovered that the wear rates on the shoulder and probe are progressive and degressive, respectively. Fig 9.10 shows that the weight loss of shoulder and probe, shoulder, and probe increases when the length of weld is increased. The cumulative mass loss with different weld length is shown in Fig 9.11.

Amirov et al. (2020) investigated the wear and durability of a tool (made of ZhS6UNi-based super alloy) used in the friction stir processing on pure titanium ZhS6U alloys. This alloy's chemical components can be found in the transfer

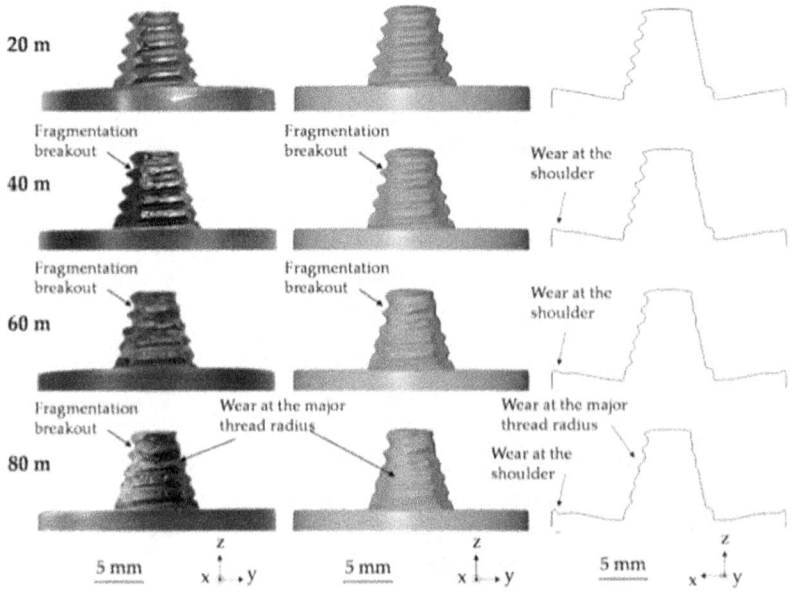

FIGURE 9.9   Image of tool after welding (a) photographic; (b) stripe light projection; (c) full section of stripe light projection (Hasieber et al., 2020).

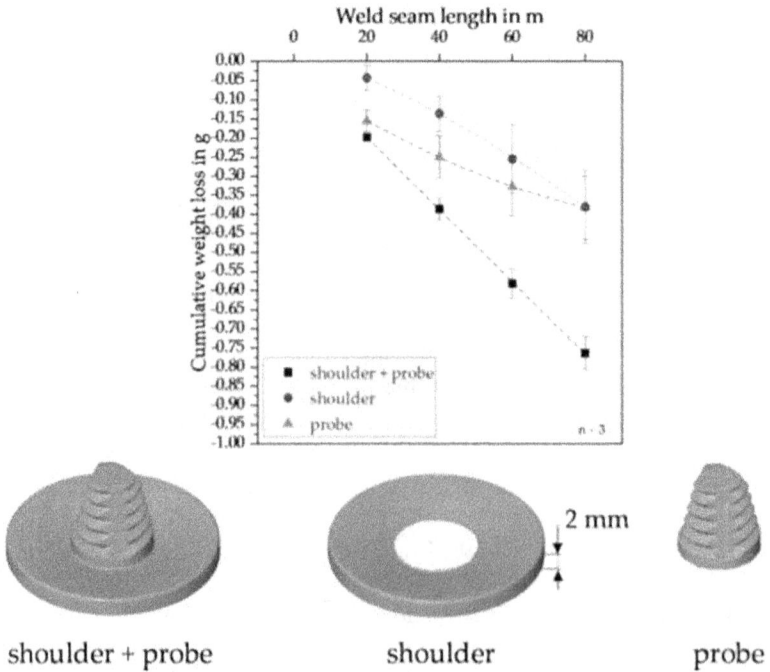

FIGURE 9.10   Cumulative weight loss with respect to weld seam length (Hasieber et al., 2020).

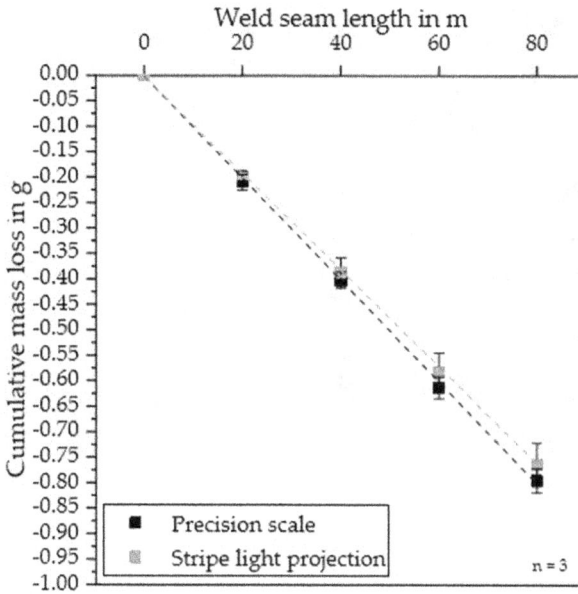

FIGURE 9.11 Cumulative mass loss with different weld length (Hasieber et al., 2020).

layer, which suggests that adhesion and diffusion are how the tool wears down. The pure titanium's tensile strength improved by 25% as a consequence of processing. The tool used in the experiment is shown in Figure 9.12(a). During this process, the tool's pin and shoulder edges became worn. The tool pin shape is still conical after processing of 1,105 mm length, and no depression is visible in Figure 9.12(b). Further processing gradually transformed the pin's conical shape into one that is virtually cylindrical (Figure 9.12(c)). While processing 2,755 mm, a depression in the shape of a circle gradually formed on the shoulder around the tool pin base, as depicted in Figure 9.12(d). The transfer layer's adherence and detachment during processing are indicated by sparse pits scattered throughout the adherent layers, which are typically aligned along the direction of rotation.

Sahlot et al. (2017) investigated the wear of a tool (made of H13 steel) during joining of CuCrZr alloys at different TRS (800, 1000, and 1,200 rpm), TTS (30, 50, and 70 mm/min) and traverse distance (300, 500, and 1,000 mm). Multiple tools with same dimensions were used in experimentation. A pilot hole with the same dimension of tool pin was made by the drill process to avoid tool wear during the plunging stage of FSW. A higher amount of total tool wear is observed at the higher TRS of 1,200 rpm (due to superior relative surface velocities), and slower traverse speed (30 mm/min). The tool wear decreases when the TTS increases (due to the reduced interaction time). The process parameters TRS and TTS significantly affect the wear rate in the initial stage of FSW.

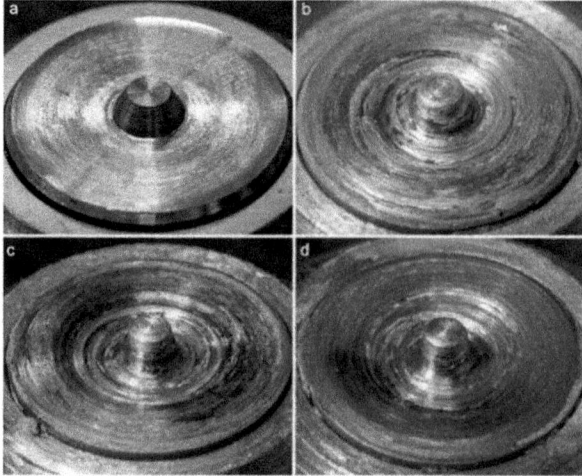

FIGURE 9.12 View of tool at different stages: (a) before joining; (b) after joining 1,105 mm; (c) 2,335 mm; (d) 2755 mm (Amirov et al., 2020).

9.5.2.9 Dissimilar Thickness of Plate

Majeed et al. (2021) joined different thickness of plates with butt joint configuration using a threaded tapered stir tool made of T4 steel at different process parameters: TTS (50,63, and 80 mm/min), TRS (560,710, and 900 rpm), surface tilt angle (10, 12, 14), and tool's shoulder diameter (18 and 22 mm). The authors observed that the greater wear appears to be the tool pin compared to the shoulder. This may be due to the large difference in thickness of plate, which offers different penetration. At the TRS of 710 rpm, TTS of 50 mm/min, surface tilt angle (12) and shoulder diameter (22 mm), during welding, the lower force generates high heat input due to low flow stress and shear stress, which leads to significant decreased tool wear and deformation.

9.6 CONCLUSIONS

The major sources of the loss of energy and failure in all different kinds of mechanical systems are wear and tribology. Tribology has a significant impact on FSW/FSP due to the interaction of the rotating tool and workpiece. In manufacturing, the tool condition plays a vital role on process dynamics and part quality. Technical support for sustaining production efficiency and quality can be obtained through effective control and implementation of tool wear deterioration. It can be hard to detect tool wear since it varies depending on the materials of the tool and the workpiece and the conditions of their joining. For friction-based procedures, tool wear is a major problem since it affects both the cost of the process and the weld quality. In this chapter, the effect of different process parameters' influences on friction stir tool wear is investigated. Different authors suggest different types of methods to reduce tool wear with significant process parameters reported in this chapter.

REFERENCES

Abdollahzadeh A., Bagheri B., Abbasi M., Sharifi F., Mirsalehi S.E., & Moghaddam A.O. (2021). A modified version of friction stir welding process of aluminum alloys: Analyzing the thermal treatment and wear behavior. *Proceedings of the Institution of Mechanical Engineers, Part L: Journal of Materials: Design and Applications*, 146442072110239. https://doi.org/10.1177/14644207211023987

Acharya U., Rou B.C., & Saha S.C. (2018). A study of tool wear and its effect on the mechanical properties of friction stir welded AA6092/17.5 sicp composite material joint. *Materials Today: Proceedings*, 5, 20371–20379.

Adesina A.Y., Iqbal Z., Al-Badour F., & Gasem Z.M. (2018). Mechanical and tribological characterization of AlCrN coated spark plasma sintered W–25%Re–Hfc Composite material for FSW tool application. *Journal of Materials Research and Technology* https://doi.org/10.1016/j.jmrt.2018.04.004

Affatato S., & Grillini L. (2013). Chapter 1, Topography in bio-tribocorrosion. *Bio-Tribocorrosion in Biomaterials and Medical Implants*, Woodhead Publishing Series in Biomaterials, 1–21, 21a–22a. https://doi.org/10.1533/9780857098603.1

Almoussawi M., Smith A.J., & Faraji M. (2018). Wear of polycrystalline boron nitride tool during the friction stir welding of steel. *Metallography, Microstructure, and Analysis*, 7(3), 252–267.

Amirafshar A., & Pouraliakbar H. (2015). Effect of tool pin design on the microstructural evolutions and tribological characteristics of friction stir processed structural steel, *Measurement*.http://dx.doi.org/10.1016/j.measurement.2015.02.05

Amirov A., Eliseev A., Kolubaev E., Filippov A., & Valery Rubtso V. (2020). Wear of ZhS6U nickel superalloy tool in friction stir processing on commercially pure titanium, *Metals*, 10, 799. https://doi.org/10.3390/met10060799

ASM International. (1992). ASM Handbook Volume 18, "Friction, Lubrication, and Wear Technology", American Society for Metals, Metals Park, Ohio, pp. 341–347.

Ayvaz S.I., Arslan D., & Ayvaz M. (2022). Investigation of mechanical and tribological behavior of SiC and B_4C reinforced Al-Zn-Mg-Si-Cu alloy matrix surface composites fabricated via friction stir processing. *Materials Today*, 31, 103419. https://doi.org/10.1016/j.mtcomm.2022.103419

Chaudhary B., Jain N.K., & Murugesan J. (2022a). Development of friction stir powder deposition process for repairing of aerospace-grade aluminum alloys. *CIRP Journal of Manufacturing Science and Technology*, 38, 252–267. https://doi.org/10.1016/j.cirpj.2022.04.016

Chaudhary S., Dangi S., Yadav V., Walia R.S., Suri N.M., & Tyagi M. (2022b). Effect of preheating on the mechanical and high temperature tribological behaviour of reusing marble dust with aluminium composite. *Advances in Materials and Processing Technologies*. https://doi.org/10.1080/2374068X.2022.2091091

Chowdhury I., Sengupta K., Maji K.K., Roy S., & Ghosal S. (2022). Experimental study of tool wears to join Al6026 aluminium alloy by ultrasonic assisted friction stir welding. *Materials Today: Proceedings*, 50, 1221–1225.

Cui Y., Li C., Li Z., Yao X., Hao W., Xing S., & Huang Y. (2022). Deformation-driven processing of CNTs/PEEK composites towards wear and tribology applications. *Coatings*, 12(7), 983.

De Castro C.C., Shen J., Plaine A. H., Suhuddin U. F., de Alcântara N. G., dos Santos J. F., & Klusemann B. (2022). Tool wear mechanisms and effects on refill friction stir spot welding of AA2198-T8 sheets. *Journal of Materials Research and Technology*. https://doi.org/10.1016/j.jmrt.2022.07.092

Du S., Liu H., Jiang M., Zhou L., Li D., & GaoY. (2022). Effects of tool wear on the microstructure evolution and mechanical properties of friction stir welded TA5 alloy. *The International Journal of Advanced Manufacturing Technology*, 119(1), 1109–1121.

Eyre T.S. (October 1976). Wear characteristics of metals. *Tribology International*, 9(5), 203–212. https://doi.org/10.1016/0301-679X(76)90077-3

Fall A., Fesharaki M.H., Khodabandeh A.R., & Jahazi M. (2016). Tool wear characteristics and effect on microstructure in Ti-6Al-4V friction stir welded joint. *Metals*, 6, 275. https://doi.org/10.3390/met6110275

Fernandez G.J., & Murr L.E. (2004). Characterization of tool wear and weld optimization in the friction-stir welding of cast aluminum 359+ 20% SiC metal-matrix composite. *Materials Characterization*, 52, 65–75.

Garg A., & Bhattacharya A. (2021). Assessing profile damage of uncoated and AlTiN coated FSW tools after successive travel on AA6061-T6 plate. *CIRP Journal of Manufacturing Science and Technology*, 35, 839–854.

Halina, G., & Motyka, M. (2019). Chapter 10, Micro and nano technologies, nanocrystalline titanium, 193–208. https://doi.org/10.1016/B978-0-12-814599-9.00010-9

Hasieber M., Grätzel M., & Bergmann J.P. (2020). A novel approach for the detection of geometricand weight-related FSW tool wear using stripe light projection. *Journal of Manufacturing and Materials Processing*, 2020(4), 60. https://doi.org/10.3390/jmmp4020060

Jin Z., & Fisher J. (2008). Chapter 2 - Tribology in joint replacement. *Joint Replacement Technology*, Woodhead Publishing Series in Biomaterials, 31–55, https://doi.org/10.1533/9781845694807.1.31

Kayaba T. (1984). The latest investigations of wear by the microscopic observations. *JSLE Ransatios*, 29, 9–14.

Khairuddin J.T., Abdullah J., Hussain Z., & Indra Putra Almanar I.P. (2012). Principles and thermo-mechanical model of friction stir welding. *Welding Processes*. https://doi.org/10.5772/50156

Kumar K., Kalyan C., Kailas Satish V., & Srivatsan T. S. (2009). An investigation of friction during friction stir welding of metallic materials, 438–445. https://doi.org/10.1080/10426910802714340

Kumar N., & Patel V.K. (2020). Effect of SiC/Si$_3$N$_4$ micro-reinforcement on mechanical and wear properties of friction stir welded AA6061-T6 aluminum alloy. *SNApplied Sciences*, 2, 1572. https://doi.org/10.1007/s42452-020-03381-y

Kumar S., Acharya U., Sethi D., Medhi T., Roy B.S., & Saha S.C. (2020b). Effect of traverse speed on microstructure and mechanical properties of friction-stir-welded third-generation Al–Li alloy. *Journal of the Brazilian Society of Mechanical Sciences and Engineering*, 42(8), 1–13. https://doi.org/10.1007/s40430-020-02509-w

Kumar S., Chaubey S.K., Sethi D., Saha S.C., & Roy B.S. (2021). Performance analysis of varying tool pin profile on friction stir welded 2050-T84Al-Cu-Li alloy plates. *Journal of Materials Engineering and Performance*, 1–12. https://doi.org/10.1007/s411665-021-06315-w

Kumar S., Choudhury S., Sethi D, Paulraj J., Bhargava M., & Saha Roy B. (2022a). Effect of Process Parameters on Third Generation of Friction Stir Welded Al-Li Alloy Plates. *CIRP Journal of Manufacturing Science and Technology*, 38, 372–385. https://doi.org/10.1016/j.cirpj.2022.05.009

Kumar S., Katiyar J.K., Acharya U., Saha S.C., & Roy B.S. (2022b). Influence of tool rotational speed on microstructure and mechanical properties of Al-Li alloy using friction stir welding. *Proceedings of the Institution of Mechanical Engineers, Part E: Journal of Process Mechanical Engineering*, 09544089221080823. https://doi. org/10.1016/j.matpr.2020.02.446

Kumar S., Sethi D., Choudhury S., Roy B.S., & Saha S.C. (2020a). An experimental investigation to the influence of traverse speed on microstructure and mechanical properties of friction stir welded AA2050-T84 Al-Cu-Li alloy plates. *Materials Today: Proceedings*, 26, 2062–2068.

Lacki P., Więckowski W, Luty G, Wieczorek P., & Maciej Motyka M. (2020). Evaluation of usefulness of AlCrN coatings for increased life of tools used in friction stir welding (FSW) of sheet aluminum alloy. *Materials*, 13, 4124. https://doi.org/10.3390/ma13184124

Lu H., Lee Y., Oguri M., & Powers J. (2006). Properties of a dental resin composite with a spherical inorganic filler. *Operative Dentistry*, 31(6), 734–740.

Madhusudan M., Shanmuganatan S.P., Kurse S. et al. (2022). Investigation on mechanical and tribological behavior of double pass friction stir welded aluminium joints. *Journal of Bio- and Tribo-Corrosion*, 8, 55. https://doi.org/10.1007/s40735-022-00654-4

Majeed T., Mehta Y., & Arshad Noor Siddiquee A.N., et al. (2021). Analysis of tool wear and deformation in friction stir welding of unequal thickness dissimilar Al alloys. *J Materials: Design and Applications*, 235(3), 501–512. https://doi.org/ 10.1177/146442072097176

Majumdar J.D., & Manna I. (2015). Laser surface engineering of titanium and its alloys for improved wear, corrosion and high-temperature oxidation resistance. *Laser Surface Engineering Processes and Applications*, Woodhead Publishing Series in Electronic and Optical Materials, 483–521. https://doi.org/10.1016/B978-1-78242-074-3.00021-0

Meshref A., Mazen A., & Ali Y. (2020). Wear behavior of hybrid composite reinforced with titanium dioxide nanoparticles. *Journal of Advanced Engineering Trends*. 39, 89–101. https://doi.org/10.21608/jaet.2020.75738

Meyghani B., Awang M.B., Poshteh R.G.M., Momeni M., Kakooei S., & Hamdi Z. (2019). The effect of friction coefficient in thermal analysis of friction stir welding (FSW). *IOP Conference Series: Materials Science and Engineering*, 495, 012102.

Michael Eff. (2012). *The Effects Of Tool Texture On Tool Wear In Friction Stir Welding of X-70 Steel*, Doctoral dissertation, The Ohio State University.

Mishra R.S., & Mahoney M.W. (2007). Friction stir welding and processing. *ASM International*, 1–333, DOI:10.1361/fswp2007p001

MollaRamezani N., Davoodi B., Aberoumand M., & RezaeeHajideh M. (2019). Assessment of tool wear and mechanical properties of Al 7075 nanocomposite in friction stir processing (FSP). *Journal of the Brazilian Society of Mechanical Sciences and Engineering*, 41(4), 1–14.

Moore M.A. (January 1974). A Review of Two-Body Abrasive Wear, *Wear*, 27(1), 1–17. https://doi.org/10.1016/0043-1648(74)90080-5

Nandan R., DebRoy T., & Bhadeshia H. (2005). Recent advances in friction-stir welding–process, weldment structure and properties. *Progress in Materials Science*, 53, 980–1023.

Nuruzzaman D.M., Chowdhury M.A., & Rahaman M.L. (2011). Effect of duration of rubbing and normal load on friction coefficient for polymer and composite materials. *Industrial Lubrication and Tribology*, 63, 320–326.

Phillips B.J., Williamson C.J., Kinser R.P., Jordon J.B., Doherty K.J., & Allison P.G. (2021). Microstructural and mechanical characterization of additive friction stir-deposition of aluminum alloy 5083 effect of lubrication on material anisotropy. *Materials*, 14, 6732. https://doi.org/10.3390/ma14216732

Phillips K. (1975). Study of the Free Abrasive Grinding of Glass and Fused Silica, Ph.D. Thesis, University of Sussex, United Kingdom.

Phulera N., Kaushik S., Kshetri R. (2021). Effects of plasma nitriding on mechanical, tribological and corrosion properties of friction stir welded joints of Al 2024. *Materials Today: Proceedings*. https://doi.org/10.1016/j.matpr.2021.04.218

Poelman L. (2011). Characterization of Tools for Friction Stir Welding of Steels, Thesis for Master of Science in Engineering Advisor: Wim De Waele, Koen Faes.

Prabhu L., & Kumar S. (2020). Tribological characteristics of FSW tool subjected to joining of dissimilar AA6061-T6 and Cu alloys. *Materials Today: Proceedings*. https://doi.org/10.1016/j.matpr.2020.06.092

Prado, R. A., Murr, L. E., Shindo, D. J., & Soto, K.F. (2001). Tool wear in the friction-stir welding of aluminum alloy 6061+ 20% Al2O3: a preliminary study. *Scripta Materialia*, 45(1), 75–80. https://doi.org/10.1016/S1359-6462(01)00994-0

Prado R.A., Murr L.E., Soto K.F., & McClure J.C. (2003). Self-optimization in tool wear for friction-stir welding of Al 6061+20% Al$_2$O$_3$ MMC. *Materials Science and Engineering A*, 349(1–2), 156–165. https://doi.org/10.1016/s0921-5093(02)00750-5

Prakash T., Sivasankaran S., & Sasikumar P. (2015). Mechanical and tribological behaviour of friction-stir-processed Al 6061 aluminium sheet metal reinforced with Al$_2$O$_3$/0.5Gr hybrid surface nanocomposite. *Arabian Journal for Science and Engineering*, 40, 559–569. https://doi.org/10.1007/s13369-014-1518-4

Prasad R, Kumar H, Kumar P, Tewari S.P., & Singh J.K. (2021). Microstructural, mechanical and tribological characterization of friction stir welded A7075/ZrB2 in situ composites. *JMEPEG*, 30, 4194–4205. https://doi.org/10.1007/s11665-021-05750-z

Prater T., Strauss A., Cook G., Gibson B., & Cox C. (2014). The effect of friction stir welding tool wear on the weld quality of aluminum alloy AMg5M. *AIP Conference Proceedings*, 1623(1), 635–638. https://doi.org/10.1063/1.4901501

Rani P., & Mishra R.S. (2022). Influence of nano-sized Al$_2$O$_3$ nanoparticles and multipass FSW on microstructure and mechanical characteristics of dissimilar welded joints of AA6061 and AA5083. *Transactions of the Indian Institute of Metals*. https://doi.org/10.1007/s12666-022-02655-w

Rani P., Misra R.S., & Mehdi H. (2022). Effect of nano-sized Al$_2$O$_3$ particles on microstructure and mechanical properties of aluminum matrix composite fabricated by multipass FSW. *Proceedings of the Institution of Mechanical Engineers, Part C: Journal of Mechanical Engineering Science*. https://doi.org/10.1177/09544062221110822

Rigney D.A. (May 1997). Comments on the sliding wear of metals. *Tribology International*, 30(5), 361–367. https://doi.org/10.1016/S0301-679X(96)00065-5

Rigney D.A., & Hirth J.P. (1979). Plastic deformation and sliding friction of metals. *Wear*, 53(2), 345–370. https://doi.org/10.1016/0043-1648(79)90087-5

Rymuza Z. (2007). *Tribology of Polymers. Archives of Civil and Mechanical Engineering*, VII(4), 177–184.

Sadooghi A., & Rahmani K. (2021). Experimental study on mechanical and tribology behaviors of Mg-SiC nano/micro composite produced by friction stir process. *Journal of Mechanical Science and Technology*, 35(3), 1121–1127. http://doi.org/10.1007/s12206-021-0225-9

Sahlot P., Jha K, Dey G.K., & Arora A. (2017). Quantitative wear analysis of H13 steel tool during friction stir welding of Cu-0.8%Cr-0.1%Zr alloy. *Wear*, 378–379, 82–89.

Schulson E.M. (2001). *Ice: Mechanical Properties, Encyclopedia of Materials: Science and Technology* (Second Edition), 4006–4018. https://doi.org/10.1016/B0-08-043152-6/00705-1

Sengupta K., Singh D.K., Mondal A.K., Bose D., Patra D., & Dhar A. (2021). Characterization of tool wear in similar and dissimilar joints of MS and SS using EAFSW. *Materials Today: Proceedings*, 44, 3967–3975.

Sharma H., Kumar T.A., & Rana S. (2022). Experimental investigation of mechanical and tribological properties of AA2024-T3 friction stir welded butt joint. *Materials Today*, 64(Part 3), 1165–1169. https://doi.org/10.1016/j.matpr.2022.03.408

Shindo D. J., Rivera A.R., Mur L.E. (2002). Shape optimization for tool wear in the friction-stir welding of cast AI359-20% SiC MMC. *Journal of Materials Science*, 37, 4999–5005.

Singh D.K., Sengupta K., Mondal A.K., Bose D., & Singh P. (2021). Characterization of tool wear in i-FSW of AISI 316L material joining. *Materials Today: Proceedings*, 46, 10628–10633.

Soundararajan R., Aravinth V., Vallarasu M. et al. (2020). Mechanical and tribological behavior of friction stir welded joint on AA 8011 at diverse strengthening condition through post processing. *Materials Today: Proceedings*. https://doi.org/10.1016/j.matpr.2020.03.035

Spikes H.A. (1998). Some challenges to tribology posed by energy efficient technology. *Tribology for Energy Conservation, Tribology Series*, 34, 35–47. https://doi.org/10.1016/S0167-8922(98)80060-6

Stachowiak, G.W., & Batchelor, A.W. (2013). *Engineering Tribology*. Butterworth-heinemann.

Tarasov S., Rubstov V.E., & Kolubaev E.A. (2014). The effect of friction stir welding tool wear on the weld quality of aluminum alloy AMg5M. Article in *AIP Conference Proceedings*. https://doi.org/10.1063/1.4901501

Thompson B.T., & Babu S.S. (2011). Application of diffusion models to predict FSW tool wear: Proceedings of XXI Int. *Offshore and Polar Engineering Conf*, Maui, Hawaii, USA, June 19–24, 2011, Maui, Hawaii, USA, p. 520.

Vicharapu B., Lemos G.V.B., Bergmann L., dos Santos J.F., De A., & Clarke T. (2021a). Probing underlying mechanisms for pcBN tool decay during friction stir welding of nickel-based alloys. *Tecnologia em Metalurgia Materiais e Mineração*, 18, e2455. http://dx.doi.org/10.4322/2176-1523.20202455

Vicharapu B., Liu H., Morisada Y., Fujii H., & De A. (2021b). Degradation of nickel-bonded tungsten carbide tools in friction stir welding of high carbon steel. *The International Journal of Advanced Manufacturing Technology*, 115(4), 1049–1061.

Vingsbo O., & Hogmark S. (1981). Wear of Steels, *ASM Materials Science Seminar on Fundamentals of Friction andWear of Materials*, 4–5 October 1980, Pittsburg, Pennsylvania, editor: D.A. Rigney, Metals Park, Ohio, Publ. ASM, 373–408.

Wang H., He D., Wu Y., & Xu S. (2021). Study on wear state evaluation of friction stir welding tools based on image of surface topography. *Measurement*, 186, 110173.

Wanga J, Su J, Mishra R.S., Xuc R., & Bauman J.A. (2014). Tool wear mechanisms in friction stir welding of Ti–6Al–4V alloy. *Wear*, 321, 25–32.

Wu L.H., Wang D., Xiao B.L., & Ma Z.Y. (2014). Tool wear and its effect on microstructure and properties of friction processed Ti-6Al-4V. *Metal Chemistry and Physics*, 146, 512–522.

Yamamoto H., Koga S., Ito K., & Mikami Y. (2022). Fatigue strength improvement due to alloying steel weld toes with WC tool constituent elements through friction stir processing. *The International Journal of Advanced Manufacturing Technology*, 119(9), 6203–6213.

Yang Y.S., Qu J.X., & Shao H.S. (1994). Mechanical–chemical effect of corrosive wear of materials. *Advanced Materials '93, Ceramics, Powders, Corrosion and Advanced Processing*, 195–198. https://doi.org/10.1016/B978-0-444-81991-8.50055-8

Zhang Y.N., Cao X., Larose S., & Wanjara P. (2012). Review of tools for friction stir welding and processing. *Article in Canadian Metallurgical Quarterly*. https://doi.org/10.1179/1879139512Y.0000000015

Zuo L., Shao W., Zhang X., & Zuo D. (2022). Investigation on tool wear in friction stir welding of SiCp/Al composites. *Wear*, 498, 204331.

Zykova A., Vorontsov A., Chumaevskii A., Gurianov D., Gusarova A., Kolubaev E., & Tarasov S. (2022). Structural evolution of contact parts of the friction stir processing heat-resistant nickel alloy tool used for multi-pass processing of Ti6Al4V/(Cu+Al) system. *Wear*, 488, 204138.

10 Tribology in Additive Manufacturing

*Y. Alex, Nidhin C. Divakaran, and
Smita Mohanty*
Central Institute of Petrochemicals Engineering and
Technology (CIPET), Bhubaneswar, India

10.1 INTRODUCTION

The rapid emergence and evolution of AM technology has created a wide range of applications in numerous industries all over the world. With this technological support, parts with complex structures can be made more quickly and easily than with traditional methods. This technology has been used primarily for prototyping, but it can also be used to create functional end-use parts in various industries, such as aerospace and medical devices. However, various barriers have emerged that prevent it from becoming widely used in the production of end-use parts. Some of these include the variability of the part's surface quality and uncertainty about the structure–property relationship. Due to the increasing number of research studies on various aspects, especially for accuracy and precision of additive manufactured parts, it is becoming more difficult to meet these standards. To overcome this, we need to consider some factors of AM machines, such as the wear, friction, and lubrication of parts; thus, the tribology community has become more interested in this technology.

Tribology is an important factor to consider in additive manufacturing. Tribology is the study of friction, lubrication, and wear between two surfaces and is critical to the success of the AM process (Czichos, 1977; Ur Rahman, Matthews, et al., 2019b). Tribology plays a major role in the printing and post-processing of 1D printed parts, as it can affect the surface finish and dimensional accuracy of parts. Improper tribology can lead to poor surface finish, warping, and even complete failure of parts. Therefore, understanding tribology is essential to achieve optimal results for additive manufacturing.

Through the identification of the process parameters that influence the tribological behavior of 1D printed parts, we can improve the performance of these components in different conditions. Due to the high degree of replicability of 1D printed parts, they are widely used in the medical sector for developing medical devices such as surgical implants and instruments (George et al., 2017;

DOI: 10.1201/9781003363576-10

Skelley et al., 2020), and fulfilling bionic requirements (Koprnicky et al., 2017; Mannoor et al., 2013). The wear and friction characteristics of 1D printed components are important factors that are considered for the design and functional performance of 1D printed parts. This technology is also widely used in the heavy manufacturing industry, where large parts are made using various 1D printing techniques (Bassoli et al., 2007). The surface of 1D printed components directly affects tribological properties. In recent years, this technology has been widely used in aerospace (Blakey-Milner et al., 2021; Pant et al., 2021). Due to its rapid production and low cost, this manufacturing process has become more attractive. In addition to mechanical strength, other factors such as thermal resistance are also considered to determine their applicability. Understanding the various aspects of 1D printing technology can help aerospace companies develop their production capabilities for harsh environments. For instance, if there is a need for spare parts on certain satellites or in outer space, this can be met by 1D printed parts ("1D Printing in Space," 2014; Ishfaq et al., 2022).

10.2 NEED FOR TRIBOLOGICAL STUDY IN ADDITIVE MANUFACTURING

Sliding can cause the loss of wear or displacement of materials from the contact surface (Azakli & Gümrük, 2021). This process consumes considerable energy and results in significant material loss. The wear caused by this process can affect the system's functionality and reliability. It can also lead to various surface fatigue and corrosion conditions. These conditions are identified due to the interaction between the medium and the contact surface (Zhu et al., 2018b). When the surfaces of various types of materials are subjected to high pressures, they can develop asperities and deformation. This can cause welding and deformation between the counter body and the base. As the plastic asperities gradually break, they can create loose debris on the two surfaces that slide (Aghababaei et al., 2016). When the solid surface is pushed against a material with a greater hardness, it can cause wear. Various mechanisms can then be used to remove the debris, some of which include plowing, micro-cutting, and fragmentation. When materials are subjected to a corrosive medium, they can develop corrosion wear. Moreover, repeated loading and unloading of materials can cause surface fatigue, as well as cracks in the subsurface or surface, resulting in the formation of large pits as large chunks of debris are removed from the subsurface or surface. A comprehensive discussion of the various wear mechanisms is provided in review papers (Khruschov, 1974; Torrance, 2005; Zmitrowicz, 2006).

Due to the increasing number of applications of AM in various fields, such as aerospace, medical, and automobile, the technology has become more prevalent (Hanon & Zsidai, 2020; Li et al., 2017; Vafadar et al., 2021). Using a 1D model of a part, a designer can create complex and personalized products. This is extremely beneficial for the development of new and innovative products. Due to its ability to customize a part for a specific patient, AM has become a widely used technology in the medical field; however, the cyclic load and wear to which these implants are subjected can cause failure. This can be caused by the release of debris (Bartolomeu et al., 2019; Roudnicka et al., 2020). The internal combustion engine of a car is powered by a series of valves that are continuously opened and closed. The wear and contact between the valve seat and the surface of the valve can lead to issues during operation (Pradeep et al., 2022; Zhou et al., 2021).

The advantages of AM are numerous, making it an ideal choice for the aerospace industry. It can be used to make components for aircraft engines, metal brackets, and structural parts (Yin et al., 2018; Zhou et al., 2021). However, wear can be a potential cause of failure in structural components and can happen as a result of the gradual removal of surface material and deformation. Before using AM components in dynamic contact applications, it is important to study their tribological properties (Ralls et al., 2021).

Due to the high velocity and temperature conditions that are commonly encountered in certain applications in the automotive and aerospace industries, the functional surfaces of these components may experience adverse tribological conditions which could lead to malfunction or loss of efficiency (Kaya et al., 2019). As the various process parameters utilized during the fabrication and processing of these components can affect the wear and tribological properties of these parts, it is important that these parameters are controlled to ensure better tribological properties of the parts. In addition to the surface quality, other factors that affect the performance of these components include density, ductility, and microhardness (Kaya et al., 2019). Creating a textured AM-built surface can help improve the surface conditions by creating micro-scale features that serve various functions under varying lubrication conditions.

Various techniques are used to create textured surfaces, such as laser etching, machining, diamond embossing, and ion beam treatments (Coblas et al., 2015; Moshkovith et al., 2007). Due to its versatility, laser texturing is commonly used to create various shapes and sizes. It can also be done with high repeatability and control. The micro-sized features created through laser texturing can act as reservoirs of lubricant and wear debris, which can influence the wear mechanism of a sliding surface., The wear rate of various textured surfaces is comparable to that of samples with protective coating. The wear debris found in micro-dimples might be caused by the material bulges created by laser texturing. This phenomenon could explain why the surface's wear resistance has improved. Understanding the various characteristics of AM components can help improve their performance and service life. This chapter focuses next on the various characteristics of 1D-printed polymers, metals, and alloys.

10.3 POLYMER AND ITS COMPOSITES FOR TRIBOLOGICAL STUDY IN AM

Under wear mechanism, the mechanical properties and morphological characteristics of various polymer composites can be exposed (Saravanan & Devaraju, 2019). The wear and friction caused by the contact substances can lead to severe damage. Polymer tribology is a completely different phenomenon from ceramics and metal. Due to their physical properties and lower melting point, polymers can overrule the laws of friction (Friedrich, 2018). The importance of lubrication is acknowledged in the design and application of polymer tribology. Aqueous-based oils were developed by combining poly-ethylenimine with reduced graphene oxide, and were utilized to reduce wear and friction between two polymer surfaces (Rahman et al., 2021; Zhang et al., 2021). The results of the water bath method were very promising. The improved performance of the aqueous-based oils can reduce the wear rate and the frictional coefficient by up to 45% and 54.6%, respectively. These factors are attributed to their low surface energy and mechanical properties (C. Liu et al., 2019). Polymer tribology is a process that involves the use of water-based and solid lubricants. These two types of products have shown exceptional performance under sliding conditions. The influence of surface forces and friction between two surfaces is known to be a crucial factor that determines the performance of polymer contact. However, these factors have not been thoroughly studied by researchers. The strong bond between the two polymer surfaces allows for an excellent adhesive interface. Different parameters that can affect the friction forces are pressure, temperature, ambient condition, and contact angle (Bernat et al., 2018; Myshkin & Kovalev, 2018).

The wear behavior of various polymer composites, such as those used for AM and electrically conductive polymers, was studied. The specific wear rate and coefficient of friction (COF) of different types of polymer composites were analyzed by comparing their contact angles, cryogenic temperatures, and lubricants (Alajmi & Shalwan, 2015; Friedrich, 2018). The analysis of the wear and friction of polymer composites will only reach an end once the erosion rate is determined. The erosion rate of polyetherimide (PEI) composites was studied using its glass fibre counterparts. The impact velocity, fibre percentage, and the angle of erosion were analyzed to determine the rate of erosion of PEI-GF composites. The impact velocity of the composites was determined by taking into account the angle of erosion between 45 and 60°. The study revealed that the maximum erosion rate was observed due to the properties of the composites' ductile character (Thakre, 2014).

The wear behavior of polyetherimide (PEI) and carbon fibre (CF) composites was studied. The objective of this study was to improve the strength of the matrix adhesion of these composites due to an increase in the oxidative functional group. The conventional method of improving the functional group was by treating the fibre surface with nitric acid. This process involved varying intervals and calculating the density and volume of the composite material. The treatment time was increased to improve the surface hardness. It was analyzed using a scanning electron microscope (SEM). The chemical properties of the different matrix structures in the composites were analyzed using the FTIR spectroscopy technique (Tiwari et al., 2011). Polymer tribology mainly focuses on the bearing industry due to how it increases the wear rate and friction on the surfaces of components. Compared to other types of coatings, aromatic thermosetting polymers offer superior performance. The properties of aromatic thermosetting polymers were studied under different operating conditions and pressures (Nunez et al., 2019). Figure 10.1 presents materials based on polymers and elastomers that could be used in tribological applications.

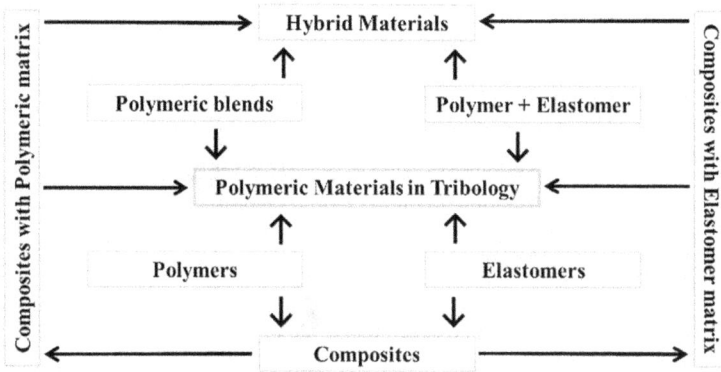

FIGURE 10.1   Materials based on polymers and elastomers, involved in tribological applications.

The properties of these polymers are studied under various conditions and pressures. Unlike chemically stable materials, semi-crystalline polymers can withstand a transition temperature. Various inorganic nano-fillers, from metals (Cu, Fe) and metallic and non-metallic oxides (CuO, Fe, TiO_2, ZrO_2, SiO_2, and ZnO), have been shown to enhance mechanical properties. They can also lower the rate of wear and the friction coefficient under different sliding conditions. Tribological applications of various polymers, such as PEEK, PPS, and PTFE, are commonly studied. They are often blended with CF fillers, TiO_2, SiC, and Si_3N_4 (Bahadur & Sunkara, 2005). Although there are various types of polymers and fillers that can improve the tribological performance of a material, they do not have the same properties in every situation. Due to the system responses generated by the materials, the wear and friction of a material can vary depending on its external environment and its intrinsic properties. A study conducted by Kurdi et al. (2019) revealed that adding a certain material to a polymeric matrix can reduce the wear and friction of the material. However, this benefit was not observed at elevated temperatures. When choosing a pair of materials for optimal tribological behavior, the functioning conditions of the materials are very important. Figure 10.2 displays the reduction in friction and wear of 5–15% of titanium dioxide at room temperature, which is not ideal for elevated temperatures. Figure 10.2(a) depicts the friction coefficients of high-performance polymers (HPPs) when subjected to different temperatures during pin-on-disk tests; Figure 10.2(b) portrays the specific wear rate of these materials (Kurdi et al., 2019). This demonstrates why it is important to consider the functioning conditions of the materials when selecting a pair.

In a study conducted on the effects of dry sliding on the wear and friction of plastic injection molded components, Hanchi and Eiss (1994) reported that the compounds' wear and friction were observed at temperatures ranging from 20 to 230 °C. The researchers noted that the tan δ peaks that were observed during transitions at the glass transition temperature were associated with the catastrophic

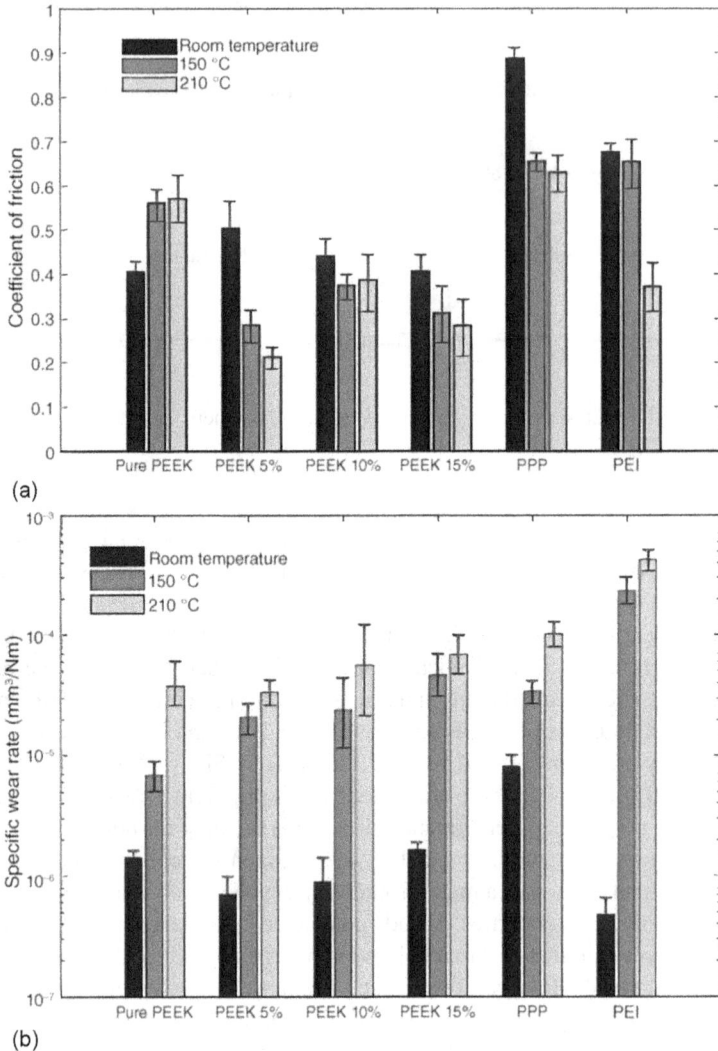

(a)

(b)

FIGURE 10.2 Influence of percentage of TiO_2 on (a) friction coefficient and (b) specific wear rate, for a pin-on-disk configuration (Kurdi et al., 2019). (Reproduced with permission, ©Elsevier. 2019.)

tribological failure of the compounds. The study revealed that the compounds, which were mainly composed of 70% PEEK and 30% PEI, exhibited significant increases in their wear and friction. The researchers noted that the absence of significant tribological failure in the compounds at the glass transition temperature was related to the transition of the lower strength. The study was conducted in a pin-on-disk tribometer test apparatus. The schematic representation of the test apparatus is displayed in Figure 10.3(a). The relationship between the changes in

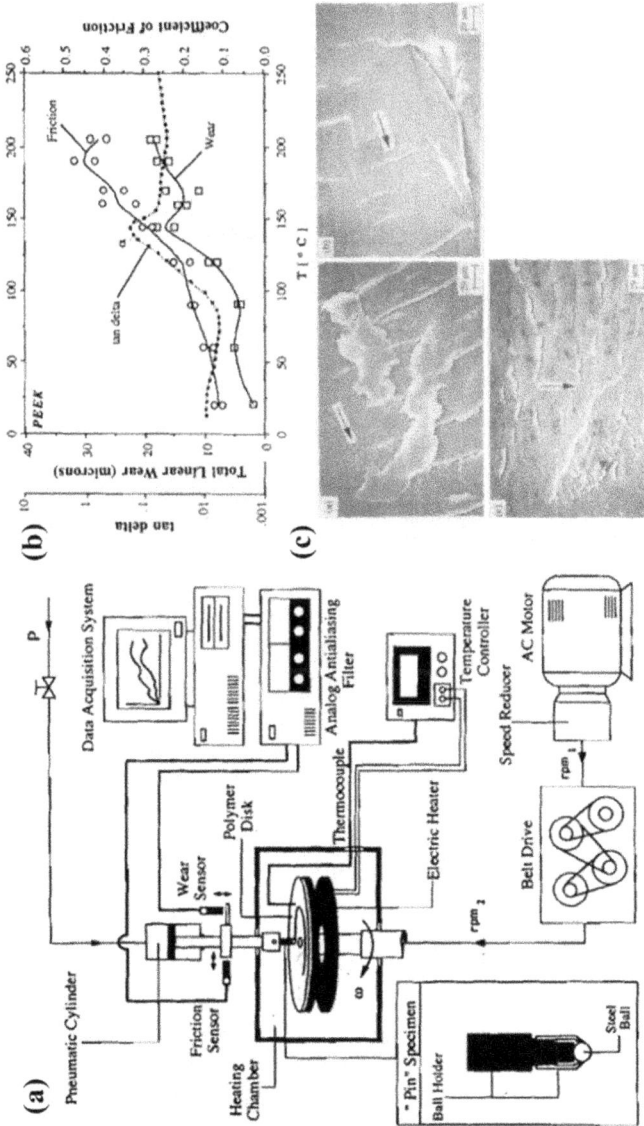

FIGURE 10.3 (a) Schematic representation of the pin-on-desk test apparatus; (b) temperature dependence of the steady-state coefficient of friction, total linear wear, and mechanical loss factor for PEEK material; (c) SEM images of PEEK on test apparatus. (Reproduced with permission, ©Taylor and Francis. 1994 Hanchi & Eiss, 1994.)

the tribological behavior and the peak in the relaxation spectrum of PEEK is shown in Figure 10.3(b). Finally, the SEM images of various polycrystalline and amorphous PEEKs, including their tribological properties, were investigated using different proportionality levels (a) PEEK/PEI 85/15 (152 °C); (b) PEEK/PEI 70/30 as-molded (159 °C); (c) PEEK/PEI 70/30 annealed (186 °C). These results reveal that the compounds exhibited severe to mild wear transitions and friction at temperatures ranging from 90 to 105 °C. The researchers noted that the significant change in the ductility of the compounds during the transitions at the glass transition temperature was associated with the development of the tribological behavior.

A study conducted by Unal and Mimaroglu (2003) revealed that the test speed and load values affected the wear and friction characteristics of polyolefin (POM, PEI, and PTFE) and pin-on-disc tribo-testers. The tests were performed at various temperature conditions, such as room temperature, under 5N, 10N, and 15N at 0.5 m/s. The results of the investigations revealed that the wear rates were influenced by mass loss. The researchers found that the COF of the tested polymers increased linearly with the increase in load. They noted that the wear rate exhibited low sensitivity to the load conditions during the test period; it also exhibited high sensitivity to the speed.

With regard to introducing polymeric materials into tribological applications, engineers need to know what they should expect from these materials. Some of the key characteristics are the addition of reinforcements and solid lubricants. A higher softening temperature can be obtained by adding glass fibres. This can result in a higher flexural modulus, which can rise to 11,000 MPa. High or acceptable wear resistance, low friction, and good strength are some of the characteristics of glass fibres that can be used to describe their properties in respect of their application. They can further resist the effects of fluids in the environment. They can be used to reinforce various materials such as solid lubricants. The processing capabilities of glass fibres are also considered to be good. These include their fast solidification and uniform flow.

Based on important works on tribology of polymer-based materials (Briscoe & Sinha, 2002; Friedrich et al., 2002; Kiran et al., 2018; Kmetty et al., 2010), their reinforcements and solid lubricants can improve their tribological properties. However, it is not a fixed rule that these materials can be used in combination with other materials to improve their performance. They should first be tested at a scale and under real-world conditions. Due to a reduction of superficial energy, some solid lubricants, such as those with sheet-like characteristics, weaken the bulk materials. Although the reinforcement of polymers makes their resistance greater, it also generates a more intense wear on the other surface. This leads to a higher friction coefficient and a worse surface quality. Even slight damage can be caused to the protective transfer films by reinforcing materials. They can also generate a sliding regime that is characterized by third body wear. For instance, glass beads reinforced with PA + 50% can exhibit third body wear at low velocities (Fusaro, 1990; Godet, 1984).

Polymer composites and hard-wearing polymers with hard components can reduce friction by adding a layer of solid lubricants that have a plaquette-like shape and a low friction coefficient. Although short fibres are commonly used for tribological applications, researchers are beginning to realize that long fibres can also be utilized for moving parts due to advancements in polymer technology. A brief discussion about the architecture of fibres is helpful here. Usually, short and tangled fibres are arranged randomly within a material, this causes an increase in their cost. A long fibre can be arranged in various ways depending on the applications. Depending on the requirements of different applications, long fibres can be arranged in multi-axial, woven, and unidirectional configurations. The wear of long fibres is usually categorized in steps depending on their properties. Since fibres with a width of 5–50 microns have a life cycle of about five to 10 years, the wear of the first two or three layers will end the tribo element's life. Some of the types of long fibres that are commonly used include carbon nanotube, glass fibres, and polymer fibres (Sharma & Kumar, 2020).

As reinforcement materials, they can have various shapes, such as almost spherical particles, beads with sheet-like structures, and plaquettes, with one dimension being very small compared to the others. When wearing polymeric blends or composites, the initial preferential wear of the latter is caused by an increase in the concentration of fibres or particles. This can then be transferred to the body through the movement of the hard materials, which can either be embedded into the matrix or torn off. The impact of the hard materials can cause the wear debris to "travel" in contact with the soft matrix, causing the friction coefficient to rise. The wear cycle continues as the hard particles are lost (Deleanu et al., 2021).

Adding materials to polymers can sometimes cause tribological behavior problems. For instance, the high concentration of glass fibres in a material can increase the wear and friction coefficient of the surface. Tribological differences between mechanical characteristics and tensile tests can be referred to as challenging. For instance, increasing the strength of a material can be done by adding reinforcements. A material's deformability can help create a fluid film when it slides. In addition, fluid lubrication can also help achieve this. In 1976, Evans and Lancaster noted that fibres in polymers can help reduce the wear and friction coefficient of a material (Evans & Lancaster, 1976).

Although some materials can be used as reinforcement materials, they can also be used to help with the heat evacuation process. A growing interest in the use of blends and polymer composites has revealed that the designer has to make compromises in order to achieve desirable results (Myshkin et al., 2015). There is some possible tribological behaviour of thermoplastic and fibre thermoplastic composites (Aldousiri et al., 2013). A summary of the thermoplastic materials' tribological behavior is explained in Figure 10.4(a). The high potential of film transfer is usually expected in the case of a neat thermoplastic. However, due to the presence of two polymers that are similar in composition, the sliding forces will lead to high heat. The presence of heat in the interface can deteriorate the soft surface. The addition of fibres or fillers can either reduce the interface's heat, or

(a)

(a)

(b)

(b)

(a)

(b)

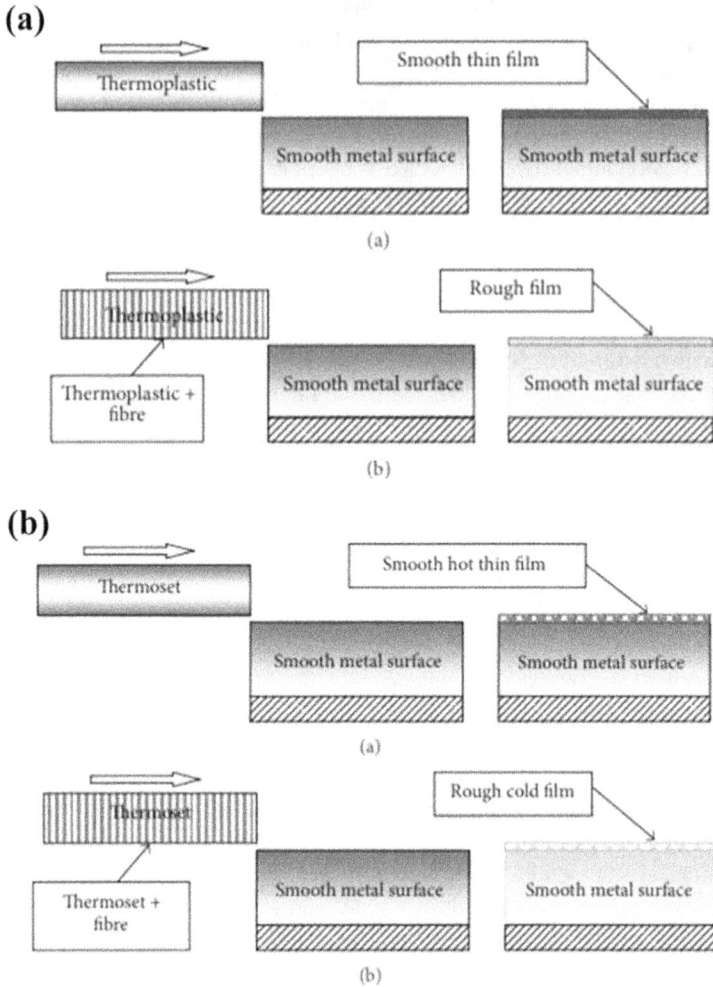

FIGURE 10.4 The possible film transfer (a) when a certain type of thermoplastic or fibre composite material slides on a metal surface, and (b) when a certain type of thermoplastic material or fibre composite material slides onto a metal surface. (Reproduced with permission, ©Taylor and Francis. 1994 Hanchi & Eiss, 1994.)

prevent detachments and the film transfer from adhering well. A material's surface may also be used to create a film transfer by combining the debris of fibre and polymer. The two properties of the surface can be described as rough or smooth; the latter surface can achieve low friction and wear, while on the other hand, the rough surface can experience a high wear rate.

Similarly, there is some possibility of tribological behaviour in thermoset and fibre thermoset composites. The information presented in Figure 10.4(b) summarizes the various characteristics of thermoset materials. In terms of film transfer, the neat thermoset is more difficult than the thermoplastic. One of the main advantages of thermoset materials is their low plasticity because they can resist plastic deformation. Most of the studies on thermoset materials mention the possibility of film transfer when the thermoset is slid against a metal surface. The film transfer characteristics of near thermoset materials are rougher than those of thermoplastics, although the interface is also cooler due to the low frictional force. The high adhesive behavior of thermoplastic on the metal surface makes it ideal for film transfer. In the case of thermoset materials, the presence of fibres can help improve the interface's cooling effect. In the case of thermoset materials with fibres, the interface's cooling effect is affected by the high degree of modification in the film transfer. This can be caused by the transition from being an adhesive to an abrasive wear (Aldousiri et al., 2013).

10.3.1 Issues Related to the Use of Polymeric Materials in Tribological Applications

Due to the increasing acceptance of plastic materials as a replacement for metallic components, manufacturers have to face the challenge of developing a good tribological behavior. This is because a reliable functioning of the material requires a set of characteristics that are favorable for its application. Besides the functional properties of the material, the dimensional stability of the polymeric materials used in tribological applications should also be considered. These materials are known to have high thermal expansion coefficients, low resistance to high temperatures, and shorter durability. Due to their low hardness, these materials are not usually used in rolling contacts. They can be utilized in various applications such as car tires and gears. While polymeric materials have many advantages in tribological applications, there are also some issues that need to be considered:

- Low load-bearing capacity (polymers generally have lower strength and stiffness compared to metals, which can limit their use in high-load applications);
- Low temperature resistance (many polymers have poor thermal stability and can become brittle at low temperatures);
- Deformation under load (polymers can experience significant deformation under load, which can lead to dimensional instability and reduced precision in certain applications);
- Creep (polymers can experience creep, which is the gradual deformation of a material under a constant load. This can lead to a loss of accuracy and precision over time in certain applications);
- Chemical degradation (some polymers can be degraded by certain chemicals, which can reduce their lifetime in certain environments);
- Wear and tear (polymers tend to wear out faster than metals and ceramics, which can limit their lifetime in certain applications);

- Surface finish (surface finish can be an issue with polymers as they tend to have a rougher surface than metals and ceramics);
- High friction and wear in dry conditions (polymers tend to have high friction and wear in dry conditions which can be an issue in certain tribological applications);
- Low thermal conductivity (polymers have low thermal conductivity, which can make them less suitable for use in high-temperature applications);
- Low impact resistance (polymers have low impact resistance, which can make them less suitable for use in applications where they are exposed to high impact loads).

10.3.2 Advantages of Using Polymeric Materials in Tribological Applications

Polymeric materials, such as composites, polymers, and their blends, have various advantages over metallic materials (Leblanc, 2009; Patnaik et al., 2010). These include their self-lubricity, low density, and resistance to tribo-corrosion. They can also be molded through injection molding (Mathew et al., 2009). There are several advantages to using polymeric materials in tribological applications, including:

- Low friction (polymeric materials have low coefficients of friction, which means they experience less resistance when in contact with other surfaces. This can help to reduce wear and tear on the components, resulting in longer lifetimes);
- Low wear (polymers are also known to have low wear rates, which means they experience less degradation when in contact with other surfaces. This can help to reduce the need for replacement parts and maintenance);
- Lightweight (polymeric materials are lightweight, which can help to reduce the overall weight of a system and increase energy efficiency);
- Low cost (polymeric materials are relatively inexpensive compared to many other materials, which can help to reduce the cost of a system);
- Good chemical resistance (many polymers have excellent chemical resistance, making them suitable for use in environments with harsh chemicals);
- Good thermal stability (many polymers have good thermal stability, which means they can maintain their properties over a wide range of temperatures);
- Easy to process (polymers are relatively easy to process, which means they can be produced in a wide range of forms, such as powders, fibres, and films);
- Can be tailored for specific tribological applications (polymers can be designed or tailored for specific tribological applications by controlling their molecular weight, molecular weight distribution and functional groups);
- Can be filled with various fillers (polymers can be filled with various fillers such as carbon, graphite, MoS2, etc. to improve their tribological properties).

10.3.3 COMMON FACTORS TO BE CONSIDERED WITH POLYMERIC MATERIALS IN TRIBOLOGICAL APPLICATIONS

One of these is that slight changes in working conditions can modify their tribological properties. For instance, low friction and wear rate are not related to one another (Myshkin et al., 2005). Polymeric materials can react differently to negative temperatures. Some become brittle, while others resist without problems. The environment and working conditions can also influence these properties. When it comes to designing polymeric components, the designer should consider the various factors that affect their longevity. For instance, if the material has good mechanical characteristics and dimensional stability, it should not be prone to aging issues. The design should also consider working conditions, which can vary in different narrow ranges. These conditions should include the temperature, load, morphology, and velocity (Briscoe & Sinha, 2014). Most of the time, tribological applications involving polymeric materials involve couples that are made up of both metallic and polymeric elements. Sometimes, a polymeric component is being used against a body that is made of the same materials; one example of this is gear transmissions.

10.4 METALS AND THEIR ALLOYS FOR THE TRIBOLOGICAL STUDY OF AM

Due to the unique properties of metal AM components, the production of complex and vital metallic parts has become more popular. Among AM components' various attributes, are their reduced weight and carbon footprint (Frazier, 2010). Metal AM is capable of creating complex parts that can be molded or fabricated using various methods. In traditional manufacturing, components are typically forged or cast (Kruth et al., 2007). Using these methods can be very challenging when creating irregular and thin-walled shapes in metal components (Cáceres et al., 1995). Metal AM offers numerous advantages, such as its ability to reduce waste and improve the design flexibility of certain parts (Atzeni & Salmi, 2012). One of the most common implementations of this technology in an industrial setting is the manufacture of fuel nozzle components for a jet engine. In this case, the use of powder-based metal AM allowed for the production of complex components much faster than traditional methods. It also reduced the weight of the parts by 75%. The most significant property of AM is its layer-by-layer process (Brandt, 2016). This allows for the creation of complex and highly detailed models, which are typically not possible using traditional methods ("Integrative Production Technology for High-Wage Countries," 2011; Zhang et al., 2014).

A study conducted by Khaing et al. revealed that through the use of powder-based AM methods, complex metal components can be created with finer details than those produced using traditional methods (2001). There are numerous advantages to metal AM components such as their strength. However, it should be noted that various printing parameters and raw material selection affect the mechanical properties of these components (Shah et al., 2021). Due to the wide variety of applications that metal AM can perform, it is widely used in various industries such as aerospace, automotive, architecture, and biomedical (Blakey-Milner et al., 2021; Zhang, Wu, et al., 2018b). Despite the technological advancements of AM, there is still much untapped potential. This technology could transform manufacturing. Through the use of additive manufacturing, components could be produced on demand and their distribution de-globalized. The reduction of manufacturing's carbon footprint could be significant. However, the layered surface quality of metal AM could prevent the industry from adopting this technology (Ford & Despeisse, 2016).

In metal AM processes, the addition of materials is achieved through repeated solidification and melting. Due to the complex thermal history of the materials involved in the AM process, the surfaces of the parts are often irregular and have a stochastic surface roughness (Kelly & Kampe, 2004). Due to the stochastic nature of the surface topography, the parts produced through traditional methods are not as efficient as they should be (Wang et al., 2015, 2016). Friction in mechanical components leads to significant energy loss. The lifespan of these components can also be reduced due to wear (Meng et al., 2020). Through tribological assessment, it is possible to reduce the wear and friction of metal AM components. This process could help improve the efficiency of the technology. Maintaining consistent mechanical properties, such as the COF, ductility, and shear strength, is a prerequisite for functional applications of AM components. A lack of familiarity with the properties of metal AM components and the anisotropic nature of their manufacturing process can limit their usefulness (Dehoff et al., 2013; Zhu et al., 2015).

Traditional techniques for the production of metal AM components cannot be replaced until the surface quality of these components can be improved with the use of different methods and materials. As-fabricated metal AM components often have higher COF and are prone to wear. This issue can be caused by their irregular surface topography (Zhu et al., 2015). The literature on the subject of metal AM components' tribological properties is very comprehensive. It covers the various aspects of the AM process, such as material selection, post-processing, and parameter optimization. Several review articles have been published on the various facets of metal AM, such as medical implants and energy applications (Shah et al., 2020; Ur Rahman, Matthews, et al., 2019b). This section aims to provide a comprehensive analysis of the various aspects of metal AM methods and their applications. It also aims to review the literature on the tribological and mechanical properties of these components.

Aluminum is a widely used material in various applications such as aerospace, automotive, and marine products. Its high strength and ductility are some of the factors that make it ideal for these industries. Aluminum components, such as cylinders, bearings, and automotive pistons can experience adverse sliding conditions during their service period. It is for this reason that much work has been done on the study of various aspects of aluminum's properties (Alidokht et al., 2011; Dai & Liu, 2008). Although studies on the tribological properties of aluminum and its alloys have not been well publicized, research on this subject is on the

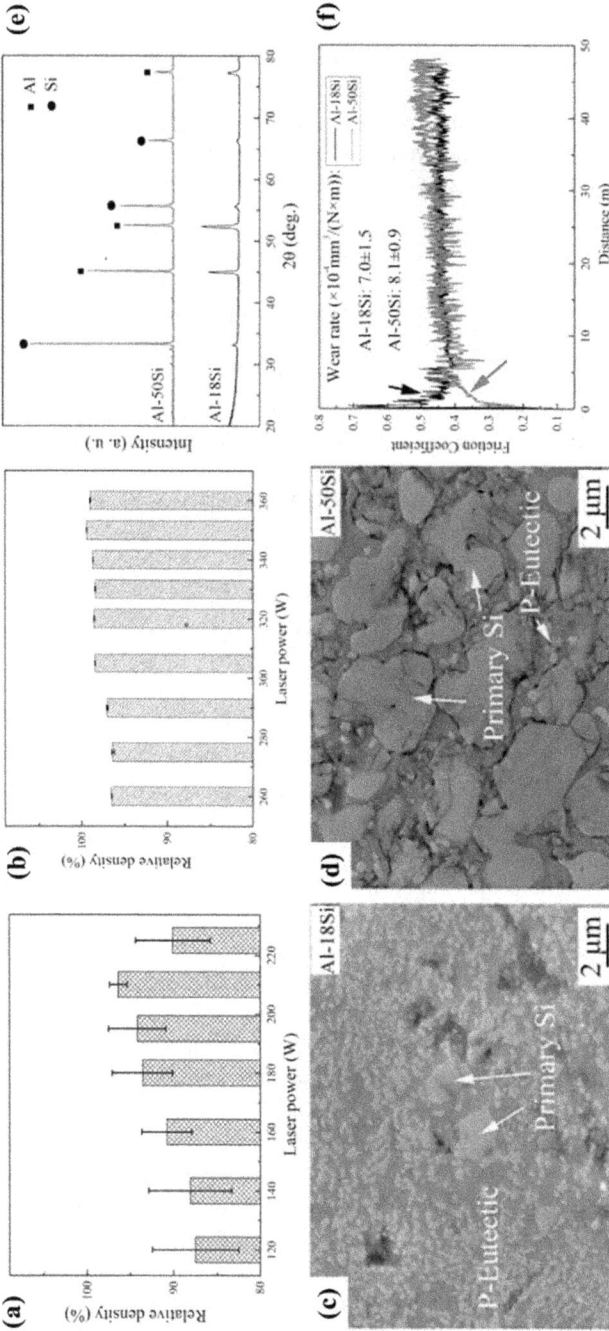

FIGURE 10.5 The relative density of SLM processed (a) Al-18Si and (b) Al-50Si on function of laser power SEM image of SLM processed; (c) Al-18Si and (d) Al-50Si alloys with several porosity; (e) XRD patterns of SLM processed Al-18Si and Al-50Si alloys with similar relative density; (f) friction coefficient and wear rate of SLM processed Al-18Si and Al-50Si alloys. Reproduced with permission, © *Tribology International* (Kang & el Mansori, 2020).

rise (Liu & Guo, 2020). A study conducted by Kang and el Mansori (2020) revealed that the wear behavior of Al-Si alloys after being processed through SLM differed from that of traditional casted materials. The microstructure of the SLMed Al-Si alloy was more extreme due to the rapid cooling and heating cycles The increase in the silicon content in Al-Si alloys from 18 to 50% during the process has led to a slight increase in the primary silicon's size (Kang & el Mansori, 2020). In SLMed Al-Si alloys, the average size of the material is around 3.12 µm. This is smaller than the average size of hundreds of micrometers in conventional casted high-Si-Al alloys.

The microstructure of Al-Si alloys subjected to SLM processing was studied to gain a deeper understanding of their tribological properties. The Al-Si alloys are composed of a pseudo eutectic structure consisting of primary silicon and super-saturated Al (Si). As the silicon content increases, the size of the primary silicon gets larger. The Al-18Si and Al-50Si alloys exhibited better wear resistance than their conventional cast counterparts. The wear rate remains the same even after the silicon content increased from 18 to 50% (Kang & el Mansori, 2020). The characteristics of the Al-Si alloy subjected to SLM processing are different to those of the casted versions. For instance, the SLM sample exhibited a simple and effective crushing-inlaying process.

10.5 TRIBOLOGY IN ADDITIVE MANUFACTURING TECHNOLOGIES

According to ISO/ASTM 52,900:2015(E), AM methods can be divided into two categories: single-step and multi-step processes (Gibson et al., 2021; Horn & Harrysson, 2012; Zhang, Jarosinski, et al., 2018a). A single-step process is commonly used to achieve the basic geometric properties of a product and its fundamental materials (Gibson et al., 2021). Multi-step AM processes are commonly used to achieve the same basic properties of a product. After providing a basic shape, the desired properties are consolidated in subsequent steps. The three different types of single-step AM processes for metallic materials are direct energy deposition (DED), powder bed fusion (PBF), and sheet lamination (SL) (Zhang et al., 2018b). The state of fusion, AM principle, form, source, and feedstock are considered when determining these processes. Most AM processes utilize either wire or powder as their feedstock material. These materials are selectively melted and deposited by a high-energy heat source. Electron beam, laser, gas metal arc, and plasma are the various types of heat sources utilized in AM processes. While laser or electron beam are commonly used for developing complex parts, plasma or arc are utilized to produce large components with better metal deposition rates. Although all AM processes utilize a layer-based principle, they have varying aspects that affect the creation and bonding of the layers (Ngo et al., 2018). The various process variables that affect the creation and bonding of a product's layers play a vital role in its overall geometric and mechanical properties (Edgar & Tint, 2015; Gibson et al., 2015).

During the early days of AM, the focus was on polymer resin systems. However, due to the technological advancements in the field, many efforts have been made to develop metal-based AM processes (DebRoy et al., 2018). Compared to traditional manufacturing methods, AM processes are more frequently utilized in certain industries, such as biomedical and aerospace. The use of AM in the aerospace industry could help improve the efficiency of the process and reduce the weight and costs of the components (Joshi & Sheikh, 2015). It is widely used in the medical sector for the production of various medical devices, such as dental implants and hearing aids.

According to Mohamed et al., (2017) the three factors that influence the wear of a component after it has been manufactured using PC-ABS are the layer thickness, the build orientation, and the raster angle. They tested the wear characteristics of a pin using a Tribometer T50 using standard ASTM G99. The pin had a diameter of 6 mm and a length of 35 mm. The pin's counterpart, which was EN 31 hardened steel, had a sliding diameter of 14 mm. Its contact force was 10 n, and it could handle a total of 10,000 cycles (Mohamed et al., 2017). The researchers found that the increase in the layer thickness of the component after manufacture can lead to higher wear rate. On the other hand, the air gap and the increase in the raster angle can improve the wear rate. The various factors that influence the wear characteristics of a component after it has been manufactured using PC-ABS. The first is the layer thickness, the second is the build orientation, and the third is the wear rate.

Almost 80% of metallic parts found in various engineering systems are made using steel. They are mainly processed through conventional means such as machining, casting, and forging. However, it is challenging to produce complex geometric shapes and functionalized materials using the traditional process. To address this issue, many researchers have developed new technologies that can be used to improve the performance of steel. Currently, around a third of the publications in the AM industry are on the subject of steel. Compared to their traditional counterparts, AM-produced steel parts are more corrosion resistant, have better wear resistance, and exhibit better strength. These characteristics can be attributed to the formation of fine precipitates and cellular sub-grains during the AM process. Besides the demand for complex parts in various industries such as aerospace, automotive, and marine, it is also important to study the tribological properties of AM steels. In addition, studies on the wear performance of 316L SS components, which are commonly used in the aerospace and automotive industries, are also important. These components usually face high velocity sliding environments. According to the literature, the high-speed sliding characteristics of certain metal components produced through conventional processes differ to those of SLM components. In this study, the influence of the build orientation on the wear behavior of 316L SS components was analyzed (Mandal et al., 2020).

In a study conducted by Hanon and Zsidai (2020) the tribological properties of samples that were manufactured using a digital light processing technique for a resin known as Wanho UV were examined. The researchers found that the steel plate samples that were not cured using UV light had a lower friction coefficient and a higher wear depth than the cured specimens. The study revealed that the cured steel plate samples had a friction coefficient ranging from 0.72 to 0.79, depending on the printing orientations.

The process of creating a part involves adding a material as wire or powder, and the surface quality can be controlled. In order to study the wear performance of the parts made using direct energy deposition of HHS alloy, Rahman et al. (Ur Rahman et al., 2019a) found that the parts produced through this process exhibited superior wear characteristics. To investigate the wear resistance of tool steels made using direct energy deposited technology, Park and colleagues conducted a test using a ball-on-disk type wear tester (Park et al., 2021). The test was performed on a plate which had a ball diameter of 3.97 mm and a material bearing ZrO_2 and steel. The tester was moved on a cyclic trajectory with a diameter ranging from 13.6 to 13.6 mm. The friction coefficient was reported to be 0.6.

Although binder jet technology has many advantages, its main disadvantage is that it produces weak and porous components. Because of this, post-treatment procedures are often utilized to improve the mechanical properties of the parts. Sintering and infiltration are the most common treatments. Depending on the process used, the final part's tribological properties can vary. In order to study the tribological properties of tungsten and stainless steel, a team of researchers (Cui et al., 2021) conducted a 5-hour treatment at a temperature of 180 °C and then infiltrated the parts with bronze. The researchers found that the values of scratch hardness exhibited by the various components were similar. However, the lower hardness of the BJP316 was more detrimental to the parts' mechanical properties. The researchers discovered that the friction coefficients varied depending on the depth of the scratch. For instance, the COF between zero and 35 nanometers can vary by as much as 0.6. The researchers conducted a friction test to determine the average coefficient of friction. They found the range to be from 0.55 to 0.75 under a contact force of 30 Newton (N). The main wear mechanisms of various components, such as the BJP316 and BJP420, are delamination, spalling, and abrasive wear. In a study on the tribological properties of 316SS/bronze, the researchers were able to conclude the same wear mechanism (Wang et al., 2021).

According to Lorusso (2019) the parts produced through the use of laser PBF have better wear resistance and a lower COF than those made using conventional technologies. The main cause of this issue is the high hardness and fine size of the grains. Porosity in the parts can also affect their wear behavior. This damages the bonding between the molten pools. The high hardness of the parts produced through laser powder bed technology makes them harder to machine. Compared to components made using traditional methods, those made using this process are more difficult to work with. It is also widely known that the surface's waviness and rough texture can affect the friction coefficient. According to Calignano (2018) the surface quality of certain parts depends on the various parameters of the printing process. These include the orientation, thermal stress, and the STL file's parameters. The roughness of a part's surface can be influenced by the partially fused materials that adhere to it. Qin et al. (2020) (Qin et al., 2020) noted that the main wear mechanisms of Fe alloy components are oxidation and plastic

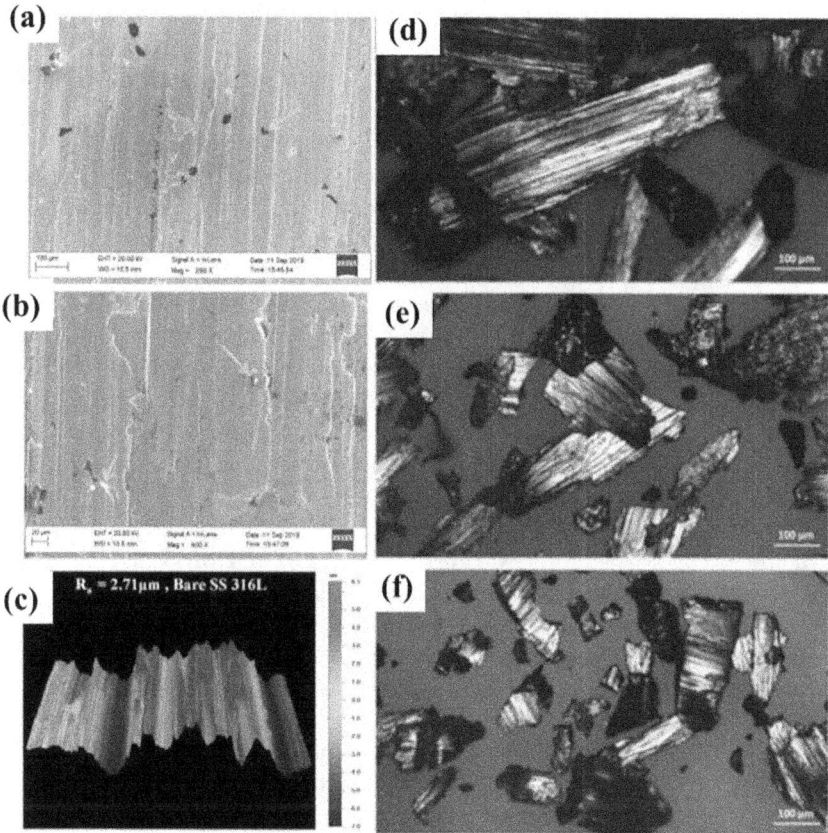

FIGURE 10.6 (a) SEM images of (a, b) the worn surface of bare SS 316L; (c) 1D surface profilometry of bare 316L; (d, e, f) debris of bare 316L tested at normal loads of 10, 8, and 6 kg, respectively, and a sliding speed of 2 m/s. Reproduced with permission, © *Tribology International* (Mandal et al., 2020).

deformation. The effects of post-treatment procedures on these components' friction and wear behavior were also studied. The graph (Qin et al., 2020) show the average wear rate and friction of various samples of Fe-Ni-Cr alloy, which were treated using HFUP or untreated. For comparison, a steel ball with a sliding speed of 8 m/s had a contact force of 400 mN. According to Qin, the treatment increased the surface hardness and reduced the friction coefficient by 11%. This is because the plastic deformation was reduced.

A post-processing step involves improving the mechanical properties of additively manufactured components. This step is usually performed after the materials have been 1D printed using PBF and binder jetting. Through the use of 1D printing technology, complex components can be manufactured with minimal reliance on the base material. The various parameters that are involved in the printing process are of great interest to researchers. The materials used in the process also play a vital role in the final components' tribological properties. Although some of the parameters that are involved in the printing process are crucial, the nature of the material used also plays a significant role in the final components' tribological properties. Through the use of 1D printing technology, the researchers can now create self-lubricating components that can be made using both non-metal and metallic base materials. For instance, the SL process can lead to the creation of steel-based components that have good wear resistance and a low friction coefficient.

The preferred processes for aerospace are direct energy deposition and powder bed fusion (Javaid & Haleem, 2018). The powdered form of various metals such as titanium, stainless steel, aluminum, copper, nickel, and aluminum alloys are available. A wide range of materials can also be utilized as wire feedstock. Despite the advantages of powder bed fusion, the deposition rates of wire feed are still unique. For instance, while powder bed technologies provide better surface finish and geometric tolerances, the wire feed processes are still capable of producing exceptional deposition rates (Dordlofva et al., 2016). The three main PBF groups that utilize a laser beam for powder bed fusion are SLM, DMLS, and SLS. They all follow a fundamental principle: The laser beam causes the powder particles on the bed to melt (Kurzynowski et al., 2020). Selective laser melting has been widely used in the production of various types of alloys, such as 316L stainless steel, Ti-6Al-4 V, Inconel 718, and CoCrMo (Amanov, 2021; Baufeld et al., 2011; Jia & Gu, 2014; Le et al., 2020). These materials are known to have excellent thermal conductivity and are capable of handling the high-speed laser processing needed for SLM. Due to the physical properties of Al-based alloys, the SLM process can be challenging. One of the main challenges that the process faces is the development of dense components. To increase the density of the materials, the process parameters can be adjusted by controlling the various aspects of the process, such as the scanning speed, power, and orientation. The scanning speed of a laser spot is measured by the speed at which it moves on a powder surface (Tran & Lo, 2019). The hatch distance is the distance between two laser beams, while the layer thickness is the amount of material that is added to each subsequent layer. Due to the varying process parameters that affect the deposition rates of SLM components, researchers are currently studying the effects of these on the densification of different metal and alloy systems (Greco et al., 2020; Zhu et al., 2018a).

10.6 CONCLUSION

The tribological behavior of materials used in AM has been a subject of ongoing research. Polymer-based materials, such as thermoplastics and thermosetting plastics, have been widely used in AM due to their low cost and ease of processing. Studies have shown that polymer-based materials exhibit lower coefficients of friction and higher wear resistance compared to metal-based materials. This is attributed to the amorphous nature of polymers, which allows for a more flexible response to deformation and a better ability to conform to asperities. Additionally, polymer-based materials can also have self-lubricating properties, which can further improve their tribological behavior. However, metal-based materials, such as titanium and aluminum, are also commonly used in additive manufacturing. These materials are known for their high strength and thermal conductivity, making them suitable for use in aerospace and medical applications. Studies have shown that the tribological behavior of metal-based materials in AM can vary depending on the specific metal used and the printing process. For example, a study by X. Zhang et al. (2018b) found that the COF of titanium printed using laser PBF was lower than that of aluminum printed using the same process. In summary, polymer-based materials have been found to have better tribological behavior than metal-based materials in additive manufacturing. Polymers exhibit lower coefficients of friction and higher wear resistance compared to metals and can also have self-lubricating properties. However, the tribological behavior of metal-based materials in AM can vary depending on the specific metal used and the printing process. Further research is needed to fully understand the tribological behavior of materials used in AM and to optimize the performance of printed parts.

REFERENCES

3D Printing in Space. (2014). In *1D Printing in Space*. https://doi.org/10.17226/18871

Aghababaei, R., Warner, D. H., & Molinari, J. F. (2016). Critical length scale controls adhesive wear mechanisms. *Nature Communications, 7.* https://doi.org/10.1038/ncomms11816

Alajmi, M., & Shalwan, A. (2015). Correlation between mechanical properties with specific wear rate and the coefficient of friction of graphite/epoxy composites. *Materials, 8*(7). https://doi.org/10.3390/ma8074162

Aldousiri, B., Shalwan, A., & Chin, C. W. (2013). A review on tribological behaviour of polymeric composites and future reinforcements. *Advances in Materials Science and Engineering, 2013.* https://doi.org/10.1155/2013/645923

Alidokht, S. A., Abdollah-Zadeh, A., Soleymani, S., & Assadi, H. (2011). Microstructure and tribological performance of an aluminium alloy based hybrid composite produced by friction stir processing. *Materials and Design, 32*(5). https://doi.org/10.1016/j.matdes.2011.01.021

Amanov, A. (2021). Effect of post-additive manufacturing surface modification temperature on the tribological and tribocorrosion properties of Co-Cr-Mo alloy for biomedical applications. *Surface and Coatings Technology, 421.* https://doi.org/10.1016/j.surfcoat.2021.127378

Atzeni, E., & Salmi, A. (2012). Economics of additive manufacturing for end-usable metal parts. *International Journal of Advanced Manufacturing Technology, 62*(9–12). https://doi.org/10.1007/s00170-011-3878-1

Azakli, Z., & Gümrük, R. (2021). Particle erosion performance of additive manufactured 316L stainless steel materials. *Tribology Letters, 69*(4). https://doi.org/10.1007/s11249-021-01503-0

Bahadur, S., & Sunkara, C. (2005). Effect of transfer film structure, composition and bonding on the tribological behavior of polyphenylene sulfide filled with nano particles of TiO_2, ZnO, CuO and SiC. *Wear, 258*(9). https://doi.org/10.1016/j.wear.2004.08.009

Bartolomeu, F., Abreu, C. S., Moura, C. G., Costa, M. M., Alves, N., Silva, F. S., & Miranda, G. (2019). Ti$_6$Al$_4$V-PEEK multi-material structures – Design, fabrication and tribological characterization focused on orthopedic implants. *Tribology International*, *131*. https://doi.org/10.1016/j.triboint.2018.11.017

Bassoli, E., Gatto, A., Iuliano, L., & Violante, M. G. (2007). 3D printing technique applied to rapid casting. *Rapid Prototyping Journal*, *13*(3). https://doi.org/10.1108/13552540710750898

Baufeld, B., van der Biest, O., Gault, R., & Ridgway, K. (2011). Manufacturing Ti-6Al-4V components by shaped metal deposition: Microstructure and mechanical properties. *IOP Conference Series: Materials Science and Engineering*, *26*(1). https://doi.org/10.1088/1757-899X/26/1/012001

Bernat, S., Brink, A., Lucas, M., & Espallargas, N. (2018). Tribological behavior of polymer seal materials in water-based hydraulic fluids. *Journal of Tribology*, *140*(6). https://doi.org/10.1115/1.4040078

Blakey-Milner, B., Gradl, P., Snedden, G., Brooks, M., Pitot, J., Lopez, E., Leary, M., Berto, F., & du Plessis, A. (2021). Metal additive manufacturing in aerospace: A review. *Materials and Design*, *209*. https://doi.org/10.1016/j.matdes.2021.110008

Brandt, M. (2016). *Laser additive manufacturing: Materials, design, technologies, and applications*. Elsevier Science.

Brecher, C., Jeschke, S., Schuh, G., Aghassi, S., Arnoscht, J., Bauhoff, F., ... & Jeschke, S. (2011). *Integrative production technology for high-wage countries* (pp. 17–76). Berlin, Heidelberg: Springer Berlin Heidelberg.

Briscoe, B. J., & Sinha, S. K. (2002). Wear of polymers. *Proceedings of the Institution of Mechanical Engineers, Part J: Journal of Engineering Tribology*, *216*(6), 401–413. https://doi.org/10.1243/135065002762355325

Briscoe, B. J., & Sinha, S. K. (2014). Tribology of polymeric solids and their composites. In *Wear – Materials, mechanisms and practice*, Elsevier Science. https://doi.org/10.1002/9780470017029.ch10

Cáceres, C. H., Davidson, C. J., & Griffiths, J. R. (1995). The deformation and fracture behaviour of an AlSiMg casting alloy. *Materials Science and Engineering A*, *197*(2). https://doi.org/10.1016/0921-5093(94)09775-5

Calignano, F. (2018). Investigation of the accuracy and roughness in the laser powder bed fusion process. *Virtual and Physical Prototyping*, *13*(2). https://doi.org/10.1080/17452759.2018.1426368

Coblas, D. G., Fatu, A., Maoui, A., & Hajjam, M. (2015). Manufacturing textured surfaces: State of art and recent developments. *Proceedings of the Institution of Mechanical Engineers, Part J: Journal of Engineering Tribology*, *229*(1). https://doi.org/10.1177/1350650114542242

Cui, S., Lu, S., Tieu, K., Meenashisundaram, G. K., Wang, L., Li, X., Wei, J., & Li, W. (2021). Detailed assessments of tribological properties of binder jetting printed stainless steel and tungsten carbide infiltrated with bronze. *Wear*, *477*. https://doi.org/10.1016/j.wear.2021.203788

Czichos, H. (1977). *Tribology – A systems approach to the science and technology of friction, lubrication and wear*. https://doi.org/10.1016/0301-679x(78)90209-8

Dai, H. S., & Liu, X. F. (2008). Refinement performance and mechanism of an Al-50Si alloy. *Materials Characterization*, *59*(11). https://doi.org/10.1016/j.matchar.2008.01.020

DebRoy, T., Wei, H. L., Zuback, J. S., Mukherjee, T., Elmer, J. W., Milewski, J. O., Beese, A. M., Wilson-Heid, A., De, A., & Zhang, W. (2018). Additive manufacturing of metallic components – Process, structure and properties. *Progress in Materials Science*, *92*. https://doi.org/10.1016/j.pmatsci.2017.10.001

Dehoff, R., Duty, C., Peter, W., Yamamoto, Y., Chen, W., Blue, C., & Tallman, C. (2013). Case study: Additive manufacturing of aerospace brackets. *Advanced Materials and Processes, 171*(3), 19–23.

Deleanu, L., Botan, M., & Georgescu, C. (2021). Tribological behavior of polymers and polymer composites. *Tribology in Materials and Manufacturing - Wear, Friction and Lubrication.* https://doi.org/10.5772/intechopen.94264

Dordlofva, C., Lindwall, A., & Törlind, P. (2016). Opportunities and challenges for additive manufacturing in space applications. *Proceedings of NordDesign, NordDesign 2016,* 1.

Edgar, J., & Tint, S. (2015). Additive manufacturing technologies: 3D printing, rapid prototyping, and direct digital manufacturing (2nd Ed.). *Johnson Matthey Technology Review, 59*(3). https://doi.org/10.1595/205651315x688406

Evans, D. C., & Lancaster, J. K. (1976). *Polymer – Fluid interactions in relation to wear.*

Ford, S., & Despeisse, M. (2016). Additive manufacturing and sustainability: An exploratory study of the advantages and challenges. *Journal of Cleaner Production, 137.* https://doi.org/10.1016/j.jclepro.2016.04.150

Frazier, W. E. (2010). Direct digital manufacturing of metallic components: Vision and roadmap. *21st Annual International Solid Freeform Fabrication Symposium – An Additive Manufacturing Conference, SFF 2010.*

Friedrich, K. (2018). Polymer composites for tribological applications. *Advanced Industrial and Engineering Polymer Research, 1*(1). https://doi.org/10.1016/j.aiepr.2018.05.001

Friedrich, K., Reinicke, R., & Zhang, Z. (2002). Wear of polymer composites. *Proceedings of the Institution of Mechanical Engineers, Part J: Journal of Engineering Tribology, 216*(6), 415–426. https://doi.org/10.1243/135065002762355334

Fusaro, R. L. (1990). Self-lubricating polymer composites and polymer transfer film lubrication for space applications. *Tribology International, 23*(2). https://doi.org/10.1016/0301-679X(90)90043-O

George, M., Aroom, K. R., Hawes, H. G., Gill, B. S., & Love, J. (2017). 3D printed surgical instruments: The design and fabrication process. *World Journal of Surgery, 41*(1). https://doi.org/10.1007/s00268-016-3814-5

Gibson, I., Rosen, D., & Stucker, B. (2015). Additive manufacturing technologies: 3D printing, rapid prototyping, and direct digital manufacturing, second edition. In *Additive manufacturing technologies: 3D printing, rapid prototyping, and direct digital manufacturing* (2nd Ed.). https://doi.org/10.1007/978-1-4939-2113-3

Gibson, I., Rosen, D., Stucker, B., & Khorasani, M. (2021). Materials for additive manufacturing. In *Additive manufacturing technologies* (pp. 379–428). Springer International Publishing. https://doi.org/10.1007/978-3-030-56127-7_14

Godet, M. (1984). The third-body approach: A mechanical view of wear. *Wear, 100*(1–3). https://doi.org/10.1016/0043-1648(84)90025-5

Greco, S., Gutzeit, K., Hotz, H., Kirsch, B., & Aurich, J. C. (2020). Selective laser melting (SLM) of AISI 316L—Impact of laser power, layer thickness, and hatch spacing on roughness, density, and microhardness at constant input energy density. *International Journal of Advanced Manufacturing Technology, 108*(5–6). https://doi.org/10.1007/s00170-020-05510-8

Hanchi, J., & Eiss, N. S. (1994). The tribological behavior of blends of polyetheretherketone (PEEK) and polyetherimide (PEI) at elevated temperatures. *Tribology Transactions, 37*(3). https://doi.org/10.1080/10402009408983322

Hanon, M. M., & Zsidai, L. (2020). Tribological and mechanical properties investigation of 3D printed polymers using DLP technique. *AIP Conference Proceedings, 2213.* https://doi.org/10.1063/5.0000267

Horn, T. J., & Harrysson, O. L. A. (2012). Overview of current additive manufacturing technologies and selected applications. *Science Progress*, *95*(3). https://doi.org/10.3184/003685012X13420984463047

Ishfaq, K., Asad, M., Mahmood, M. A., Abdullah, M., & Pruncu, C. I. (2022). Opportunities and challenges in additive manufacturing used in space sector:a comprehensive review. *Rapid Prototyping Journal*, *28*(2). https://doi.org/10.1108/RPJ-04-2021-0091

Javaid, M., & Haleem, A. (2018). Additive manufacturing applications in medical cases: A literature based review. *Alexandria Journal of Medicine*, *54*(4). https://doi.org/10.1016/j.ajme.2017.09.003

Jia, Q., & Gu, D. (2014). Selective laser melting additive manufacturing of Inconel 718 superalloy parts: Densification, microstructure and properties. *Journal of Alloys and Compounds*, *585*. https://doi.org/10.1016/j.jallcom.2013.09.171

Joshi, S. C., & Sheikh, A. A. (2015). 3D printing in aerospace and its long-term sustainability. *Virtual and Physical Prototyping*, *10*(4). https://doi.org/10.1080/17452759.2015.1111519

Kang, N., & el Mansori, M. (2020). A new insight on induced-tribological behaviour of hypereutectic Al-Si alloys manufactured by selective laser melting. *Tribology International*, *149*. https://doi.org/10.1016/j.triboint.2019.04.035

Kaya, G., Yildiz, F., & Hacisalihoğlu, A. (2019). Characterization of the structural and tribological properties of medical Ti_6Al_4V alloy produced in different production parameters using selective laser melting. *3D Printing and Additive Manufacturing*, *6*(5). https://doi.org/10.1089/3dp.2019.0017

Kelly, S. M., & Kampe, S. L. (2004). Microstructural evolution in laser-deposited multilayer Ti-6Al-4V builds: Part 1. Microstructural characterization. *Metallurgical and Materials Transactions A: Physical Metallurgy and Materials Science*, *35*. https://doi.org/10.1007/s11661-004-0094-8

Khaing, M. W., Fuh, J. Y. H., & Lu, L. (2001). Direct metal laser sintering for rapid tooling: Processing and characterisation of EOS parts. *Journal of Materials Processing Technology*, *113*(1–3). https://doi.org/10.1016/S0924-0136(01)00584-2

Khruschov, M. M. (1974). Principles of abrasive wear. *Wear*, *28*(1). https://doi.org/10.1016/0043-1648(74)90102-1

Kiran, M. D., Govindaraju, H. K., Jayaraju, T., & Kumar, N. (2018). Review-effect of fillers on mechanical properties of polymer matrix composites. *Materials Today: Proceedings*, *5*(10). https://doi.org/10.1016/j.matpr.2018.06.611

Kmetty, Á., Bárány, T., & Karger-Kocsis, J. (2010). Self-reinforced polymeric materials: A review. *Progress in Polymer Science (Oxford)*, *35*(10). https://doi.org/10.1016/j.progpolymsci.2010.07.002

Koprnicky, J., Najman, P., & Safka, J. (2017). 3D printed bionic prosthetic hands. *Proceedings of the 2017 IEEE International Workshop of Electronics, Control, Measurement, Signals and Their Application to Mechatronics, ECMSM 2017*. https://doi.org/10.1109/ECMSM.2017.7945898

Kruth, J. P., Levy, G., Klocke, F., & Childs, T. H. C. (2007). Consolidation phenomena in laser and powder-bed based layered manufacturing. *CIRP Annals – Manufacturing Technology*, *56*(2). https://doi.org/10.1016/j.cirp.2007.10.004

Kurdi, A., Kan, W. H., & Chang, L. (2019). Tribological behaviour of high performance polymers and polymer composites at elevated temperature. *Tribology International*, *130*. https://doi.org/10.1016/j.triboint.2018.09.010

Kurzynowski, T., Pawlak, A., & Smolina, I. (2020). The potential of SLM technology for processing magnesium alloys in aerospace industry. *Archives of Civil and Mechanical Engineering*, *20*(1). https://doi.org/10.1007/s43452-020-00033-1

Le, V. T., Mai, D. S., & Hoang, Q. H. (2020). Effects of cooling conditions on the shape, microstructures, and material properties of SS308L thin walls built by wire arc additive manufacturing. *Materials Letters*, *280*. https://doi.org/10.1016/j.matlet.2020.128580

Leblanc, J. L. (2009). Filled polymers: Science and industrial applications. In *Filled Polymers: Science and Industrial Applications, CRC Press.*

Li, F., Wang, Z., & Zeng, X. (2017). Microstructures and mechanical properties of Ti_6Al_4V alloy fabricated by multi-laser beam selective laser melting. *Materials Letters*, *199*. https://doi.org/10.1016/j.matlet.2017.04.050

Liu, C., Guo, Y., & Wang, D. (2019). PEI-RGO nanosheets as a nanoadditive for enhancing the tribological properties of water-based lubricants. *Tribology International*, *140*. https://doi.org/10.1016/j.triboint.2019.105851

Liu, S., & Guo, H. (2020). A review of SLMed magnesium alloys: Processing, properties, alloying elements and postprocessing. *Metals*, *10*(8). https://doi.org/10.3390/met10081073

Lorusso, M. (2019). Tribological and wear behavior of metal alloys produced by laser powder bed fusion (LPBF). *Friction, Lubrication and Wear*. https://doi.org/10.5772/intechopen.85167

Mandal, A., Tiwari, J. K., AlMangour, B., Sathish, N., Kumar, S., Kamaraj, M., Ashiq, M., & Srivastava, A. K. (2020). Tribological behavior of graphene-reinforced 316L stainless-steel composite prepared via selective laser melting. *Tribology International*, *151*. https://doi.org/10.1016/j.triboint.2020.106525

Mannoor, M. S., Jiang, Z., James, T., Kong, Y. L., Malatesta, K. A., Soboyejo, W. O., Verma, N., Gracias, D. H., & McAlpine, M. C. (2013). 3D printed bionic ears. *Nano Letters*, *13*(6). https://doi.org/10.1021/nl4007744

Mathew, M. T., Srinivasa Pai, P., Pourzal, R., Fischer, A., & Wimmer, M. A. (2009). Significance of tribocorrosion in biomedical applications: Overview and current status. *Advances in Tribology*. https://doi.org/10.1155/2009/250986

Meng, Y., Xu, J., Jin, Z., Prakash, B., & Hu, Y. (2020). A review of recent advances in tribology. *Friction*, *8*(2). https://doi.org/10.1007/s40544-020-0367-2

Mohamed, O. A., Masood, S. H., Bhowmik, J. L., & Somers, A. E. (2017). Investigation on the tribological behavior and wear mechanism of parts processed by fused deposition additive manufacturing process. *Journal of Manufacturing Processes*, *29*. https://doi.org/10.1016/j.jmapro.2017.07.019

Moshkovith, A., Perfiliev, V., Gindin, D., Parkansky, N., Boxman, R., & Rapoport, L. (2007). Surface texturing using pulsed air arc treatment. *Wear*, *263*, 7–12. https://doi.org/10.1016/j.wear.2006.11.043

Myshkin, N., & Kovalev, A. (2018). Adhesion and surface forces in polymer tribology—A review. *Friction*, *6*(2). https://doi.org/10.1007/s40544-018-0203-0

Myshkin, N. K., Pesetskii, S. S., & Grigoriev, A. Y. (2015). Polymer tribology: Current state and applications. *Tribology in Industry*, *37*(3), 284–290.

Myshkin, N. K., Petrokovets, M. I., & Kovalev, A.V. (2005). Tribology of polymers: Adhesion, friction, wear, and mass-transfer. *Tribology International*, *38*, 11–12. https://doi.org/10.1016/j.triboint.2005.07.016

Ngo, T. D., Kashani, A., Imbalzano, G., Nguyen, K. T. Q., & Hui, D. (2018). Additive manufacturing (3D printing): A review of materials, methods, applications and challenges. *Composites Part B: Engineering*, *143*, 172–196. https://doi.org/10.1016/j.compositesb.2018.02.012

Nunez, E. E., Gheisari, R., & Polycarpou, A. A. (2019). Tribology review of blended bulk polymers and their coatings for high-load bearing applications. *Tribology International*, *129*. https://doi.org/10.1016/j.triboint.2018.08.002

Pant, M., Pidge, P., Nagdeve, L., & Kumar, H. (2021). A review of additive manufacturing in aerospace application. *Revue des Composites et des Materiaux Avances, 31*(2). https://doi.org/10.18280/RCMA.310206

Patnaik, A., Satapathy, A., Chand, N., Barkoula, N. M., & Biswas, S. (2010). Solid particle erosion wear characteristics of fibre and particulate filled polymer composites: A review. *Wear, 268*(1). https://doi.org/10.1016/j.wear.2009.07.021

Pradeep, G. V. K., Duraiselvam, M., & Sivaprasad, K. (2022). Tribological behavior of laser surface melted γ-TiAl fabricated by electron beam additive manufacturing. *Journal of Materials Engineering and Performance, 31*(2). https://doi.org/10.1007/s11665-021-06278-y

Qin, H., Xu, R., Lan, P., Wang, J., & Lu, W. (2020). Wear performance of metal materials fabricated by powder bed fusion: A literature review. *Metals, 10*(3). https://doi.org/10.3390/met10030304

Ralls, A. M., Kumar, P., & Menezes, P. L. (2021). Tribological properties of additive manufactured materials for energy applications: A review. *Processes, 9*(1). https://doi.org/10.3390/pr9010031

Roudnicka, M., Bayer, F., Michalcova, A., Kubasek, J., Hamed Alzubi, E. G., & Vojtech, D. (2020). Biomedical titanium alloy prepared by additive manufacturing: Effect of processing on tribology. *Manufacturing Technology, 20*(6). https://doi.org/10.21062/MFT.2020.112

Saravanan, I., & Devaraju, A. (2019). Wear mechanism of UHMWPE polymer composites for bio medical applications. *Materials Research Express, 6*(10). https://doi.org/10.1088/2053-1591/ab3ed9

Shah, R., Gashi, B., Hoque, S., Marian, M., & Rosenkranz, A. (2021). Enhancing mechanical and biomedical properties of protheses – Surface and material design. *Surfaces and Interfaces, 27*. https://doi.org/10.1016/j.surfin.2021.101498

Shah, R., Woydt, M., Huq, N., & Rosenkranz, A. (2020). Tribology meets sustainability. *Industrial Lubrication and Tribology, 73*(3). https://doi.org/10.1108/ILT-09-2020-0356

Sharma, R. P., & Kumar, M. (2020). Mechanical and tribological performance of polymer composite materials: A review. *Journal of Physics: Conference Series, 1455*(1). https://doi.org/10.1088/1742-6596/1455/1/012033

Skelley, N. W., Hagerty, M. P., Stannard, J. T., Feltz, K. P., & Ma, R. (2020). Sterility of 3D-printed orthopedic implants using fused deposition modeling. *Orthopedics, 43*(1). https://doi.org/10.3928/01477447-20191031-07

Thakre, A. A. (2014). Prediction of erosion of polyetherimide and its composites using response surface methodology. *Journal of Tribology, 137*(1). https://doi.org/10.1115/1.4028267

Tiwari, S., Bijwe, J., & Panier, S. (2011). Tribological studies on polyetherimide composites based on carbon fabric with optimized oxidation treatment. *Wear, 271*(9–10). https://doi.org/10.1016/j.wear.2010.11.052

Torrance, A. A. (2005). Modelling abrasive wear. *Wear, 258*, 1–4. https://doi.org/10.1016/j.wear.2004.09.065

Tran, H. C., & Lo, Y. L. (2019). Systematic approach for determining optimal processing parameters to produce parts with high density in selective laser melting process. *International Journal of Advanced Manufacturing Technology, 105*(10). https://doi.org/10.1007/s00170-019-04517-0

Unal, H., & Mimaroglu, A. (2003). Influence of test conditions on the tribological properties of polymers. *Industrial Lubrication and Tribology, 55*(4). https://doi.org/10.1108/00368790310480362

Ur Rahman, N., de Rooij, M. B., Matthews, D. T. A., Walmag, G., Sinnaeve, M., & Römer, G. R. B. E. (2019a). Wear characterization of multilayer laser cladded high speed steels. *Tribology International, 130*. https://doi.org/10.1016/j.triboint.2018.08.019

Ur Rahman, N., Matthews, D. T. A., de Rooij, M., Khorasani, A. M., Gibson, I., Cordova, L., & Römer, G. W. (2019b). An overview: laser-based additive manufacturing for high temperature tribology. *Frontiers in Mechanical Engineering, 5*. https://doi.org/10.3389/fmech.2019.00016

Vafadar, A., Guzzomi, F., Rassau, A., & Hayward, K. (2021). Advances in metal additive manufacturing: A review of common processes, industrial applications, and current challenges. *Applied Sciences, 11*(3), 1213. https://doi.org/10.3390/app11031213

Wang, L., Tieu, A. K., Lu, S., Jamali, S., Hai, G., Zhu, Q., Nguyen, H. H., & Cui, S. (2021). Sliding wear behavior and electrochemical properties of binder jet additively manufactured 316SS/bronze composites in marine environment. *Tribology International, 156*. https://doi.org/10.1016/j.triboint.2020.106810

Wang, P., Nai, M. L. S., Sin, W. J., & Wei, J. (2015). Effect of building height on microstructure and mechanical properties of big-sized Ti-6Al-4V plate fabricated by electron beam melting. *MATEC Web of Conferences, 30*. https://doi.org/10.1051/matecconf/20153002001

Wang, Z., Palmer, T. A., & Beese, A. M. (2016). Effect of processing parameters on microstructure and tensile properties of austenitic stainless steel 304L made by directed energy deposition additive manufacturing. *Acta Materialia, 110*. https://doi.org/10.1016/j.actamat.2016.03.019

Yin, S., Chen, C., Yan, X., Feng, X., Jenkins, R., O'Reilly, P., Liu, M., Li, H., & Lupoi, R. (2018). The influence of aging temperature and aging time on the mechanical and tribological properties of selective laser melted maraging 18Ni-300 steel. *Additive Manufacturing, 22*. https://doi.org/10.1016/j.addma.2018.06.005

Zhang, S., Wei, Q., Cheng, L., Li, S., & Shi, Y. (2014). Effects of scan line spacing on pore characteristics and mechanical properties of porous Ti_6Al_4V implants fabricated by selective laser melting. *Materials and Design, 63*. https://doi.org/10.1016/j.matdes.2014.05.021

Zhang, Y., Jarosinski, W., Jung, Y. G., & Zhang, J. (2018a). Additive manufacturing processes and equipment. In *Additive Manufacturing: Materials, Processes, Quantifications and Applications*. https://doi.org/10.1016/B978-0-12-812155-9.00002-5

Zhang, Y., Wu, L., Guo, X., Kane, S., Deng, Y., Jung, Y. G., Lee, J. H., & Zhang, J. (2018b). Additive manufacturing of metallic materials: A review. *Journal of Materials Engineering and Performance, 27*(1). https://doi.org/10.1007/s11665-017-2747-y

Zhou, H., Wu, C., Tang, D. yan, Shi, X., Xue, Y., Huang, Q., Zhang, J., Elsheikh, A. H., & Ibrahim, A. M. M. (2021). Tribological performance of gradient Ag-multilayer graphene/TC4 alloy self-lubricating composites prepared by laser additive manufacturing. *Tribology Transactions, 64*(5). https://doi.org/10.1080/10402004.2021.1922789

Zhu, Y., Lin, G., Khonsari, M. M., Zhang, J., & Yang, H. (2018a). Material characterization and lubricating behaviors of porous stainless steel fabricated by selective laser melting. *Journal of Materials Processing Technology, 262*. https://doi.org/10.1016/j.jmatprotec.2018.06.027

Zhu, Y., Tian, X., Li, J., & Wang, H. (2015). The anisotropy of laser melting deposition additive manufacturing Ti-6.5Al-3.5Mo-1.5Zr-0.3Si titanium alloy. *Materials and Design, 67*. https://doi.org/10.1016/j.matdes.2014.11.001

Zhu, Y., Zou, J., & Yang, H. Yong. (2018b). Wear performance of metal parts fabricated by selective laser melting: a literature review. *Journal of Zhejiang University: Science A, 19*(2). https://doi.org/10.1631/jzus.A1700328

Zmitrowicz, A. (2006). wear patterns and laws of wear—A review. *journal of Theoretical and Applied Mechanics, 44*, 219–253.

11 Parametric Optimization of Steel Turning Variables under Metal and Metal Oxide Nanofluid Cooling Environment with MCDM Hybrid Method

Anup A. Junankar and Yashpal Kaushik
Poornima University Jaipur, Jaipur, Rajasthan, India

Jayant K. Purohit
Banasthali Vidyapith Tonk, Tonk, India

11.1 INTRODUCTION

In the metal-cutting industry, the process evaluation of machining operation is evaluated based on product surface quality and the tool's lifespan. For manpower in the metal-cutting industry, improving metal-cutting performance without compromising the machined workpiece product surface quality and tool life is the critical parameter. Over the last few years, eminent researchers have shown sustainable results in solving the difficulties related to metal-cutting performance. A unique alternate cooling system for conventional cooling has been presented during a metal-cutting operation (Katiyar et al., 2022). In previously published research, minimum quantity lubrication (MQL) was noted as an effective, sustainable technique out of all the cooling systems. Due to less usage of coolant in MQL, machine operators' exposure to suspended particles of coolant and unnecessary gases is reduced. The accurate penetration of coolant at the cutting zone is possible using an MQL system.

The 'nanofluid' was supplied as a coolant using the MQL system to improve the metal-cutting performance and showed excellent results. The synthesised

DOI: 10.1201/9781003363576-11

TABLE 11.1

Literature for the Utilisation of Nanofluid for the Turning Process

Ref.	Nanofluid	Surface Roughness	Tool Wear	Cutting Temperature	Cutting Forces
Vasu and Pradeep (2011)	Al_2O_3	✔	✔	✔	
Khandekar et al. (2012)	Al_2O_3	✔	✔		✔
Sharma et al. (2016a)	Al_2O_3	✔	✔		✔
Behera et al. (2017)	Al_2O_3		✔		
Mahboob Ali (2017)	Al_2O_3	✔	✔		
Faheem et al. (2020)	Al_2O_3	✔			
Abbas et al. (2019)	Al_2O_3	✔			
Kumar et al. (2019)	Al_2O_3	✔	✔	✔	
Duc et al. (2020)	Al_2O_3	✔			✔
Ghalme et al. (2020)	Al_2O_3		✔		
Das et al. (2021)	Al_2O_3		✔		✔
Sharma et al. (2016b)	TiO_2	✔	✔		✔
Kumar et al. (2019)	TiO_2	✔	✔	✔	
Nune and Chaganti (2020)	SiO_2	✔	✔	✔	
Junankar et al. (2021)	CuO and ZnO	✔		✔	
Elsheikh et al. (2021)	Al_2O_3 and CuO	✔	✔		✔
Junankar et al. (2020)	Cu	✔		✔	
Mushtaq and Hanief (2021)	GnP	✔	✔		
Javid et al. (2021)	SiO_2	✔	✔		
Ramanan et al. (2021)	Al_2O_3	✔	✔		✔

nanoparticles are mixed with base fluid in a standard proportion, and the generated fluid is called a 'nanofluid'. Nanofluids were prepared with the help of the one-step or two-step technique. Eminent researchers used pure metal, oxides, carbonic, and carbides as nanoparticles. The nanofluid with an MQL combination enhanced the specific thermophysical properties of the coolant. The detailed literature presented relate to the effective utilisation of nanofluid for the turning process, the output variables considered and the method of optimisation as shown in Table 11.1.

After the literature review, it was noted that the eminent investigators implemented nanofluids as a cutting fluid for the turning process. Very few studies are reported by investigators for the comparative evaluation of metal oxide and metal-based nanofluid for the turning process. This motivated us to perform a comparative review to study the influence of copper nanofluid and silicon dioxide nanofluid during the turning operation of bearing steel. For the comparative evaluation, surface roughness and cutting temperature were selected as the response parameters. Multi-criteria decision-making hybrid method was implemented to optimise the machining variables under two MQL cooling conditions – Cu nanofluid and SiO_2 nanofluid.

11.2 EXPERIMENTAL SETUP AND PROCESS PARAMETERS

The experimental setup and process parameters are shown in Table 11.2. The investigation was conducted on a computer numerical control lathe machine with an external MQL setup. The detailed experimental setup is presented in Figure 11.1. The workpiece was cylindrical bearing steel selected (70 mm length and 23 mm diameter) with hardness 55 HRBW. The surface roughness and cutting temperature were quantified with the help of a roughness tester and non-contact thermometer, respectively.

11.3 PREPARATION OF CU AND SIO$_2$ NANOFLUIDS

The synthesised Cu nanoparticles were purchased and characterised using SEM and EDAX tests at SAIF-IIT Madras. The outcomes of SEM and EDAX tests are

TABLE 11.2
Experimental Setup and Process Parameters

Parameters	Specification
Input Parameters	Cutting Speed (140, 170, 200) (m/min)
	Feed Rate (0.2, 0.3, 0.4) (mm/rev)
	Depth of Cut (0.15, 0.25, 0.35) (mm)
Response Parameters	Surface Roughness (SR) (um)
	Cutting Temperature (CT) (°C)
Cooling Condition	MQL + Cu Nanofluid (6 bar, 150 ml/hr)
	MQL + SiO$_2$ Nanofluid (6 bar, 150 ml/hr)
Cutting Tool	1108E-TMT9125 DNMG

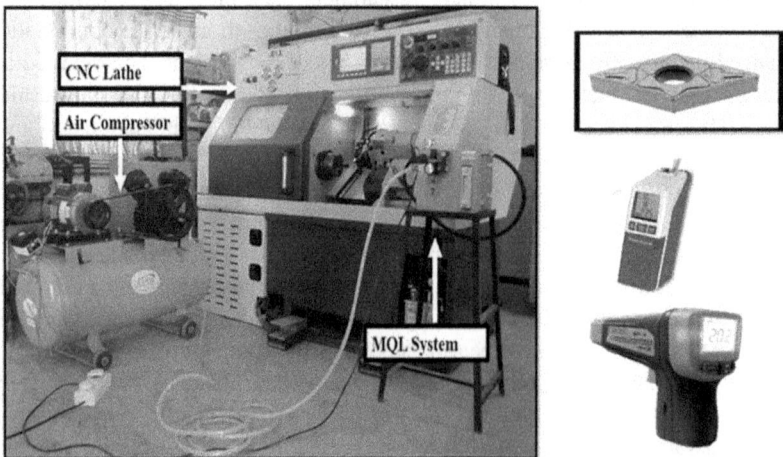

FIGURE 11.1 Experimental setup and response parameter's respective devices.

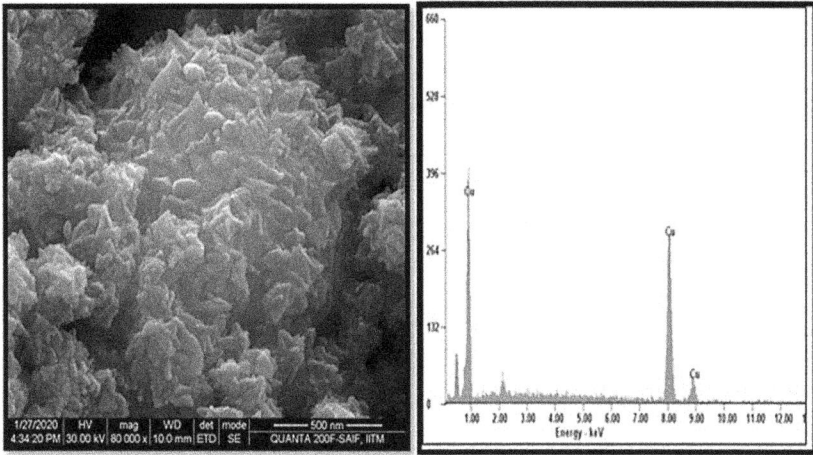

FIGURE 11.2   Cu nanoparticle SEM and EDAX test results.

shown in Figure 11.2. The nanoparticles are 500 nm in size. After characterisation testing, Cu nanoparticles were added to biodegradable vegetable oil along with ethylene glycol and butenol to ensure the proper dispersion of the nanoparticles. In addition, ultrasonication followed by magnetic stirring of the prepared mixture was carried out, and finally, it was noted that Cu nanoparticles were appropriately scattered throughout.

The synthesised SiO_2 nanoparticles were purchased and characterised using SEM and EDAX tests at SAIF-IIT Madras. The results of SEM and EDAX tests are shown in Figure 11.3. The nanoparticles are 500 nm in size. After characterisation

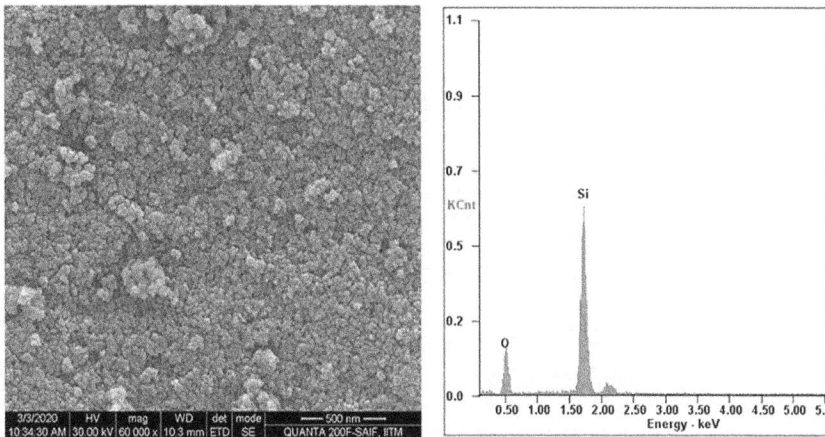

FIGURE 11.3 SiO_2 nanoparticle SEM and EDAX test results.

testing, SiO_2 nanoparticles were added to biodegradable vegetable oil along with ethylene glycol and butenol to ensure the proper dispersion of the nanoparticles. In addition, ultrasonication and magnetic stirring of the prepared mixture were carried out, and finally, it was noted that SiO_2 nanoparticles were appropriately scattered throughout.

11.4 BEARING STEEL TURNING PROCESS ASSESSMENT

The MCEM hybrid method was implemented to find the optimal process condition. In the hybrid method, two stages are involved – (a) entropy weighted technique (EWT) and (b) weighted aggregated sum product assessment (WASPAS). The first (EWT) determines the weights for each response parameter, and the optimal process conditions are estimated on the specified weights.

11.4.1 ENTROPY WEIGHT TECHNIQUE

Shannon and Weaver (in (Junankar et al. 2021) jointly presented the entropy weight technique to determine the weights of the response parameters. Probability theory is the primary basis for the entropy weight technique.

Step a) Collection of response parameters.
Considering the specific evaluation criteria, the machining experiments were decided and recorded data are presented in Table 11.3.

TABLE 11.3
Decision Matrix

Experiments	Cooling Condition	Surface Roughness	Cutting Temperature
01	Cu Nanofluid	2.08	28.0
02	Cu Nanofluid	2.93	31.5
03	Cu Nanofluid	2.98	32.6
04	Cu Nanofluid	2.10	32.4
05	Cu Nanofluid	1.48	29.6
06	Cu Nanofluid	2.66	30.8
07	Cu Nanofluid	1.89	31.2
08	Cu Nanofluid	1.96	31.7
09	Cu Nanofluid	2.18	30.6
10	SiO_2 Nanofluid	2.00	28.9
11	SiO_2 Nanofluid	2.63	31.0
12	SiO_2 Nanofluid	2.60	31.6
13	SiO_2 Nanofluid	2.12	31.4
14	SiO_2 Nanofluid	1.60	29.8
15	SiO_2 Nanofluid	2.33	30.4
16	SiO_2 Nanofluid	1.95	30.9
17	SiO_2 Nanofluid	1.96	30.2
18	SiO_2 Nanofluid	2.20	31.1

Step b) Decision matrix
This matrix is presented in Equation 11.1. The rows of the decision matrix are allotted to each trial (experiment), and the column is given to response parameters. The data presented in Table 11.3(a) is considered a decision matrix.

$$
DM = \begin{bmatrix}
q_{11} & q_{12} & q_{1j} & q_{1m} \\
q_{21} & q_{22} & q_{2j} & q_{2m} \\
q_{i1} & q_{i2} & q_{ij} & q_{im} \\
.. & .. & .. & .. \\
q_{n1} & q_{n2} & q_{nj} & q_{nm}
\end{bmatrix}
\tag{11.1}
$$

Step c) Data normalization
The response parameters were normalised by considering two criteria such as beneficial and non-beneficial criteria. For the current investigation, non-beneficial criteria were decided for both response parameters and presented by the following equation

$$
NDM_{ij} = \frac{\min q_{ij}}{q_{ij}}
\tag{11.2}
$$

Step d) Probability and entropy assessment
Using Equations 11.3 and 11.4, the probability and entropy of response parameters were evaluated, and data is presented in Table 11.4.

$$
Pr_{ij} = \frac{NDM_{ij}}{\displaystyle\sum_{i=1}^{n} NDM_{ij}}
\tag{11.3}
$$

$$
En_j = -Y \sum_{i=1}^{n} Pr_{ij} \log_s \left(Pr_{ij} \right)
\tag{11.4}
$$

Step e) Divergence and entropy weights
Divergence and entropy were estimated using Equations 11.5 and 11.6. The estimated data is presented in Table 11.5.

$$
Div_j = \left| 1 - n \right|
\tag{11.5}
$$

$$
Ew_j = \frac{Div_j}{\displaystyle\sum_{j=1}^{m} Div_j}
\tag{11.6}
$$

TABLE 11.4
Weight Estimation

Expt No.	Normalization SR	Normalization CT	Total Probability SR	Total Probability CT	Entropy SR	Entropy CT	Divergence SR	Divergence CT	Weights SR	Weights CT
					Cu Nanofluid					
1	0.7115	1.0000	−2.1743	−2.1962	0.98953	0.9999492	0.0105	0.0005	0.9537	0.0463
2	0.5051	0.8889								
3	0.4966	0.8589								
4	0.7048	0.8642								
5	1.0000	0.9459								
6	0.5564	0.9091								
7	0.7831	0.8974								
8	0.7551	0.8833								
9	0.6789	0.9150								
					SiO₂ Nanofluid					
10	0.8000	1.0000	−2.1862	−2.1969	0.994945	0.999793	0.0051	0.0002	0.9606	0.0394
11	0.6084	0.9323								
12	0.6154	0.9146								
13	0.7547	0.9204								
14	1.0000	0.9698								
15	0.6867	0.9507								
16	0.8205	0.9353								
17	0.8163	0.9570								
18	0.7273	0.9293								

TABLE 11.5
Rank Estimation

	Cu Nanofluid					
	TRI (Q^1)		TRI (Q^2)		FRI	
Expt. No.	SR	MT	SR	MT	Q_i	Rank
1	0.6786	0.0463	0.7228	1.0000	0.7239	4
2	0.4817	0.0412	0.5213	0.9946	0.5207	8
3	0.4736	0.0398	0.5130	0.9930	0.5114	9
4	0.6721	0.0400	0.7163	0.9933	0.7118	5
5	0.9537	0.0438	1.0000	0.9974	0.9975	1
6	0.5306	0.0421	0.5717	0.9956	0.5710	7
7	0.7468	0.0416	0.7920	0.9950	0.7882	2
8	0.7201	0.0409	0.7650	0.9943	0.7608	3
9	0.6475	0.0424	0.6912	0.9959	0.6891	6
	SiO$_2$ Nanofluid					
10	0.7685	0.0394	0.8071	1.0000	0.8075	4
11	0.5844	0.0367	0.6204	0.9972	0.6199	9
12	0.5911	0.0360	0.6273	0.9965	0.6261	8
13	0.7250	0.0363	0.7631	0.9967	0.7609	5
14	0.9606	0.0382	1.0000	0.9988	0.9988	1
15	0.6596	0.0375	0.6969	0.9980	0.6963	7
16	0.7882	0.0368	0.8269	0.9974	0.8249	2
17	0.7842	0.0377	0.8229	0.9983	0.8217	3
18	0.6986	0.0366	0.7365	0.9971	0.7348	6

11.4.2 WEIGHTED AGGREGATED SUM PRODUCT ASSESSMENT

This assessment was designed by Zavadskas and followed a stepwise approach to determine the optimal machining conditions for both cooling conditions.

Step a) Decision matrix and data normalization
The prepared decision matrix and normalised data using Equations 11.1 and 11.2 were utilised for WASPAS.

Step b) Total relative Importance for weighted sum technique
Using Equation 11.7, the total relative importance of WST was estimated.

$$Q_i^{(1)} = \sum_{j=1}^{n} x_{ij}.w_j \tag{11.7}$$

Step c) Total relative Importance for weighted product technique
By using Equation 11.8, the total relative importance of WPT was estimated.

$$Q_i^{(2)} = \prod_{j=1}^{n} x_{ij}^{w_j} \tag{11.8}$$

Step d) Final relative importance (FRI)

FRI (Q_i) was estimated to improve the ranking accurateness and assessed with Equation 11.9.

$$Q_i = \lambda.Q_i^1 + (1-\lambda)Q_i^2 \qquad (11.9)$$

For the present experimentation, L9 OA (Orthogonal Array) was utilised for DOE, and the response parameters recorded are presented in Table 11.4(a).

The main effect plot for Cu and SiO$_2$ nanofluid cooling environment is presented in Figures 11.4 and 11.5. The outcomes for effective machining parameters and optimal machining conditions are shown in Tables 11.6 and 11.7, respectively.

11.5 RESULTS

The surface quality of the product is defined based on the SR parameter. For the present study, the minimum SR is 1.48 um and 1.60 um, noted under Cu nanofluid and SiO$_2$ nanofluid cooling environment, respectively. The comparative analysis of SR measurement is shown in Figure 11.6. It can be seen that the cutting temperature directly affects the cutting tool's lifespan. The lowest cutting temperature, 28 °C and 28.9 °C, are noted under Cu nanofluid and SiO$_2$ nanofluid cooling environments, respectively. A comparative analysis of cutting temperature measurement is shown in Figure 11.7. The Cu nanoparticles generated a protective film to

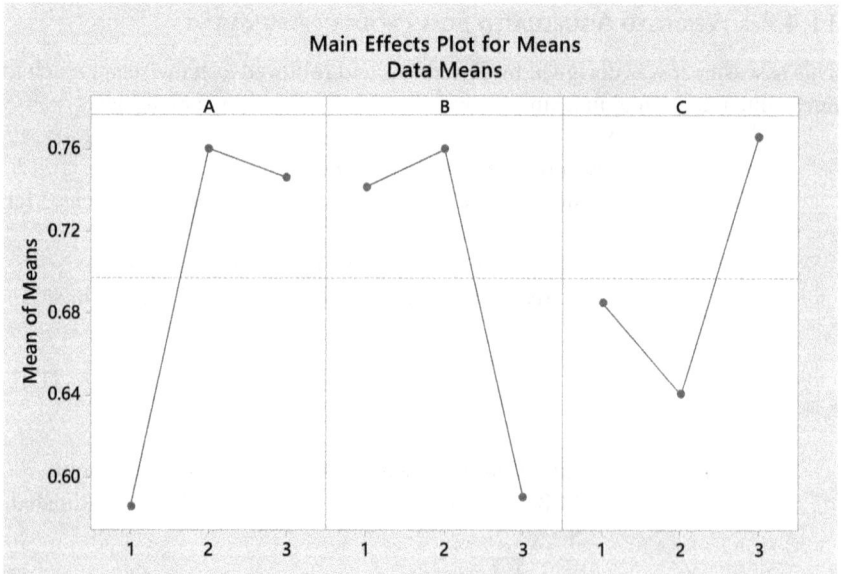

FIGURE 11.4 Main effects plot – Cu nanofluid cooling environment.

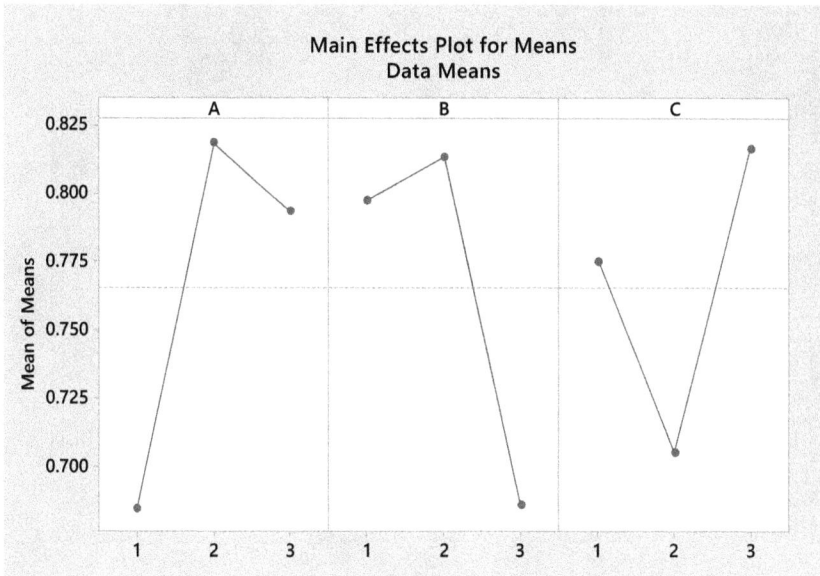

FIGURE 11.5 Main effects plot – SiO$_2$ Nanofluid cooling environment.

TABLE 11.6
Effective Machining Parameters as per FRI

Parameters	L1	L2	L3	Δ	Rank
		Cu Nanofluid			
CS	0.5853	0.7601	0.7460	0.1748	1
FR	0.7413	0.7597	0.5905	0.1692	2
DOC	0.6852	0.6405	0.7657	0.1252	3
		SiO$_2$ Nanofluid			
CS	0.6845	0.8187	0.7938	0.1342	1
FR	0.7978	0.8135	0.6857	0.1277	2
DOC	0.7752	0.7052	0.8166	0.1114	3

TABLE 11.7
Optimal Machining Conditions

Variables	Cu Nanofluid	SiO$_2$ Nanofluid
CS	170 m/min	170 m/min
FR	0.2 mm/rev	0.2 mm/rev
DOC	0.35 mm	0.35 mm

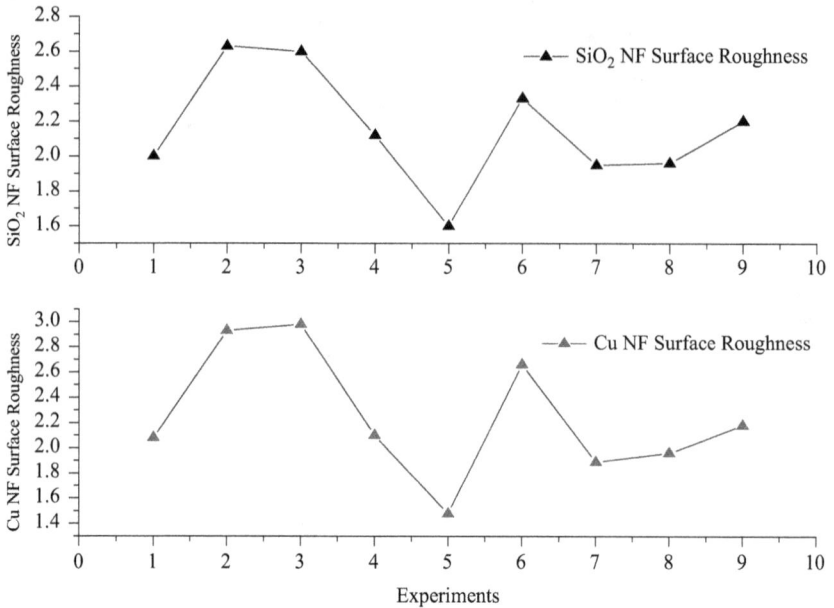

FIGURE 11.6   Surface roughness for Cu and SiO_2 nanofluid.

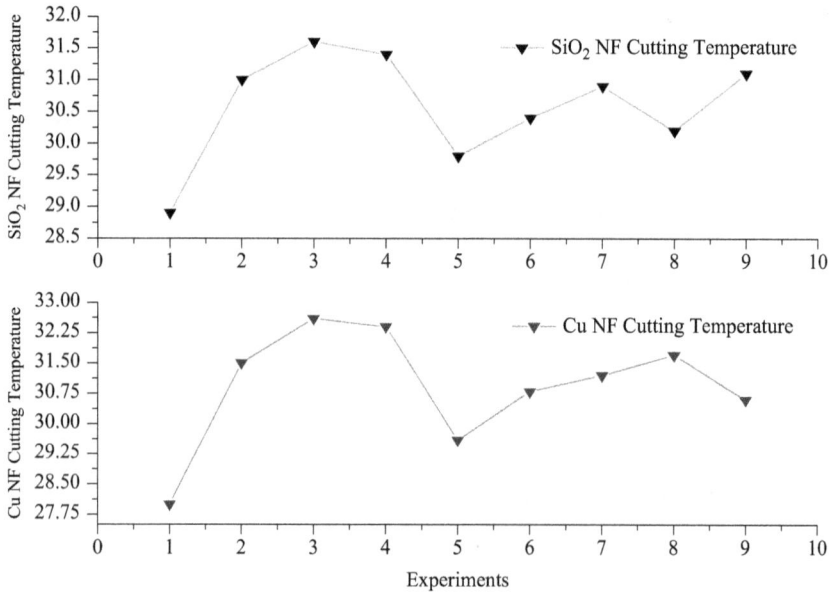

FIGURE 11.7   Cutting temperature for Cu and SiO_2 nanofluid.

reduce the friction between sliding surfaces. Furthermore, the elasticity decreased along with wear reduction. Similarly, Cu nanofluid possesses excellent thermal conductivity compared to SiO_2 nanofluid. Therefore, significant outcomes were observed under Cu nanofluid compared to SiO_2 nanofluid cooling environments during the bearing steel turning process.

11.6 CONCLUSION

The experiment of the bearing steel turning under Cu nanofluid and SiO_2 nanofluid cooling environments was centred on optimising response parameters. The following are the experimental outcomes:

- For Cu nanofluid cooling conditions, cutting speed was noted as the most significant parameter, followed by feed rate and depth of cut.
- Similarly, for SiO_2 nanofluid cooling conditions, cutting speed was noted to be an effective parameter, followed by feed rate and depth of cut.
- Cu nanofluid's protective film and thermal conductivity decreased the coefficient of friction at the cutting zone, resulting in the effective reduction of SR and cutting temperature compared to SiO_2 nanofluid.
- The Cu nanofluid with the MQL system capably advanced the turning process performance of bearing steel compared to other cooling conditions.
- Finally, combining the biodegradable vegetable oil and metal (Cu)-based nanofluid promotes green machining and sustainability.

REFERENCES

Abbas, A.T., Gupta, M.K., Soliman, M.S., Mia, M., Higab, H., Luqman, M., Pimenov, D.Y., 2019, Sustainability assessment associated with surface roughness and power consumption characteristics in nanofluid MQL-assisted turning of AISI 1045 steel. *International Journal of Advanced Manufacturing Technology*, 1–17. https://doi.org/10.1007/s00170-019-04325-6

Behera, B.C., Chetan Setti, D., Ghosh, S., 2017, Spreadability studies of metal working fluids on tool surface and its impact on minimum amount cooling and lubrication turning. *Journal of Materials Processing Technology*, 244, 1–16.

Das, A., Patel, S.K., Biswal, B.B., Sahoo, N., Pradhan, A., 2021, Performance evaluation of various cutting fluids using MQL technique in hard turning of AISI 4340 alloy steel. *Measurement*, 150, 1–28.

Duc, T.M., Long, T.T., Chien, T.Q., 2020, Performance evaluation of MQL parameters using Al_2O_3 and MoS_2 nanofluids in hard turning 90CrSi steel. *Lubricants*, 40(7), 1–17. https://doi.org/10.3390/lubricants7050040

Elsheikh, A.H., Elaziz, M.A., Das, S.R., Muthuramalingam, T., Lu, S., 2021, A new optimized predictive model based on political optimizer for eco-friendly MQL-turning of AISI 4340 alloy with nano-libricants, *Journal of Manufacturing Processes*, 67, 562–578.

Faheem, A., Husain, T., Hasan, F., Murtaza, Q., 2020, Effect of nanoparticles in cutting fluid for structural machining of Inconel 718. *Advances in Materials and Processing Technologies*, 1–20. https://doi.org/10.1080/2374068X.2020.1802563

Ghalme, S., Koinkar, P., Bhalerao, Y., 2020, Effect of aluminium oxide nanoparticles addition into lubricating oil on tribological performance. *Tribology in industry*, 42, 194–502.

Javid, H., Jahanzaib, M., Jawad, M., Ali, M.A., Farooq, M.U., Pruncu, C.I., Hussain, S., 2021, Parametric analysis of turning HSLA steel under minimum quantity lubrication and nanofluid-based minimum quantity lubrication: A concept of one-step sustainable machining, *The International Journal of Advanced Manufacturing Technology*, 117, 1915–1934.

Junankar, A.A., Purohit, J.K., Gohane, G.M., Pachbhai, J.S., Gupta, P.M., Sayed, A.R., 2020, Performance evaluation of Cu nanofluid in bearing steel MQL based turning operation, *Materials Today: Proceedings*, 44(6), 4309–4314.

Junankar, A.A., Yashpal, Y., Purohit, J.K., 2021, Experimental investigation to study the effect of synthesized and characterized monotype and hybrid nanofluids in minimum quantity lubrication assisted turning of bearing steel. *Journal of Engineering Tribology*, 236(9), 1794–1813.

Katiyar, J.K., Sahu, R.K., Gupta, TCSM, 2022, *Sustainable Lubrication*, 1st Edition, CRC Press, Boca Raton UK.

Khandekar, S., Sankar, M.R., Agnihotri, V, 2012, Nano-cutting fluid for enhancement of metal cutting performance. *Materials and Manufacturing Processes*, 27(9), 963–967.

Kumar, R., Sahoo, A.K., Mishra, P.C., Das, R.K., 2019, Influence of Al_2O_3 and TiO_2 nanofluid on hard turning performance. *International Journal of Advanced Manufacturing Technology*, 1–16. https://doi.org/10.1007/s00170-019-04754-3

Mahboob Ali, M.A., Azmi, A.I., Khalil, A.N., 2017, Experimental study on minimal nanolubrication with surfactant in the turning of titanium alloys. *International Journal of Advanced Manufacturing Technology*, 92(1–4), 117–127.

Mushtaq, Z., Hanief, M., 2021, Enhancing the tribological characteristics of Jatropha oil using graphene nanoflakes. *Jurnal Tribologi*, 28, 129–143.

Nune, M.M.R., Chaganti, P.K., 2020, Performance evaluation of novel developed biodegradable metal working fluid during turning of AISI 420 material. *Journal of the Brazilian Society of Mechanical Sciences and Engineering*, 42(319), 1–16.

Ramanan, K.V., Babu, S.R., Jebraj, M., Ross, K.N.S., 2021, Face turning of Incoloy 800 under MQL and nano-MQL environments. *Materials and Manufacturing Processes*, 36(15), 1769–1780.

Sharma, A.K., Singh, R.K., Dixit, A.R., 2016a, Characterization and experimental investigation of Al_2O_3 nanoparticle based cutting fluid in turning of AISI 1040 steel under minimum quantity lubrication (MQL). *Materials Today: Proceedings* 3, 1899–1906.

Sharma, A.K., Tiwari, A.K., Dixit, A.R., 2016b, Effects of Minimum Quantity Lubrication (MQL) in machining processes using conventional and nanofluid based cutting fluids. *Journal of Cleaner Production*, 127, 1–18.

Vasu, V., Pradeep, R.G., 2011, Effect of minimum quantity lubrication with Al_2O_3 nanoparticles on surface roughness, tool wear and temperature dissipation in machining Inconel 600 alloy. *Proceedings of the Institution of Mechanical Engineers*, 225(1), 3–16.

12 Edge Band Defect in Cold-Rolled IF Grade Steel and Its Remedy
A Case Study

Subho Chakraborty and Ashwin Pandit
Tata Steel Limited, Jamshedpur, India

12.1 INTRODUCTION TO STEEL

Steel is an alloy of iron and carbon with varying percentages of Mn, Si, P, S and O. Pure iron is quite ductile and easy to form. In the case of steel, the inclusions work as hardening agents to prevent dislocations [1]. It is a go-to material when engineering and construction application is considered. Steel plays a pivotal role in our daily lives starting with providing a roof over our head, to white goods, mobility, infrastructure development and medical appliances. Steel is safer to use because it has the highest strength-to-weight ratio. It is economical compared to its peers in the market.

12.1.1 STEEL IN THE AUTOMOTIVE SECTOR

The automotive sector is a key area for steel consumption. On average, 900 kg of steel is used per vehicle. Around 60% of advanced high strength steel is used to make today's vehicles' body structure, which imparts enhanced safety, makes the car lighter in weight and improves overall fuel efficiency [2, 3]. Steel sheet finds its application in body structure, door parts and trunks. These steel sheets are either used as CRCA or zinc-coated steel. Coatings improve the corrosion resistance as well as the aesthetic value. The metal sheet surface needs to be defect free, irrespective of its final finish.

12.1.2 COLD-ROLLING MANUFACTURING PROCESS

Before any coating application, the process of steel sheet manufacturing is important to understand. Steel slabs processed through casters are hot rolled in a hot-rolling mill. The hot-rolled coils have a final thickness between 2.0 and 3.0 mm.

DOI: 10.1201/9781003363576-12

The hot-rolled steel is poor in terms of surface finish, aesthetics and desired properties. To improve its surface finish, the coils are subjected to further rolling. Hot rolling occurs at very high temperatures that are above the recrystallization temperature of steel. This high temperature operation results in rapid oxidation of substrate leading to primary and secondary scale formation. The hot-rolled coils are acid pickled to remove primary and secondary oxide scales. A process flow chart is provided in Figure 12.1. The removal of scale is very important before the next process commences. Any remaining loose scales would lead to surface defects. The pickled coils are next rolled in a reversing or tandem cold-rolling mill. The coils are cold rolled to a final thickness of 0.1 to 0.4 mm. Cold rolling is performed below the recrystallization temperature of steel and is employed to change the material properties of sheets [5–8]. Cold rolling increases yield strength, tensile strength and material hardness and imparts a superior surface finish.

12.1.3 FRICTION IN COLD ROLLING

The first law of friction proposed by Amonton and Coulomb [9] is expressed in Equation 12.1, where P is normal load and F is the tangential force.

$$F = \mu P \qquad (12.1)$$

μ represents the coefficient of friction (CoF). It was suggested that in the case of dry friction, the frictional force is caused by mechanical interlocking of asperities of two solid surfaces. At the contact position, the asperities of the two contact surfaces remain in touch against each other. The real contact A_r is shown by Equation 12.2,

$$A_r = P / p_m \qquad (12.2)$$

where P is normal load, p_m is plastic flow stress = $3Y$, Y is yield stress.

This contact model was further explained for relative motion V by Bowden and Taylor in Equation 12.3 where real contact area A_r is expressed as:

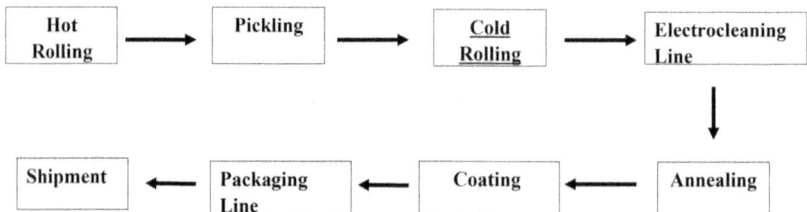

FIGURE 12.1 Process flow chart of steel sheet manufacturing.

$$A_r = P / p \qquad (12.3)$$

where p is normal stress and τ is tangential stress. As,

$$F = \tau A_r \qquad (12.4)$$

coefficient of friction (μ) in dry condition can be derived as:

$$\mu = \frac{F}{P} = \tau A_r / p A_r = \tau / p \qquad (12.5)$$

which explains why CoF is independent of apparent contact area. The CoF in dry condition is always expected to be higher.

12.1.4 TRIBOLOGY IN COLD ROLLING

The lubrication regime in cold rolling is close to the boundary and mixed-lubrication regimes. It involves a rolling and sliding motion. At the roll bite region (contact point between work roll and steel sheet), the contact pressure is expected to be much higher than the yield stress of the steel sheet resulting in plastic deformation. During cold rolling the roll bite temperature can reach up to 200 °C. The rise in temperature can be attributed to plastic deformation energy and friction energy [10]. The mean temperature rise T_{mw} by plastic deformation can be expressed as:

$$T_{mw} = \sigma_m \varepsilon_m / \rho \qquad (12.6)$$

where σ_m is average stress, ε_m is average strain and ρ is density. During cold rolling if V_a is the roll speed, V_b is the inlet strip speed and V_c is the outlet strip speed, the relation can be expressed as:

$$V_a \leq V_b \leq V_c \qquad (12.7)$$

At the neutral point, the strip and roll speed are equal, hence the relative sliding speed becomes 0.

12.1.5 STRIBECK CURVE

The Stribeck curve mentioned in Figure 12.2 represents CoF in the Y axis and the parameter $\eta V/P$ in X axis, where η represents viscosity of lubricant, V represents velocity of lubricant and P represents the normal load. The CoF is highest in the boundary region and maintains a constant value. In this region the asperity to asperity contact between the work roll and the steel surface is highest, that is, the film thickness of boundary film is negligible and tends to 0. In the mixed-lubrication regime, the CoF decreases abruptly with increasing parameter and thickness of lubricant film almost equal to surface roughness ($h \approx R$). In the

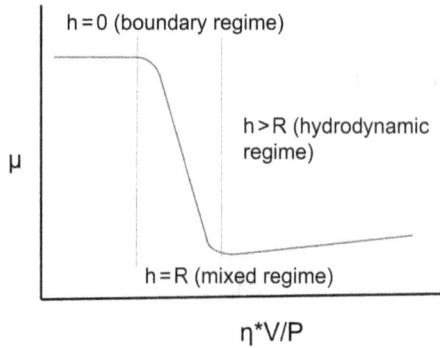

FIGURE 12.2   Stribeck curve.

elasto-hydrodynamic regime the lubricant layer forms a thick film between the asperities and the film thickness is greater than the surface roughness ($h \gg R$). The CoF value in the boundary regime remains closer to 0.1, while in the elasto-hydrodynamic region it is closer to 0.02. The mixed-lubrication regime remains between these two values [11]. During cold rolling in a 5-stand tandem mill, a gradual transition from boundary to mixed and finally to EHL is observed from the initial stands to the final stands.

12.2 FILM FORMATION OF LUBRICANT DURING COLD ROLLING

Phosphorous and sulfur chemistry present in rolling oil formulation acts as anti-wear additives [12–15]. These additives react with iron surface initially forming a tribo-chemical film which is in the form of a monolayer. In addition, some fatty acids present in rolling oil also assist in the boundary film formation through physio-adsorption techniques. These films act as a boundary layer which lays the foundation for the film formation by reducing friction and wear. The thickness of the boundary lubrication layer is only a few nano-metres. As cold rolling progresses from the initial stands to subsequent stands, a thicker film formation is required. The formation of a robust thicker lubricant film prevents contact of the asperities between the work roll and steel, and the mill is able to run at higher speeds with lower roll forces. The formation of thicker lubricant film in the mixed and EHL regimes is ensured by the correct choice of esters and triglycerides present in rolling oil chemistry. This EHL film can be a few hundred nano-metres thick [10]. The relation between mill speed and film thickness is depicted in Figure 12.3.

12.2.1 LUBRICATION IN A COLD-ROLLING MILL

Lubrication during cold rolling is provided by cold-rolling oil. It is applied in the form of an emulsion at concentration of 3–5% [16, 17]. It is diluted with demineralized water and sprayed at the rolling mill through flat jet nozzles. The

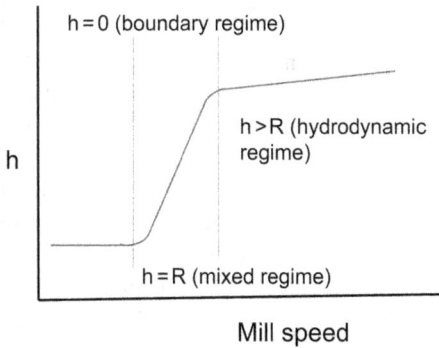

FIGURE 12.3   Relation between film thickness and mill speed.

cold-rolling emulsion is prepared in a premix tank at higher concentration and its final concentration is adjusted at the clean oil tank (COT) using further addition of demineralized water. The working temperature of emulsion is maintained in the range 50–55 °C through the use of heat exchangers. The emulsion is pumped from COT to various stands of the rolling mill. The emulsion provides the required lubrication during reduction at roll bite through plate-out formation [18–21]. The excess fluid washes the iron fines generated during rolling and is collected in the dirty oil tank (DOT). This fluid is filtered through a vacuum filter bed and magnetic separator to arrest the excess dirt and iron fines generated during rolling. The filtered rolling oil emulsion next passes to the COT and is then recirculated back to the mill [22–25]. A detailed process flow diagram is shown in Figure 12.4.

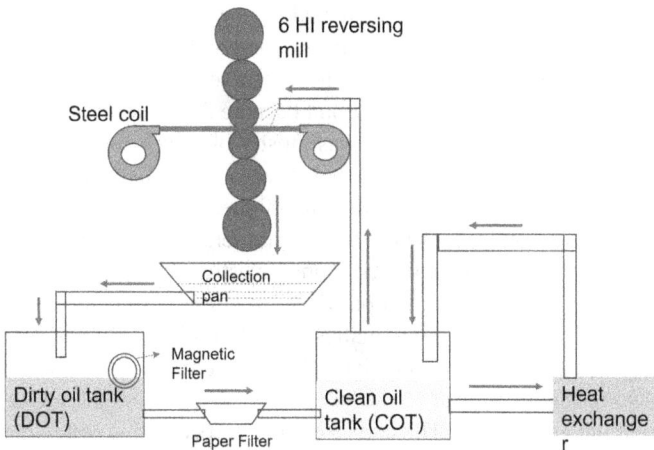

FIGURE 12.4   Process flow chart.

12.2.2 DAILY MAINTENANCE PRACTICE OF COLD-ROLLING EMULSION

The cold-rolling emulsion is maintained at a temperature band between 50 and 55 °C. There is a continuous process of water evaporation along with the oil losses encountered during the cold-rolling process. These losses are countered by periodic replenishment of cold-rolling oil and water, which helps to maintain the required concentration of oil in water emulsion [5]. Most cold-rolling emulsion has a service period of one year before being dumped and a fresh emulsion made. Assessing the health of emulsion is critical and has a direct impact on the performance of lubricants. Many parameters such as pH, conductivity, oil concentration, iron content, chloride content, etc. are checked every day and any deviation outside the control limits are immediately addressed [16, 26].

12.2.3 COLD-ROLLING DEFECTS

Blemish-free surface finish is the utmost criteria for any cold-rolled coil. There are many defects which can arise from a slight change in operational parameters. Similarly, the role of a cold-rolling lubricant is also very important, for a defect-free surface to be delivered to customers for further processing. Since this chapter is focused on lubricants and their impact on tribology, we will be concentrating on the defects arising out of lubricants. Patches due to emulsion carryover have been reported in earlier works. Corrosion problems on steel sheet, machinery and pipelines due to malfunctioning of cold rolled lubricants have also been reported in previous works by Chakraborty et al. [27–31].

12.3 INTRODUCTION TO A CASE STUDY

In one of the six-high reversing cold-rolling mills of Tata Steel, a yellowish band formation suddenly started appearing post cold rolling throughout the length of the coil (Figure 12.5). This defect was observed during 5-pass rolling. The reduction pattern followed in the cold-rolling mill is shown in Table 12.1. The intensity of the defect formation was much lighter when observed immediately during rolling after the 5th pass. The intensity of the band became darker as the coil aged before the next processing step. This defect was mostly observed on the top surface of the coil and was seen on all coils, irrespective of coil chemistry. Furthermore, during cold rolling, the mill also encountered higher roll forces, especially during the 3rd pass and above. The input process parameters during the hot rolled coil and pickling processes were backtracked and investigated to identify if there were any connection in the defect formation. Detailed study of the process parameters indicated no process changes or anomalies in the hot rolled coils and pickling section. During the next process of alkaline degreasing the coils were passed through alkaline cleaners maintained at 75–80 °C. There was a very slight reduction in the intensity of band appearance. The coils were next processed at annealing furnace, but this defect was still observed on the edges. Post coating, the edge band area looked dull compared to the centre of the coil. This led to rejections and down

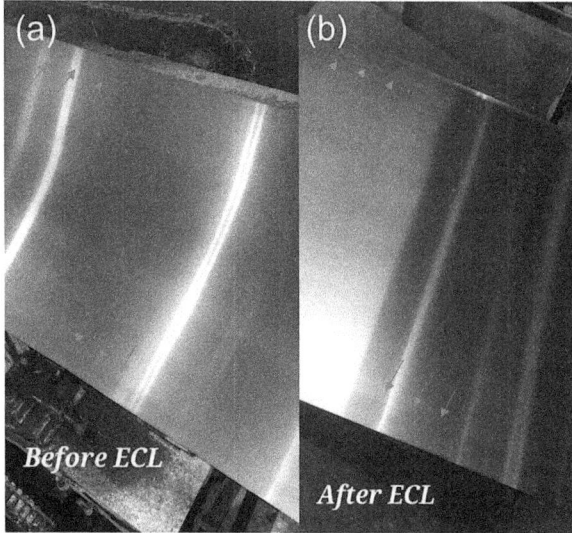

FIGURE 12.5   Yellow band defect; (a) after cold rolling; (b) after ECL.

TABLE 12.1
Typical Pass Reduction During 5-Pass Rolling

	1st Pass	2nd Pass	3rd Pass	4th Pass	5th Pass
% Reduction	16–22	16–23	17–23	17–23	17–23

gradation of almost 600 tonnes of coils. Some significant observations were noted during the course of the investigation which would guide us to solve the problem:

a. The mill speed was increased from 1,050 mpm to 1,200 mpm to increase productivity. This increase in mill speed was made just one week before the problem started.

b. The yellowish band problem reduced significantly if the reduction pattern was followed in six passes instead of five passes. However, this step had a negative impact on productivity and was not a feasible solution.

c. Increasing the alkaline cleaner solution concentration by 1% and 10 °C had little/no impact on the intensity of the yellowish band.

d. The rolling emulsion bath was only one month old, so the ageing effect could be ruled out.

e. The rolling emulsion was based on cationic technology. During the problematic period, a change in emulsion colour was noted from greyish to off-white and some scum layers were observed floating on the top.

f. The coil temperature reached 160 °C during the 5-pass rolling, which was considerably on the higher side.

In this study, the authors tried to identify and understand the reason behind the defect formation. Various aspects of the cold-rolling mill processes and rolling oil emulsion parameters were studied to find the probable reason behind defect formation. Various characterization techniques were employed to understand the nature of the defect formation.

12.4 MATERIALS AND METHODS

12.4.1 MATERIALS

The defect sample with yellowish band defect on the edge was collected from the cold-rolling mill. The typical composition of the steel sample is depicted in Table 12.2. Tables 12.3 and 12.4 describe the operating parameters of the cold-rolling mill and the operational parameters of cold-rolling emulsion.

The defect sample was subjected to various characterization techniques. The emulsion used during rolling was also subjected to complete analysis and the data compared against that of a good period.

12.4.2 SCANNING ELECTRON MICROSCOPY (SEM) AND ENERGY DISPERSIVE SPECTROSCOPY (EDS)

Surface morphology of the affected area was studied using a Zeiss scanning electron microscope (SEM). The defect and non-defect area was ultrasonicated using alcohol to remove traces of rolling oil from the substrate. The microscopy

TABLE 12.2
Chemical Composition of the Affected Steel Sample

	C	Mn	Si	P	S
Specification	0.1 max.	0.7 max.	0.03 max.	0.02 max.	0.02 max.

TABLE 12.3
Typical Operational Parameters of a Cold-rolling Mill

Parameters	Value
Speed	1,200 max.
Emulsion flow rate (l pm)	5,000–7,000
Avg. roll force (ton)	700–900
Final pass tension (kg)	4,500–5,500

TABLE 12.4
Typical Operational Parameters of Cold-Rolling Emulsion

Parameters	Value
Temperature (°C)	50–60
Oil conc. (%)	2–4
pH	5.0–6.0
Conductivity (µS/cm)	300 max.
Iron content (ppm)	250 max.

study was performed in secondary electron mode with a 15 kV electron beam. Elemental composition of defect and non-defect area was captured using energy dispersive spectroscopy (EDS).

12.4.3 PHYSICO-CHEMICAL ANALYSIS OF THE ROLLING EMULSION SAMPLE

The pH and conductivity of the rolling oil emulsion was monitored regularly using a calibrated pH meter (model – pH700 from Eutech Instruments) and conductivity meter (model – Cond 7110 from Xylem Analytics). The pH meter was calibrated against a pH buffer of 4.0 and 7.0, respectively, whereas the conductivity meter was calibrated against 0.01 (M) KCl with the calibrated reading of 1,409 µS/cm at 25 °C. Iron content in emulsion was estimated titrimetrically through the acid digestion method. The acid digested liquor was titrated against 0.1(M) EDTA using sulfosalicylic acid as indicator. The acid number was also estimated titrimetrically. The oil was extracted from the rolling oil emulsion using a solvent extraction method. The extracted oil was filtered and a known weight of oil sample was titrated against a 0.1(N) KOH solution using phenolphthalein as the indicator. Oil concentration of the emulsion was checked instrumentally every day using moisture balance (model-MA35 from Sartorius).

12.4.4 X-RAY FLUORESCENCE

The X-ray fluorescence (XRF) technique was used to analyse the elemental composition of rolling oil emulsion and water sample. It is a good analytical tool based on a non-destructive technique. The elemental composition is ascertained by measuring the energy (keV) and intensity of the generated X-rays. A Shimadzu make (model-EDX7000) energy dispersive X-ray fluorescence spectrometer was used for the study.

12.4.5 FOURIER TRANSFORM INFRARED SPECTROSCOPY (FTIR)

The Fourier transform infrared spectroscopy (FTIR) technique was employed to compare the differences in functional groups of the rolling oil samples arising out of contamination or deterioration. The oil samples studied were a fresh rolling oil supply and the oil under service. The oil in service was extracted through a similar method used for performing acid number study. A Bruker made FTIR was used for the comparative study. Scanning was performed between 4,000–400 cm^{-1} wavenumbers at 4 cm^{-1} resolution.

12.4.6 RAMAN STUDY

Renishaw made Invia Raman spectroscopy was used for the comparative analysis of the cold rolled defect and unaffected area. The spectra were acquired using a laser source of 785 nm and magnification lens of 50×. Previously, Raman has proved to be a very useful tool in decoding the various compounds and reaction intermediates formed on the steel substrate during rolling.

12.4.7 TRIBO STUDIES

A Plint TE77 tribology tester was used to study and understand the comparative tribological behavior between the freshly supplied oils and the extracted oil. A flat steel specimen sample of BS4659-1971 with a dimension of 58 mm × 38 mm × 4 mm along with a cylinder (BS1804), 16 mm long and 6mm diameter was used for the study. The average surface roughness (Ra) of the flat steel specimen was 0.4. The flat steel sample present in the specimen bath was completely immersed in the oil to be evaluated. The test was performed under two different set of conditions. One set of tests was performed at lower temperature (100 °C) and at lower frequency to understand the role at boundary conditions, and the second set was performed at higher temperatures (150 °C) and at higher frequency. Constant load was set for all the experiments.

12.5 RESULTS

The as-received defect sample shown in Figure 12.5a shows yellowish band mark of 50–80 mm at the edges along the length of the coil. The intensity of the yellow band became darker with time. Alkaline electro degreasing technique was also unable to clean the substrate completely (Figure 12.5b). Closer inspection showed a shiny area adjacent to the band area. The nature of defect was similar for all the affected coils.

12.5.1 MICROSCOPY AND EDS ANALYSIS

The roughness of the band and non-band sample was measured through OLS spectroscopy. Average roughness measurement data shows roughness (Ra) varying in the range 0.27–0.32 in both the defect and non-defect areas, indicating no major variation in surface roughness. The SEM images of both the defect and non-defect sample show wear marks in the form of ploughing along the direction of cold rolling which is a normal phenomenon and has also been reported in earlier studies (Figure 12.6a and b). Under microscopy, no variation in surface

Non-defect   Defect
2000X   2000X

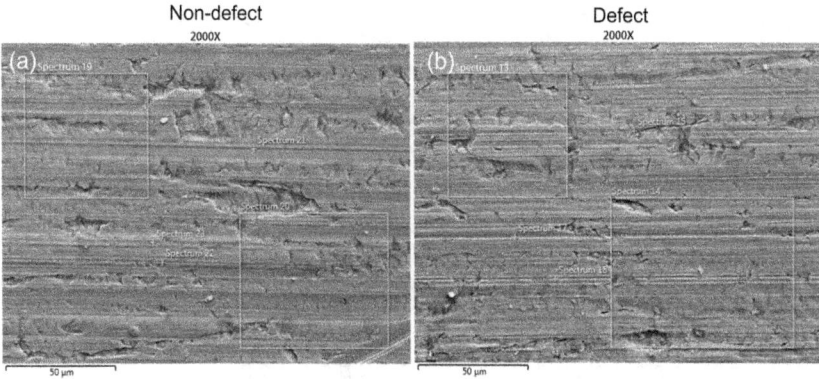

FIGURE 12.6   SEM image of (a) defect and (b) non-defect surface.

TABLE 12.5
Comparative Elemental Analysis of Defect and Non-defect Area

Parameters	Defect (%)	Non-defect (%)
C	15.2	8.2
O	2.7	0.5
Fe	82.1	91.3
Total	100	100

morphology could be observed between the band and the non-band surfaces. Elemental analysis (average data, Table 12.5) indicates a higher proportion of carbonaceous and oxide resides on the band surface. Band surface was reported at around 2.7% of oxygen compared to 0.5% against the non-band surface, which is an indication of higher surface oxidation of the band area compared to non-band area. The higher % of carbon in the defect area could be attributed to the presence of a higher amount of carbonaceous deposits arising out of the rolling oil residues.

12.5.2   PHYSICO-CHEMICAL PROPERTIES OF DEMINERALIZED WATER

The demineralized water used in the dilution of rolling emulsion was subjected to detailed investigation. The data was monitored one month before the problem started. No major deviation in pH of the input DM water was reported (Figure 12.7a). The conductivity of the water sample was also within specification, but an increasing tendency in the conductivity of water was observed for one month leading to the band problem (Figure 12.7b).

Detailed analysis of the water samples based on conductivity revealed some interesting findings. Though the conductivity of the DM water was within the specified limit, an increase in Ca hardness as reported by XRF studies was found to be responsible for the rise in conductivity to $10\mu S/cm$ (Table 12.6).

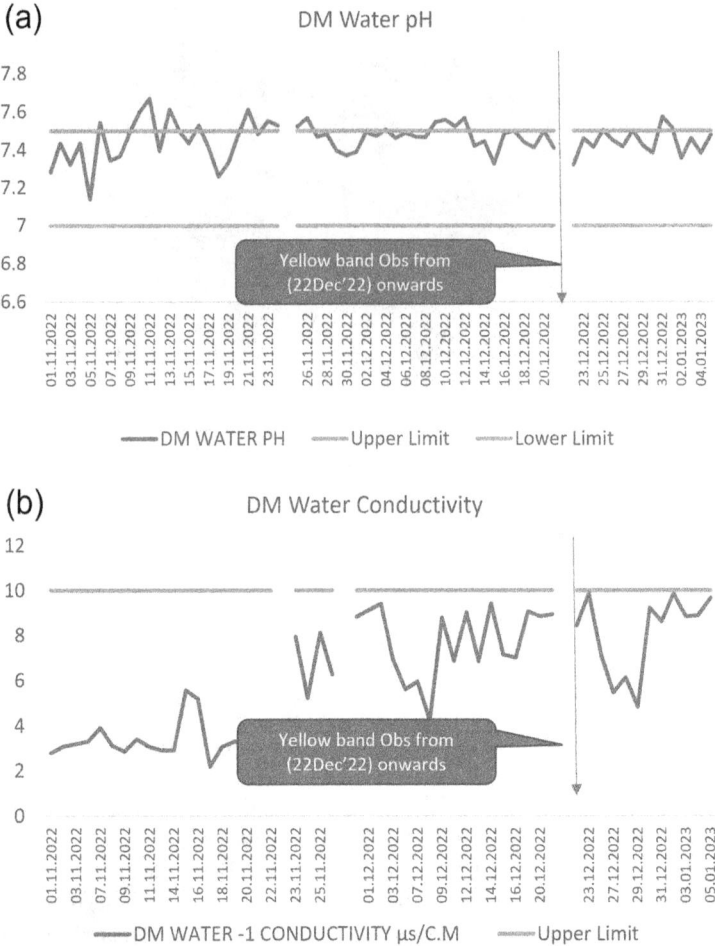

FIGURE 12.7   (a and b) pH and conductivity of demineralized water.

TABLE 12.6
Comparative Water Analysis of Varying Conductivity

Parameters	Conductivity (4 μS/cm)	Conductivity (10 μS/cm)
Free alkalinity (ppm)	0	0
Total alkalinity (ppm)	2	2
Total hardness (ppm)	2	8
Chloride (ppm)	nil	nil
Calcium (ppm, XRF)	nil	6

12.5.3   Physico-chemical Properties of Rolling Oil Emulsion

The emulsion parameters were studied for a period of one month prior to the start of the problematic period. Oil concentration was observed to be within the specified limit. The pH and conductivity of the emulsion was also within the operating range. A change in coloration of the emulsion was noticed during the defect period. During the good period, the appearance of the emulsion was greyish in nature, while during the problematic period the colour of the emulsion suddenly became off-white. Some scum formation was also observed floating on top of the emulsion. No change in chloride content of the emulsion was noted. A build-up in iron content in the emulsion bath was observed, simultaneously with the advent of the problem (Figure 12.8b). Build-up in iron fines indicates more wear particles was being generated during cold rolling. Additionally, an abrupt increase in acid number was observed 10 days before the start of the problem (Figure 12.8a).

12.5.4   FTIR Studies

An FTIR study was performed with the solvent-extracted oil collected from rolling oil emulsion during the problematic period. This spectrum was compared against the fresh supplied oil and the old oil sample extracted during the non-problematic period. The comparative spectra showed similar peaks for all the oil samples. (Figure 12.9). The oil sample collected during the problematic period showed a reduction in ester peak at $1,740$ cm^{-1} and formation of a new peak at $1,710$ cm^{-1}. The peak at $1,710$ cm^{-1} indicates the formation of free fatty acid within the oil sample, which is undesirable during cold rolling [26]. The presence of $1,710$ cm^{-1} peak in FTIR also confirms the finding on the higher acid number as depicted in Figure 12.8a.

12.5.5   Raman Studies

The band and non-band areas were subjected to Raman studies (Figure 12.10). The cold-rolled surface showed the formation of iron oxide in the form of hematite and iron phosphate. During cold rolling, extreme pressure and anti-wear agents in the form of phosphorous additives react with steel substrate forming iron phosphate ($1,000$ cm^{-1}) [5]. The formation of iron oxides during cold rolling has been reported in earlier studies. In addition, a formation of carbonaceous residues in the form of D and G bands was reported in the Raman spectroscopy [5]. A careful analysis of Raman spectra indicates iron oxides of higher intensity at the band area. A higher intensity of carbonaceous peaks was also observed in the band area compared to the non-defect area.

12.6   DISCUSSION

The speed of the six-high reversing mill was increased from $1,050$ mpm to $1,200$ mpm to increase productivity. This did not have immediate implications during

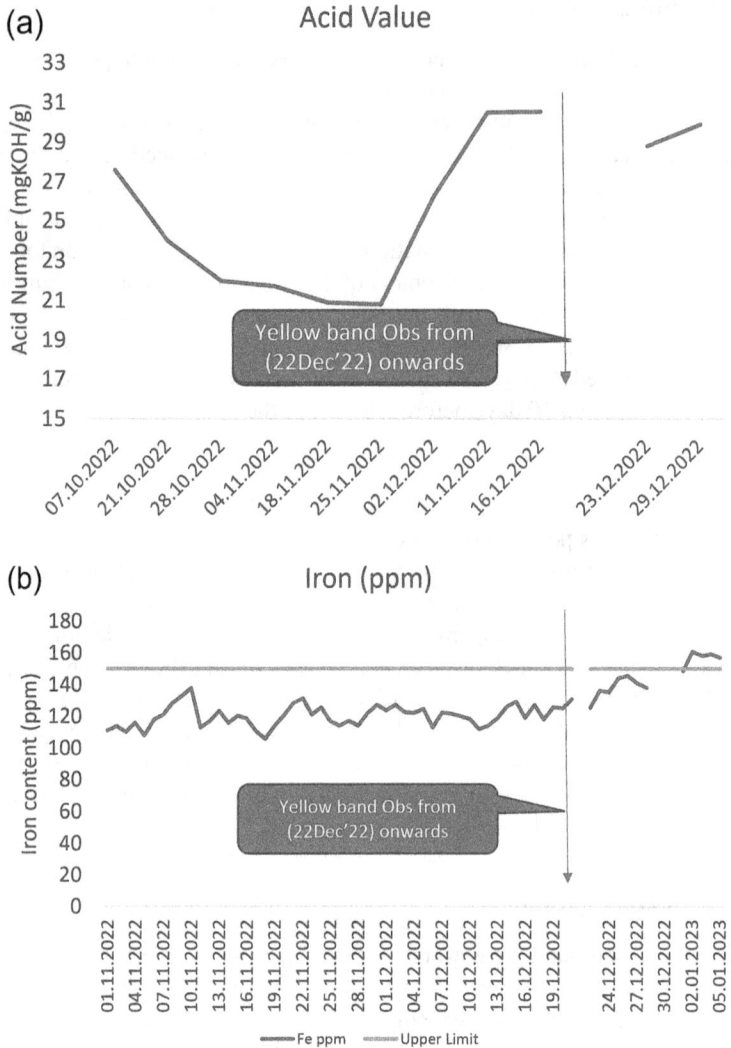

FIGURE 12.8 Variation in (a) acid number; (b) iron content in rolling oil emulsion.

cold-rolling process parameters but in due course, within a span of one week, problems related to yellowish edge band problem started appearing. This was accompanied by higher roll forces during later passes. The coil temperature touched 150 °C and above after the 5th pass which was 20–30 °C higher than normal conditions.

The increase in coil temperature during the later passes could be attributed to the higher frictional forces encountered during the rolling process. To understand and correlate with the mill conditions, a simulative study was conducted on a

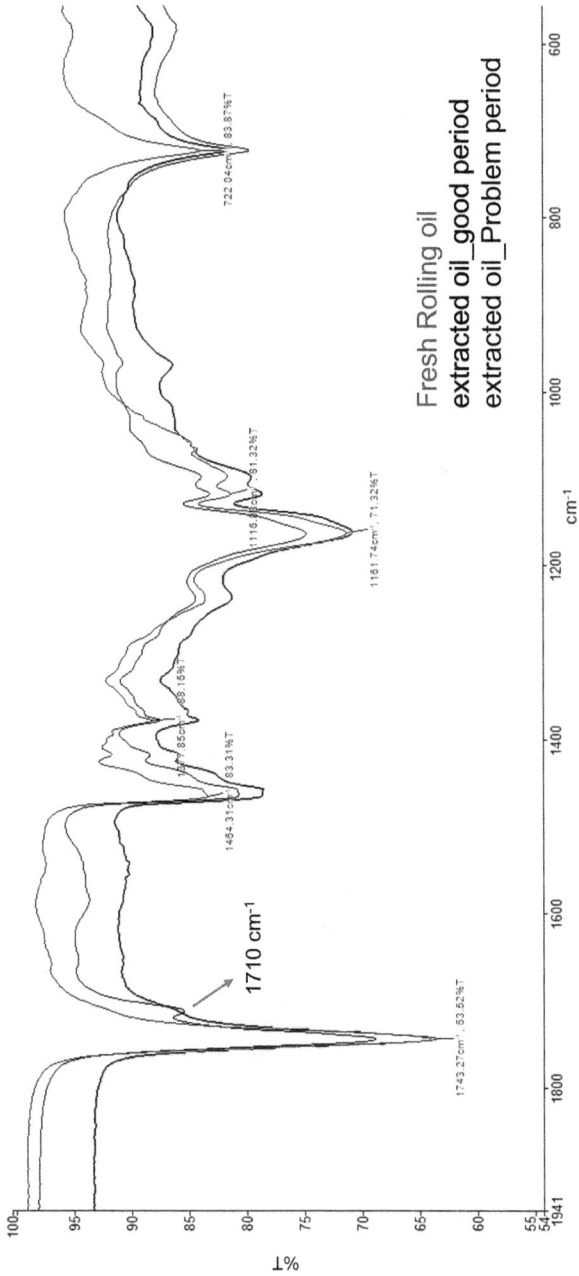

FIGURE 12.9 Comparative FTIR spectra of fresh, extracted oil in the problematic and non-problematic periods.

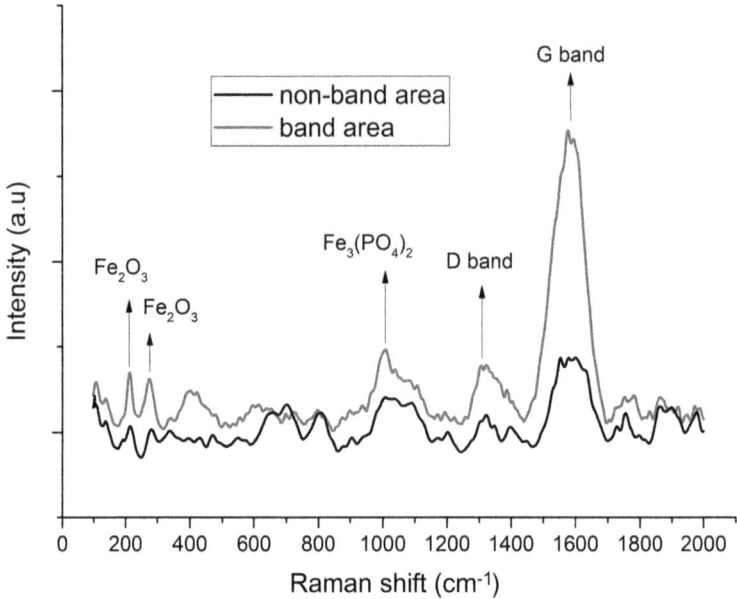

FIGURE 12.10   Comparative Raman spectra for band and non-band surface.

tribological rig equipment. Two oil samples, viz. freshly supplied oil and MEK-extracted oil from emulsion were subjected to a reciprocating wear test at lower frequency and fixed load to simulate the boundary lubrication conditions. The test was conducted at 100 °C. The test results did not show any difference in CoF between the two oils, indicating no major issues in boundary lubrication regime (Figure 12.11a).

Similarly, to simulate the mixed-lubrication regime conditions at the mill, a reciprocating wear test at higher frequency, fixed load and higher temperature of 150 °C was conducted. The test results showed higher CoF with the extracted oil used during the problem period compared to the supplied oil. This indicates that the extracted oil during the problematic period showed problems in the mixed-lubrication regime. (Figure 12.11b).

The higher CoF in the mixed-lubrication regime leads to higher friction leading to increased coil temperature. The higher frictional conditions also lead to the increased generation of iron fines which could be confirmed and reported by the earlier findings (Figure 12.8b).

The roll bite temperature could be considered to be much higher compared to the final coil temperature of 150 °C and above. This instantaneous rise in roll bite temperature often leads to incomplete burning of rolling oil residues at the roll bite. This in turn leads to the formation of carbonaceous residues which could be confirmed through Raman studies and EDS findings. This increase in temperature also leads to rapid oxidation of the residual rolling oil and the metal substrate at coil edges.

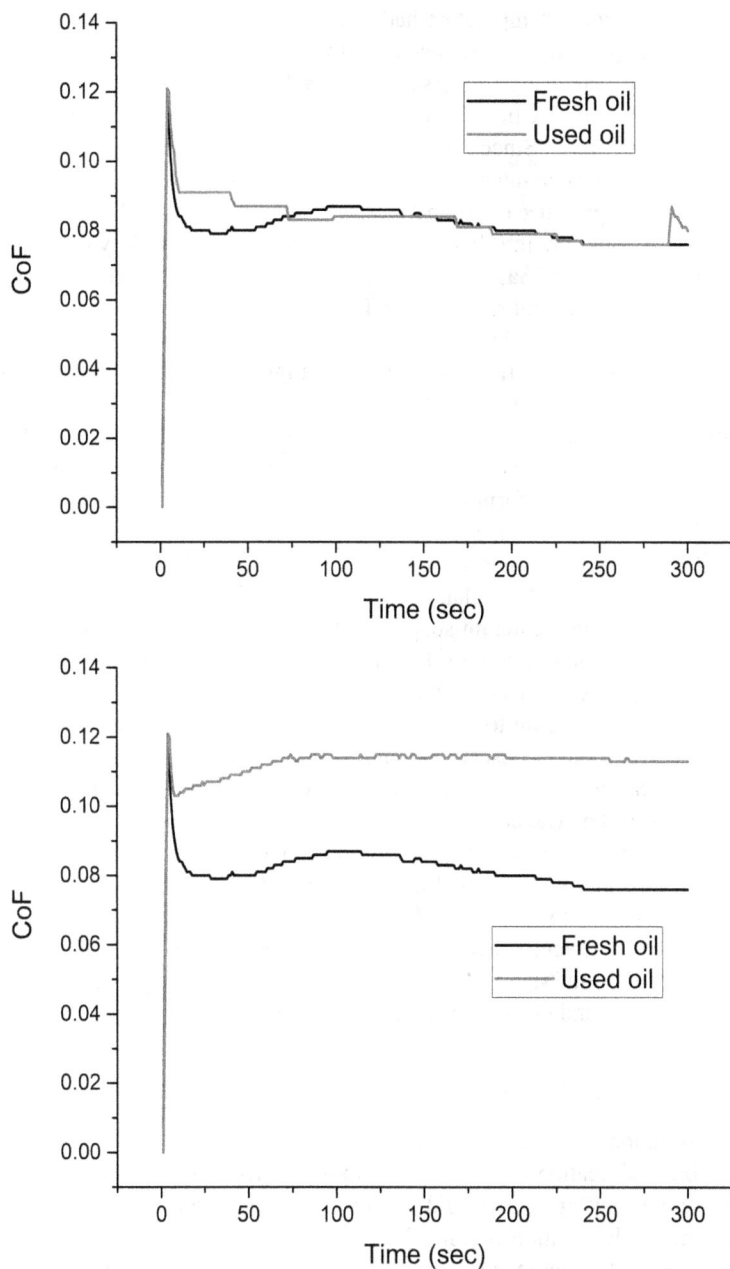

FIGURE 12.11   (a) Comparative CoF at low frequency and 100 °C; (b) comparative CoF at high frequency and 150 °C.

The increase in coil temperature had a direct impact on the emulsion temperature. The emulsion temperature increased from 55 °C to 59 °C. The increase in emulsion temperature has an adverse impact on the cooling efficiency of the cold-rolled coil. Moreover, the increase in emulsion temperature also increases the rate of emulsion hydrolysis. Since rolling oil constitutes around 50–70% of triglycerides or esters, there is an inherent tendency of ester hydrolysis with an increase in temperature. The presence of around 95% water makes the hydrolysis convenient. The hydrolysis of esters into free fatty acids could be confirmed by the increased acid number (Figure 12.8a). The breakdown of esters to form free fatty acid in extracted oil could be confirmed from FTIR studies which showed a reduction in intensity of ester peak at 1,740 cm^{-1} and the formation of a new peak at 1,710 cm^{-1} representing formation of free fatty acid [26] (Figure 12.9). This change was not visible in fresh oil and extracted oil during the good period.

During this period, an increase in Ca hardness was observed in the demineralized water. The free fatty acid formed due to hydrolysis of esters reacts with the calcium present in water forming calcium soaps which are insoluble in emulsion and lead to increased scum formation, which was observed during the defect period. The role of ester hydrolysis on the tribological or frictional properties on the cold-rolling oil could be related. The hydrolysis of esters into free fatty acids and its conversion into calcium soap leads to a deterioration in the properties of esters which plays an important role in the mixed-lubrication regime. The loss of esters had a negative impact during the transition from the boundary to the mixed-lubrication regime leading to an increase in CoF. This could be translated to the sudden rise in roll forces observed in later passes of the rolling operation.

Post root cause analysis, the action was taken on rising calcium hardness in demineralized water. The conductivity of the water was restored back to 5 µS/cm and below. The temperature of rolling oil emulsion was brought down below 55 °C. Some fresh rolling oil was added in the system. This action led to lowering of scum formation, and the lowering of emulsion temperature led to a better cooling effect. This had a direct impact on the lowering of the acid number. With time the conditions of the mill stabilized and roll forces were once again under control. The yellow edge band problem as finally resolved.

12.7 CONCLUSIONS

The deterioration of esters into fatty acids was mainly responsible for the failure in the mixed-lubrication regime, leading to higher roll forces in the final passes. The reduction in ester- content did not allow a smooth transition from a boundary to a mixed-lubrication regime. This led to higher asperity contact between the work roll surface and the steel sheet, resulting in higher frictional forces. The high friction increased the CoF, whereas higher wear rate led to greater generation of iron fines. An increase in friction resulted in increase of coil temperature by 20–30 °C against the normal rolling period. An increase in coil temperature also increased the emulsion temperature to 59 °C, which deteriorated the cooling activity of the emulsion during the rolling process. A higher emulsion temperature increased the rate of hydrolysis of the emulsion resulting in higher acid number. The free fatty acid formed further reacted with the calcium ion present in demineralized water resulting in calcium soap formation which is a sticky material and led to the generation of scum. The oxidation or hydrolysis of oil accompanied with the higher coil temperature accelerated the oxidation of the steel substrate resulting in the yellow edge band problem. Reducing the calcium content in water, accompanied with lowering of the emulsion temperature resulted in the lowering of acid number and Ca-soap formation. These steps led to the mill's stabilization and the yellow edge band problem could be eliminated.

ACKNOWLEDGEMENTS

The authors would like to thank Mr. Suvendu Sekhar Giri from Technosoft Services Private Limited for performing the tribological studies. The authors are also grateful to Tata Steel for providing the necessary support during the course of the investigation.

REFERENCES

1. About Steel, World Steel Association, (2022).
2. Steel in Automotive, World Steel Association, (2019).
3. T. Team, *Trends in Steel Usage In The Automotive Industry*, Forbes, New York, (2015).
4. N.K. Akafuah, S. Poozesh, A. Salaimeh, G. Patrick, K. Lawler, K. Saito, Evolution of the Automotive Body Coating, *Coatings*, 6(24), 24, (2016).
5. S. Chakraborty, R. Prakash, A.N. Bhagat, M. Dutta, S.S. Giri, Influence of Surface Competition Between Lubricant Additives on the Defect Formation in Cold Rolled Steel, *J. Mater. Eng. Perform*, 30(11), 8652–8662, (2021), DOI: 10.1007/s11665-021-06041-3
6. Stephen Hsu, Richard Gates, *Boundary Lubrication and Boundary Lubricating Films*, (2000), DOI: 10.1201/9780849377877.ch12
7. Michael Sutcliffe, Surface Finish and Friction in Cold Metal Rolling, Chapter 4, *Metal Forming Science and Practice* (pp. 19–59), (Dec 2002), DOI: 10.1016/B978-008044024-8/50004-7
8. S. Chakraborty, S.S. Giri, A.N. Bhagat, S. Singh, J. Ghori, Effect of Tramp Oil Ingress in Cold Rolling Oil Bath Leading to White Spot Defect on CRCA at Tinplate Mill, *Eng. Fail. Anal.*, 140, 106601, (2022), DOI: 10.1016/j.engfailanal.2022.106601
9. A. Azushima, Fundamentals of Tribology, Chapter-1, *Tribology in Sheet Rolling Technology*, DOI: 10.1007/978-3-319-17226-2_1
10. A. Azushima, Tribology in Metal Forming, Chapter-2, *Tribology in Sheet Rolling Technology*, DOI: 10.1007/978-3-319-17226-2_2
11. A. Azushima, Tribology in Cold Sheet Rolling, Chapter-4, *Tribology in Sheet Rolling Technology*, DOI: 10.1007/978-3-319-17226-2_4
12. I. Minami, S. Mori, Concept of Molecular Design Towards Additive Technology for Advanced Lubricants, *Lubr. Sci.*, 19(2), 127–149, (2007), DOI: 10.1002/ls.37
13. Aldara Naveira Suárez, *The Behaviour of Antiwear Additives in Lubricated Rolling-Sliding Contacts*, Lulea University of Technology, (2011), https://www.diva-portal.org/smash/get/diva2:998822/FULLTEXT01.pdf. Accessed 12th Jan 2023

14. R.D. Evans, H.P. Nixon, C.V. Darragh et al., Effects of Extreme Pressure Additive Chemistry on Rolling Element Bearing Surface Durability, *Tribol. Int.*, 40, 1649–1654, (2007), DOI: 10.1016/j.triboint.2007.01.012

15. P.A. Willermet et al., Mechanism of Formation of Antiwear Films from Zinc Dialkyldithiophosphates, *Tribol. Int.*, 28(3), 177–187, (1995).

16. T. Radu, A. Ciocan, Behavior of Cold Rolling Emulsions in the Obtaining Process of Steel Strips the Annals of "Dunarea de Jos" University of Galati. Fascicle IX, *Metall. Mater. Sci.*, 37, 57–63, (2014).

17. P. Vergne, M. Kamel, M. Querry, Behavior of Cold-Rolling Oil-in-Water Emulsions: A Rheological Approach, *ASME. J. Tribol.*, 119(2), 250–258, (1997), DOI: 10.1115/1.2833173

18. W.L. Roberts, *Friction and lubrication in metal processing*, New York: ASME, pp. 103–110, (1966).

19. N. Fujita, Y. Kimura, Influence of Plate-out Oil Film on Lubrication Characteristics in Cold Rolling, *ISIJ Int.*, 52, 850–857, (2012), DOI: 0.2355/isijinternational.52.850

20. W.R.D. Wilson, Y. Sakaguchi, S.R. Schmid, A Dynamic Concentration Model of Emulsions, *Wear*, 161, 207, (1993), DOI: 10.1016/0043-1648(93)90471-W

21. A. Azushima, K. Noro, Y. Iyanagi, Surface Qualities after Rolling and Oil Film Introduced during Rolling in Emulsion Lubrication, *J. Jpn. Soc. Tribologists*, 34, 879–886, (1989).

22. Y. Kimura, N. Fujita, Y. Matsubara, et al., High-Speed Rolling by Hybrid-Lubrication System in Tandem Cold Rolling Mills, *J. Mater. Process. Technol.*, 216, 357–368, (2015), DOI: 10.1016/j.jmatprotec.2014.10.002

23. H Dong, Y Zhang, W Hu, Z Ba, Y Zhang. Design and Engineering Application of Direct Mixing Lubrication System for Emulsion Pipeline in Secondary Cold Rolling Mill, *Mech. Eng. Sci.*, 2(1), 36–47, (2020), DOI: 10.33142/mes.v2i1.2619

24. K.F. Karhausen, O. Seiferth. Aluminium hot and cold rolling. In: Mang T, ed. *Encyclopedia of Lubricants and Lubrication*, Springer, (2014).

25. S Chakraborty, S.S. Giri, A. Pandit, A. Bhagat, A.K. Jha, Kinetics Study of Cold Rolling Lubricant Degradation Through Advanced Instrumental Techniques, *Lubr. Sci.*, 1–12, (2022), DOI: 10.1002/ls.1630

26. S. Chakraborty, S.S. Giri, K. Mondal, A.N. Bhagat, M. Dutta, Generation of Free Fatty Acid during Lockdown and its Effect on the Corrosion in Rolling Emulsion Tank, *Eng. Fail. Anal.*, 129, 105685, (2021), ISSN 1350-6307, DOI: 10.1016/j.engfailanal.2021.105685

27. S. Pawar, S. Chakraborty, H. Jugade, G. Mukhopadhyay, Study of White Patch Defect in Automotive Grade Interstitial Free Steel, *J. Fail. Anal. Prev.*, 20(6), 1819–1824, (2020), DOI: 10.1007/s11668-020-01016-3

28. R. Smits, J. Smeulders, Surface Chemistry-Based Defects on Steel Sheet, Investigated by SEM/EDS, *Iron Steel Technol.*, 14(2), 46–57, (2017).

29. R. Prakash, B.N. Roy, Quality Improvement in Finished Cold Rolled Sheet by Reducing the Defect, *Int. J. Sci. Eng. Res.*, 7(3), 595–602, (2016).

30. K. Kenmochi, I. Yarita, H. Abe, A. Fukuhara, T. Komatu, H. Kaito, Effect of Micro-Defects on the Surface Brightness of Cold-Rolled Stainless-Steel Strip, *J. Mater. Process. Technol.*, 69(1–3), 106–111, (1997), DOI: 10.1016/S0924-0136(97)00003-4

31. Y. Li, J.-L. Sun, J.-Y. Chen, Analysis of White-Stripe Defects of Non-Oriented Silicon Steel Rolled Surface with Emulsion, *Metall. Res. Technol.*, 112(5), 507, (2015), DOI: 10.1051/metal/2015035

13 Case Study
Analysis of Multi-Recessed Spherical Hybrid Journal Bearings Lubricated with Power-Law Lubricant

Adesh Kumar Tomar
Graphic Era (Deemed to be University), Dehradun, India

Krishnkant Sahu
Sharad Institute of Technology College of Engineering,
Ichalkaranji, India

13.1 INTRODUCTION

In the present era of technological advancement, multi-recess journal bearing configurations have gained significant attention from researchers due to their excellent lubricating bearing performance [1–5]. This class of bearings provides excellent performance characteristics, such as the larger value of lubricant film stiffness as well as damping coefficients, and also provides smooth operations. Multi-recess journal bearing configurations support the external load due to the generation of lubricant film pressure by the pumping and wedging action of the lubricant film in the clearance gap among journal and bearing surfaces via control flow devices or restrictors, including orifice, capillary, membrane restrictor, constant flow valve, etc. Several researchers have investigated the behavior of multi-recessed hydrostatic/hybrid journal bearings. The lubricating performance characteristics of multi-recessed journal bearing systems have been analyzed, taking into account the effect of the pockets' geometric shape and the recess number [6–8]. Recently, Zhang et al. [9] scrutinized the lubricating behavior of the multi-recessed hydrostatic journal bearing configuration using the lumped-parameter technique. Newer bearing configurations have been developed in order to improve their performance. Spherical journal bearing has a spherical tribo-pair which produces lesser noise and vibration and provides longer life. Hybrid spherical bearing configurations have various advantages compared to conventional circular bearings, such as spherical bearing configurations can adjust the shaft misalignment and can sustain both radial and axial loading [10]. There are few studies that have been performed related to spherical bearing configuration. Chiang et al. [11] performed an early work related to spherical bearing. The perturbation method was used in their study to theoretically investigate the behavior of spherical hybrid squeeze bearing configuration. The results presented were only for the small radial displacement. Goenka and Booker [12] conducted pioneering work in the area of spherical bearing configuration. They theoretically investigated the behavior of hydrodynamic spherical bearing configuration using the finite element technique. They reported that the spherical bearing configuration could be used as an alternative to a circular bearing configuration, especially under the condition where both radial and axial loadings act on the bearing. The influence of fluid inertia effect on hydrostatic thrust spherical bearing was investigated by Dowson et al. [13]. They found that the performance of the bearing improves under the influence of fluid inertia effects. Andres [14] analyzed the hydrostatic journal bearing performance using a CFD approach and solved the equations of mass and momentum for the flow of barotropic liquid in the clearance domain of the bearing. A newer configuration of the hydrostatic spherical hinge was developed by Xu et al. [15] in their recent work. They applied the perturbation technique for the solution of the governing Reynolds equation for the developed bearing configurations.

Commonly, while designing a film bearing system, the behavior of lubricants is considered to be the same as a Newtonian fluid, but most of the lubricant available in reality shows non-Newtonian behavior due to the existence of some polymer additives in the lubricant for improving its performance. Further, it is a recognized fact that the operations of fluid film bearing systems are greatly dependent on the lubricating characteristics of the lubricant. Every lubricant has different lubricating characteristics and is selected according to its particular applications. The main purpose of the lubricants in this class of bearing systems is to support the externally applied load deprived of degradation in the strength or significant deformation. The lubricant properties, that is, volatility, oxidation resistance, viscosity index, etc. are improved by blending the specific additives, for example, polymethacrylate (PMA) or polyisobutylene (PIB) in the Newtonian lubricant. These additives enhanced the lubricant's viscosity and flow resistance. Owing to this behavior, the lubricant acts as a non-Newtonian lubricant. The lubricant's viscosity considerably influenced the performance of the fluid film journal bearings. The various rheological models – namely the cubic shear model, couple stress fluid model, Carreau fluid model, power-law model, etc. [16–21] – presented empirical relationships among the shear stress of the lubricant and shear strain for examining the lubricating performance of non-linear flow behavior of the lubricants. The power-law model provides the relationship of the power-law index (n) which determines the flow behavior of the lubricant, and the consistency index (m) which refers to the lubricant viscosity with respect to the shear stress. For Newtonian lubricant, the power-law or flow behavior index is $n = 1$, while $n <$ 1 means the lubricant is a pseudoplastic or shear thinning lubricant, and $n > 1$ implies the lubricant is a shear thickening or dilatant lubricant. Many studies are available in the open literature of fluid film journal bearings lubricating with non-linear lubricants. Nevertheless, some prominent studies concerning the non-Newtonian rheology of lubricating oils are available in the literature [22–25]. Tayal et al. [24, 26] examined the hydrodynamic circular bearing considering operating with the power-law lubricant and cubic law model. Sinhasan and Sah [27] scrutinized the lubricating performance indicators of the hydrostatic journal bearing using the cubic law model. Prashad [28] presented a method to analyze the performance parameters of the hydrodynamic bearing configuration considering the maximum lubricant temperature in the clearance region of the bearing for a range of clearance ratio and viscosity of the lubricant. Wu and Dareing [29] studied the hydrostatic journal bearing system using the power-law lubricant model, which consists of the powder lubricant slurries and predicated the fluid film pressure in the hydrostatic journal bearing system. Rahmatabadi et al. [30, 31] theoretically examined the non-circular/circular bearing considering the effect of micropolar fluid on performance indicators. Dowson [32] developed a generalized Reynolds equation for a hydrostatic journal bearing system, which considered the change of physical variables along and across the lubricating film.

   A thorough review of the existing literature demonstrates that the flow behavior of the lubricant has a substantial influence on the lubricating performance indicators of fluid film journal bearings. Nevertheless, to the best of the author's knowledge, no study has yet been carried out regarding four-pocketed hybrid spherical journal bearing systems compensated by capillary compensator lubricating with power lubricant. Therefore, the current study planned to examine the performance characteristics parameters of capillary compensating four-pocketed spherical hybrid journal bearing using power-law lubricant.

13.2 MATHEMATICAL MODELLING

The schematic diagram of four-pocketed spherical hybrid journal bearing using the capillary compensator is represented in Figure 13.1. The governing Reynolds

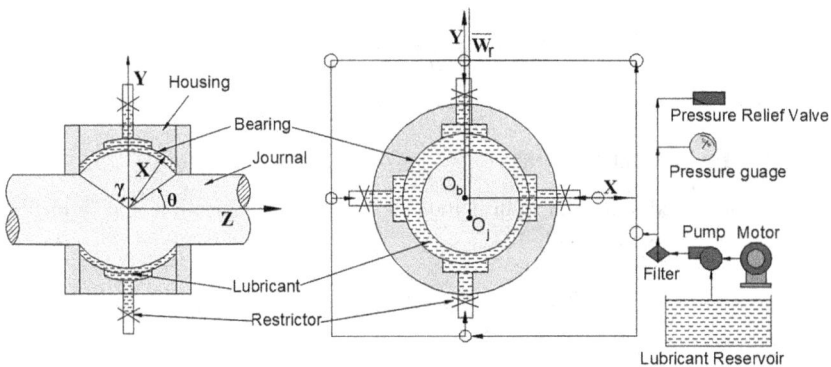

FIGURE 13.1 Schematic illustration of four-pocketed spherical hybrid journal bearing configuration.

equation in the dimensionless form of four-pocketed spherical hybrid journal bearings, considering the non-Newtonian and laminar flow of the lubricant in the bearing clearance may be given as [11, 12]:

$$\frac{1}{R^2}\left[\frac{1}{\sin\theta}\frac{\partial}{\partial\theta}\left(F_2\sin\theta\frac{\partial p}{\partial\theta}\right)+\frac{1}{\sin^2\theta}\frac{\partial}{\partial\phi}\left(F_2\frac{\partial p}{\partial\phi}\right)\right]$$
$$=\omega_j\left[\frac{\partial}{\partial\phi}\left(h-\frac{F_1}{F_0}\right)\right]+\frac{\partial h}{\partial t} \tag{13.1}$$

where, $F_0=\int_0^h\frac{1}{\mu}dr, F_1=\int_0^h\frac{r}{\mu}dr, F_2=\int_0^h\left(\frac{r^2}{\mu}-\frac{r}{\mu}\frac{F_1}{F_0}\right)dr$

The aforementioned equation can be expressed in dimensionless form using the non-dimensionalized terms given below:

$$\bar{h}=\frac{h}{c};\bar{\mu}=\frac{\mu}{\mu_r};\bar{p}=\frac{p}{p_s};\bar{t}=\frac{t}{\left[\frac{\mu R^2}{c^2 p_s}\right]};\Omega=\frac{\omega_j}{\left[\frac{c^2 p_s}{\mu R^2}\right]};$$

$$\bar{F}_0=F_0\left(\mu_r/h\right);\bar{F}_1=F_1\left(\mu_r/h^2\right);\bar{F}_2=F_2\left(\mu_r/h^3\right)$$

$$\frac{1}{\sin\theta}\frac{\partial}{\partial\theta}\left(\bar{h}^3\bar{F}_2\sin\theta\frac{\partial\bar{p}}{\partial\theta}\right)+\frac{1}{\sin^2\theta}\frac{\partial}{\partial\phi}\left(\bar{h}^3\bar{F}_2\frac{\partial\bar{p}}{\partial\phi}\right)$$
$$=\Omega\left[\frac{\partial}{\partial\phi}\left\{\left(1-\frac{\bar{F}_1}{\bar{F}_0}\right)\bar{h}\right\}\right]+\frac{\partial\bar{h}}{\partial t} \tag{13.2}$$

Here, $\bar{F}_0, \bar{F}_1$ and $\bar{F}_2$ denotes the cross-film viscosity integrals, which are defined as:

$$\bar{F}_0=\int_0^1\frac{1}{\bar{\mu}}d\bar{r};\bar{F}_1=\int_0^1\frac{\bar{r}}{\bar{\mu}}d\bar{r};\bar{F}_2=\int_0^1\left(\frac{\bar{r}^2}{\bar{\mu}}-\frac{\bar{r}}{\bar{\mu}}\frac{\bar{F}_1}{\bar{F}_0}\right)d\bar{r}$$

13.2.1 FLUID FILM THICKNESS

For four-pocketed spherical hybrid journal bearing configuration, the fluid film thickness is expressed as [11, 12]:

$$\bar{h}=1-\bar{X}_j\sin\theta\cos\phi-\bar{Y}_j\sin\theta\sin\phi-\bar{Z}_j\cos\theta \tag{13.3}$$

13.2.2 FE ANALYSIS

The finite element (FE) technique has been used to solve the non-dimensional modified governing Reynolds equation. The finite element method (FEM) has the capability to handle the irregular shape of the fluid region and provide accurate results. To obtain the unknown value of the nodal pressure, the discretization of the lubricant domain has been carried out by using 4-noded quadrilateral isoparametric elements. The Lagrange polynomial function has been incorporated to interpolate the unknown value of lubricant film pressure within an element as:

$$\bar{p} = \sum_{j=1}^{\eta^e} \bar{p}_j N_j$$

where, η^e and N_j refer to the number of nodes over an element and interpolation function, respectively.

The eth element equation is obtained using Galerkin's method as given below:

$$\left[\bar{F}_{ij}\right]^e \{\bar{p}\}^e = \{\bar{Q}_j\}^e + \Omega\{\bar{R}_{H_j}\}^e + \bar{X}_j\{\bar{R}_{X_j}\}^e + \bar{Y}_j\{\bar{R}_{Y_j}\}^e + \bar{Z}_j\{\bar{R}_{Z_j}\}^e \quad (13.4)$$

where,

$$\left[\bar{F}_{ij}\right]^e = \iint_{A^e} \bar{h}^3 \bar{F}_2 \left[\sin\theta \frac{\partial N_i}{\partial\theta}\frac{\partial N_j}{\partial\theta} + \frac{1}{\sin\theta}\frac{\partial N_i}{\partial\phi}\frac{\partial N_j}{\partial\phi}\right] d\theta d\phi$$

$$\{\bar{Q}_j\}^e = \int_{\Gamma^e} \left\{\left(\frac{1}{\sin\theta}\bar{h}^3\bar{F}_2\frac{\partial\bar{p}}{\partial\phi} - \sin\theta\left(1-\frac{\bar{F}_1}{\bar{F}_0}\right)\bar{h}\Omega\right)\eta_\phi + \left(\sin\theta\bar{h}^3\bar{F}_2\frac{\partial\bar{p}}{\partial\theta}\right)\eta_\theta\right\} N_i d\Gamma^e$$

$$\{\bar{R}_{H_j}\}^e = \iint_{A^e} \bar{h}\sin\theta\left(1-\frac{\bar{F}_1}{\bar{F}_0}\right)\frac{\partial N_i}{\partial\phi} d\theta d\phi$$

$$\{\bar{R}_{X_j}\}^e = \iint_{A^e} N_i \sin^2\theta\cos\phi \, d\theta d\phi$$

$$\{\bar{R}_{Y_j}\}^e = \iint_{A^e} N_i \sin^2\theta\sin\phi \, d\theta d\phi$$

$$\{\bar{R}_{Z_j}\}^e = \iint_{A^e} N_i \sin\theta\cos\theta \, d\theta d\phi$$

where η_ϕ and η_θ refers to the direction cosines, and e^{th} element boundaries and domain are denoted by Γ^e and A^e, respectively.

By applying the usual procedure of the globalization of FE analysis, the global system equation is attained in the following algebraic form.

$$\left[\bar{F}_{ij}\right]\{\bar{p}\} = \{\bar{Q}_j\} + \Omega\{\bar{R}_{H_j}\} + \bar{X}_j\{\bar{R}_{X_j}\} + \bar{Y}_j\{\bar{R}_{Y_j}\} + \bar{Z}_j\{\bar{R}_{Z_j}\} \quad (13.5)$$

13.2.3 Power-Law Model

This model governs the lubricating performance of the mostly available polymeric-thickener-based mineral oil. This model presents the mathematical expression which relates to shear stress $(\bar{\tau})$ with strain rate $(\bar{\dot{\gamma}})$, as given below [24, 26, 27]:

$$\bar{\tau} = \bar{m}\left(\bar{\dot{\gamma}}\right)^n \tag{13.6}$$

Here, $\bar{m}$ and n represent the indexes of the lubricant, that is, consistency and flow behavior, respectively.

The lubricant's apparent viscosity can be expressed as:

$$\bar{\mu}_a = \bar{\tau} / \bar{\dot{\gamma}} \tag{13.7}$$

The shear strain rate $(\bar{\dot{\gamma}})$ of the power-law lubricant is expressed as:

$$\bar{\dot{\gamma}} = \left[\left(\frac{\bar{h}}{\bar{\mu}\sin\theta}\frac{\partial\bar{p}}{\partial\phi}\left(\bar{r}-\frac{\bar{F}_1}{\bar{F}_0}\right)+\frac{\Omega\sin\theta}{\bar{h}\,\bar{\mu}\,\bar{F}_0}\right)^2+\left(\frac{\bar{h}}{\bar{\mu}}\frac{\partial\bar{p}}{\partial\theta}\left(\bar{r}-\frac{\bar{F}_1}{\bar{F}_0}\right)\right)^2\right]^{\frac{1}{2}} \tag{13.8}$$

13.2.4 Lubricant Flow

The lubricant flow equation in the dimensionless form of a capillary compensator is expressed as given below [1, 33]:

$$\bar{Q}_R = C_{s2}\left(1-\bar{p}_c\right) \tag{13.9}$$

13.2.5 Performance Indicators

The expressions for bearing lubricating performance indicators of four-pocketed spherical hybrid journal bearings are defined in the following sections.

13.2.5.1 Fluid Film Reactions

The fluid film pressure acting on the surface of a four-pocketed spherical hybrid journal bearing is integrated for computing the fluid film reactions as given below [12, 33]:

$$\left.\begin{aligned}\bar{F}_x &= \iint_A \bar{p}\sin^2\theta\cos\phi\,d\theta d\phi \\ \bar{F}_Y &= \iint_A \bar{p}\sin^2\theta\sin\phi\,d\theta d\phi \\ \bar{F}_z &= \iint_A \bar{p}\cos\theta\sin\theta d\theta d\phi\end{aligned}\right\} \tag{13.10}$$

where, radial load $\overline{F}_R = \sqrt{\left(\overline{F}_x^{\,2} + \overline{F}_y^{\,2}\right)}$ and axial load $\overline{F}_A = \overline{F}_z$

The result of fluid film reactions of four-pocketed spherical hybrid journal bearings is expressed as:

$$\overline{F}_{RES} = \sqrt{\left(\overline{F}_R^{\,2} + \overline{F}_A^{\,2}\right)} \tag{13.11}$$

13.2.5.2 Rotor Dynamic Coefficients

The lubricant film stiffness and dynamic coefficients are the bearing rotor dynamic coefficients as discussed below.

13.2.5.2.1 Lubricant Film Stiffness Coefficients

For four-pocketed spherical hybrid journal bearings, lubricant film stiffness coefficients are given as:

$$\overline{S}_{ij} = -\frac{\partial \overline{F}_i}{\partial \overline{q}_j} \left(i, j = x, y, z;\ \text{and}\ \overline{q}_j = \overline{X}_j, \overline{Y}_j, \overline{Z}_j\right) \tag{13.12}$$

Further, these coefficients may also be presented in the form of a matrix as given below:

$$
\begin{vmatrix}
\overline{S}_{xx} & \overline{S}_{xy} & \overline{S}_{xz} \\
\overline{S}_{yx} & \overline{S}_{yy} & \overline{S}_{yz} \\
\overline{S}_{zx} & \overline{S}_{zy} & \overline{S}_{zz}
\end{vmatrix}
= -
\begin{bmatrix}
\iint_A \dfrac{\partial \overline{p}}{\partial \overline{X}_j} \sin^2 \theta \cos\phi\, d\theta d\phi & \iint_A \dfrac{\partial \overline{p}}{\partial \overline{Y}_j} \sin^2 \theta \cos\phi\, d\theta d\phi & \iint_A \dfrac{\partial \overline{p}}{\partial \overline{Z}_j} \sin^2 \theta \cos\phi\, d\theta d\phi \\
\iint_A \dfrac{\partial \overline{p}}{\partial \overline{X}_j} \sin^2 \theta \sin\phi\, d\theta d\phi & \iint_A \dfrac{\partial \overline{p}}{\partial \overline{Y}_j} \sin^2 \theta \sin\phi\, d\theta d\phi & \iint_A \dfrac{\partial \overline{p}}{\partial \overline{Z}_j} \sin^2 \theta \sin\phi\, d\theta d\phi \\
\iint_A \dfrac{\partial \overline{p}}{\partial \overline{X}_j} \cos\theta \sin\theta\, d\theta d\phi & \iint_A \dfrac{\partial \overline{p}}{\partial \overline{Y}_j} \cos\theta \sin\theta\, d\theta d\phi & \iint_A \dfrac{\partial \overline{p}}{\partial \overline{Z}_j} \cos\theta \sin\theta\, d\theta d\phi
\end{bmatrix}
$$

$$\tag{13.13}$$

13.2.5.2.2 Lubricant film damping coefficients

For four-pocketed spherical hybrid journal bearings, lubricant film damping coefficients are given as:

$$\overline{C}_{ij} = -\frac{\partial \overline{F}_i}{\partial \dot{\overline{q}}_j} \left(i, j = x, y, z\ \text{and}\ \dot{\overline{q}}_j = \dot{\overline{X}}_j, \dot{\overline{Y}}_j, \dot{\overline{Z}}_j\right) \tag{13.14}$$

Further, these coefficients may be written in the form of a matrix as given below:

$$
\begin{vmatrix} \bar{C}_{xx} & \bar{C}_{xy} & \bar{C}_{xz} \\ \bar{C}_{yx} & \bar{C}_{yy} & \bar{C}_{yz} \\ \bar{C}_{zx} & \bar{C}_{zy} & \bar{C}_{zz} \end{vmatrix} = - \begin{bmatrix} \iint_A \dfrac{\partial \bar{p}}{\partial \dot{X}_j} \sin^2\theta \cos\phi\, d\theta d\phi & \iint_A \dfrac{\partial \bar{p}}{\partial \dot{Y}_j} \sin^2\theta \cos\phi\, d\theta d\phi & \iint_A \dfrac{\partial \bar{p}}{\partial \dot{Z}_j} \sin^2\theta \cos\phi\, d\theta d\phi \\ \iint_A \dfrac{\partial \bar{p}}{\partial \dot{X}_j} \sin^2\theta \sin\phi\, d\theta d\phi & \iint_A \dfrac{\partial \bar{p}}{\partial \dot{Y}_j} \sin^2\theta \sin\phi\, d\theta d\phi & \iint_A \dfrac{\partial \bar{p}}{\partial \dot{z}_j} \sin^2\theta \sin\phi\, d\theta d\phi \\ \iint_A \dfrac{\partial \bar{p}}{\partial \dot{X}_j} \cos\theta \sin\theta\, d\theta d\phi & \iint_A \dfrac{\partial \bar{p}}{\partial \dot{Y}_j} \cos\theta \sin\theta\, d\theta d\phi & \iint_A \dfrac{\partial \bar{p}}{\partial \dot{z}_j} \cos\theta \sin\theta\, d\theta d\phi \end{bmatrix}
$$

(13.15)

i, $\bar{q}_j$ and $\dot{q}_j$ represent the force direction, displacement of journal center, and velocities components.

13.2.6 BOUNDARY CONDITIONS

The suitable boundary conditions required for solving the modified Reynolds equation are written below:

(i) Nodes positioned on the bearing's exterior boundary have null fluid film pressure.
(ii) Nodes situated in a recess have the same nodal pressure.
(iii) The input flow to the bearing and the restrictor flow is the same in order to maintain the continuity of the flow.
(iv) The value of pressure gradient is null in circumferential direction, i.e., $\bar{p} = \partial \bar{p} / \partial \phi = 0$.

13.3 COMPUTATIONAL SCHEME

The capillary compensated multi-recess hybrid spherical journal bearing operating with the power-law model needs the simultaneous solution of governing modified Reynolds equation, and flow equation of the restrictors, along with applying the necessary boundary conditions. The Gauss-Seidel technique was used to solve the governing equations, and the value of fluid film pressures computed. The power-law model mathematical equations have been solved using the Newton-Raphson method to compute the value of shear strain rate, equivalent shear stress, and lubricant's viscosity, obtained using Eqn. (13.7). Continue the iterations until the below convergence requirements are met in order to reach an equilibrium position.

$$
\left| \frac{\left[\left(\Delta \bar{X}_j^i \right)^2 + \left(\Delta \bar{Y}_j^i \right)^2 \right]^{1/2}}{\left[\left(\bar{X}_j^i \right)^2 + \left(\bar{Y}_j^i \right)^2 \right]^{1/2}} \right| \times 100 \le 0.001
$$

The performance indicators of four-pocketed spherical hybrid journal bearing configurations are examined once the convergence criterion is determined.

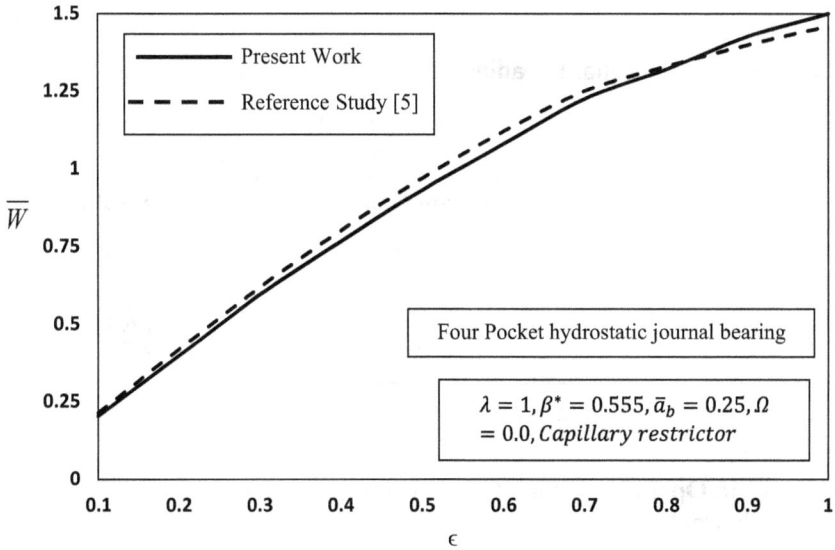

FIGURE 13.2 Loading capacity $(\overline{W})$ versus eccentricity ratio (ϵ).

13.4 VALIDATION

A source code in MATLAB software was developed to consider the non-Newtonian behavior of power-law lubricant on four-pocketed spherical hybrid journal bearings. The current model has been verified with an earlier published study by Metman et al. [4] to authenticate the methodology adopted in this work. The source code has been changed for the case of capillary compensating four-pocketed hydrostatic journal bearing configuration by selecting the same operating and geometrical parametric values as used in the reference work of Metman et al. [4]. The fluid film reaction computed from the developed source code is compared with the reference study as depicted in Figure 13.2. It may be noticed that the results obtained from the present model show close agreement with the results of the reference study. The current numerical model has also been verified with the existing results of Goenka and Booker [12] for spherical journal bearing configuration. The variation of the results obtained from the current model with the reference work is shown in Table 13.1. The results show close agreement, and a maximum variation of the order of 5% is observed. The accuracy of the present numerical model is ensured by these verifications.

13.5 RESULTS AND DISCUSSION

This study uses the power-law model to analyze the lubricating performance characteristics of capillary compensating four-pocketed hybrid spherical journal bearing configuration, such as min. fluid film thickness, lubricant flow rate,

TABLE 13.1

Comparison of Resultant Loading Capacity Versus Eccentricity Ratio

Eccentricity Ratio	Dimensionless Resultant Loading Capacity with Span Angle ($\gamma = 90°$)	
	Current Study	Goenka et al. [12]
0.25	0.183	0.184
0.5	0.477	0.456
0.75	1.161	1.11
0.9	2.741	2.614
0.925	3.431	3.292

TABLE 13.2

Geometric/Operating Parameters of Multi-Recess Spherical Hybrid Journal Bearings

Parameters	Value/Range
Land width ratio $\left(\bar{a}_b \right)$	0.14
Bearing radius (R)	50 mm
Concentric design pressure ratio (β^*)	0.5
Clearance (c)	50.2 μm
Speed of journal (N)	2500 rpm
Span angle (γ)	90°
Supply pressure (p_s)	8.96 MPa
No. of pockets	4
Inter-recess angle (φ)	18°
Radial load $\left(\bar{W}_r \right)$	0.4–1
Power-law index (n)	0.5,0.7,1,1.3,1.5
Lubricant viscosity (μ)	0.0345 Pa.s
Speed (Ω)	1
Restrictor	Capillary

and bearing dynamic coefficients. Table 13.2 shows the geometric specifications and operating parameters of the multi-recessed spherical hybrid journal bearing employed in the current work. These values of the parameters have been selected based on the earlier reported works stated in the literature [1, 4, 11, 12, 32].

13.5.1 Influence on Min. Fluid Film Thickness $\left(\bar{h}_{min} \right)$

Figure 13.3 shows the variation of min. fluid film thickness $\left(\bar{h}_{min} \right)$ for four-pocketed spherical hybrid journal bearing versus the external radial load ($\bar{W}_r$). It may be noticed that there is a reduction in the value of $\left(\bar{h}_{min} \right)$ with an increment in the external radial loading. Such a phenomenon can be due to the reason that the value of $\bar{h}_{min}$ is reduced in order to equilibrate the increment in the external loading. Further, it can also be noticed that a larger value of the power-law index

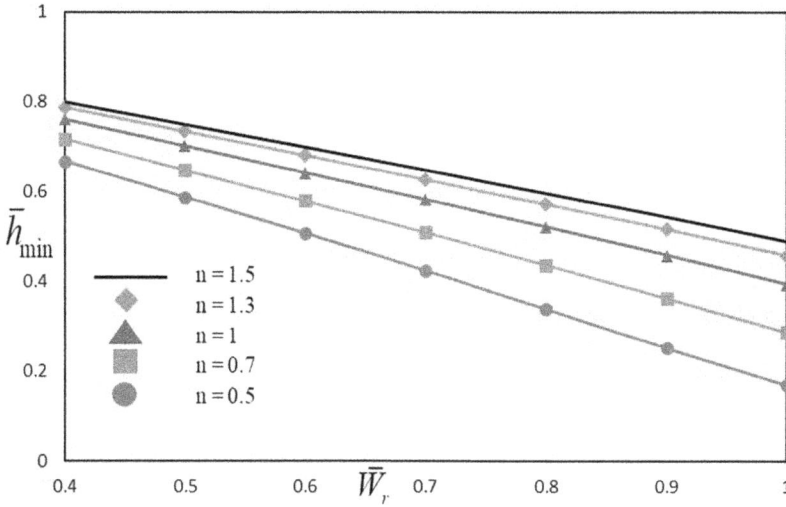

FIGURE 13.3 Min. fluid film thickness $\left(\bar{h}_{min}\right)$ versus radial load $\left(\bar{W}_r\right)$.

(n) offers an improved value of the min. fluid film thickness. Therefore, dilatant lubricant provides a larger value of the min. fluid film thickness as compared to pseudoplastic and Newtonian lubricants. At a chosen value $\bar{W}_r = 1$, the percentage difference in the value of $\bar{h}_{min}$ is obtained to be the order of 24.07% for the dilatant lubricant ($n = 1.5$) with respect to the Newtonian lubricant. The useful trends obtained for the value of $\bar{h}_{min}$ from the results of the numerical simulation are given below:

$$\bar{h}_{min}\left[\left(n = 1.5\right) > \left(n = 1.3\right) > \left(n = 1\right) > \left(n = 0.7\right) > \left(n = 0.5\right)\right]$$

$$\left[\bar{h}_{min\binom{Dilatant}{Lubricant}} > \bar{h}_{min\binom{Newtonian}{lubricant}} > \bar{h}_{min\binom{Pseudoplastic}{lubricant}}\right]$$

13.5.2 Influence on the Lubricant Flow Rate $\left(\bar{Q}\right)$

The variation of lubricant flow $\left(\bar{Q}\right)$ corresponding to the externally applied radial load ($\bar{W}_r$) for four-pocketed spherical hybrid journal bearing configuration is shown in Figure 13.4. It can be seen from Figure 13.4 that the demand for the lubricant flow decreases for a higher value of the power-law index (n). Dilatant lubricants require a lower value of lubricant flow rate, whereas pseudoplastic lubricants require a higher lubricant flow rate. Therefore, pseudoplastic lubricant is suitable from the viewpoint of lubricant rate requirement. The following useful trends are obtained for the value of $\bar{Q}$ from the results of the numerical simulation:

$$\bar{Q}\left[\left(n = 0.5\right) > \left(n = 0.7\right) > \left(n = 1\right) > \left(n = 1.3\right) > \left(n = 1.5\right)\right]$$

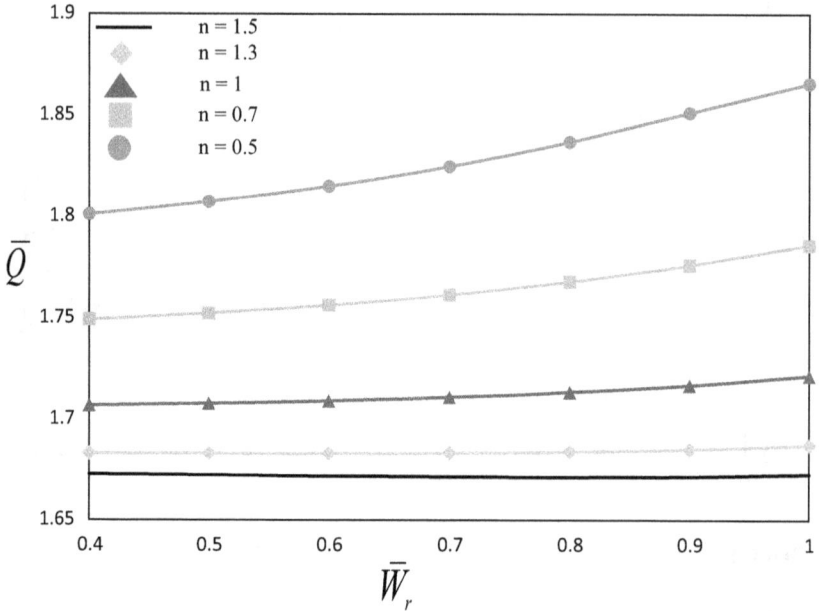

FIGURE 13.4   Lubricant flow rate $\left(\bar{Q}\right)$ versus radial load $\left(\bar{W}_r\right)$.

$$\left[\bar{Q}_{\substack{\text{Pseudoplastic}\\\text{lubricant}}} > \bar{Q}_{\substack{\text{Newtonian}\\\text{lubricant}}} > \bar{Q}_{\substack{\text{Dilatant}\\\text{Lubricant}}}\right]$$

13.5.3   Influence on the Lubricant Film Stiffness Coefficient $\left(\bar{S}_{yy}\right)$

The comparison of the lubricant film stiffness coefficient $\left(\bar{S}_{yy}\right)$ versus the externally applied radial load $\left(\bar{W}_r\right)$ is shown in Figure 13.5. It may be seen from the Figure 13.5 that as the power-law index (n) value of the lubricant decreases, the value of the lubricant film stiffness coefficient $\left(\bar{S}_{yy}\right)$ is enhanced. Therefore, pseudoplastic lubricants offer a higher value of lubricant stiffness coefficient $\left(\bar{S}_{yy}\right)$ versus Newtonian and dilatant lubricants. At a given value $\bar{W}_r = 1$, the percentage difference in the value of $\bar{h}_{\min}$ is obtained to be the order of 12.86% for the pseudoplastic lubricant ($n = 0.5$) with respect to the Newtonian lubricant. The useful trends obtained for the value of $\bar{S}_{yy}$ from the results of the numerical simulation are given below:

$$\bar{S}_{yy}\left[\left(n=0.5\right)>\left(n=0.7\right)>\left(n=1\right)>\left(n=1.3\right)>\left(n=1.5\right)\right]$$

$$\left[\bar{S}_{yy\substack{\text{Pseudoplastic}\\\text{lubricant}}} > \bar{S}_{yy\substack{\text{Newtonian}\\\text{lubricant}}} > \bar{S}_{yy\substack{\text{Dilatant}\\\text{Lubricant}}}\right]$$

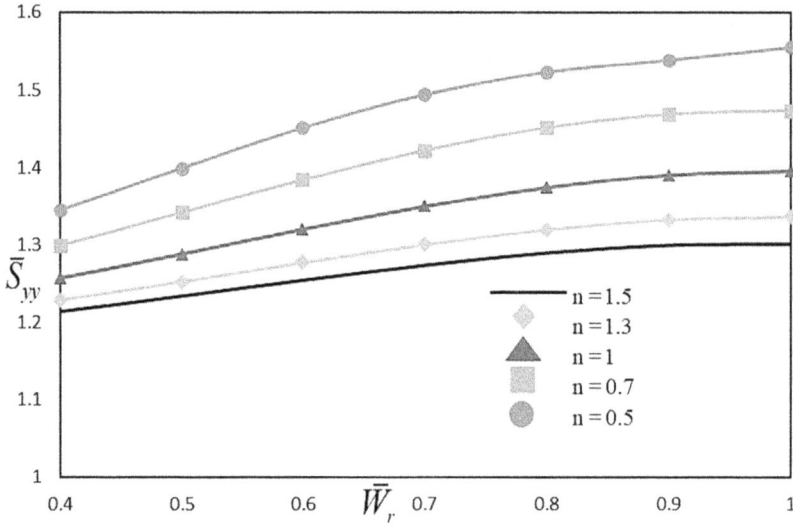

FIGURE 13.5 Stiffness coefficient $\left(\bar{S}_{yy}\right)$ versus radial load $\left(\bar{W}_r\right)$.

13.5.4 INFLUENCE ON THE LUBRICANT FILM DAMPING COEFFICIENT $\left(\bar{C}_{yy}\right)$

Figure 13.6 depicts the variation of the lubricant film damping coefficients corresponding to the externally applied radial load $\left(\bar{W}_r\right)$. It can be seen in Figure 13.6 that as the power-law index (n) of the lubricant increased, the value of the

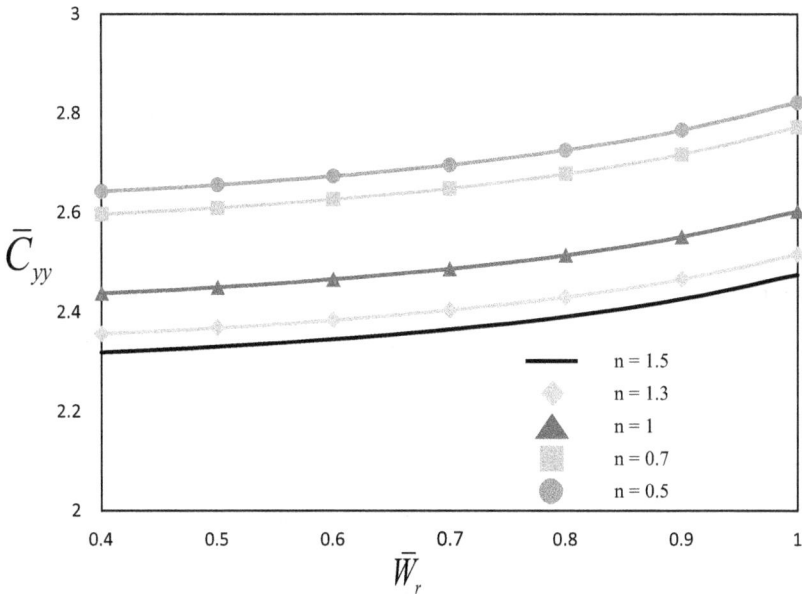

FIGURE 13.6 Damping coefficient $\left(\bar{C}_{yy}\right)$ versus radial load $\left(\bar{W}_r\right)$.

lubricant film damping coefficient $\left(\bar{C}_{yy}\right)$ reduced. Therefore, pseudoplastic lubricants offer a higher value of stiffness coefficient $\left(\bar{C}_{yy}\right)$ as compared to Newtonian and dilatant lubricants. At a chosen value $\bar{W}_r = 1$, the percentage variation in the value of $\bar{C}_{yy}$ is obtained to be the order of 19.67% for the pseudoplastic lubricant $(n = 0.5)$ with respect to the Newtonian lubricant. The other bearing rotor dynamic coefficients have also been computed. However, for the brevity of the study, these coefficients have not been presented. The useful trends obtained for the value of $\bar{C}_{yy}$ from the numerically simulated results are given below:

$$\bar{C}_{yy}\left[\left(n = 0.5\right) > \left(n = 0.7\right) > \left(n = 1\right) > \left(n = 1.3\right) > \left(n = 1.5\right)\right]$$

$$\left[\bar{C}_{yy\left(\substack{\text{Pseudoplastic}\\ \text{lubricant}}\right)} > \bar{C}_{yy\left(\substack{\text{Newtonian}\\ \text{lubricant}}\right)} > \bar{C}_{yy\left(\substack{\text{Dilatant}\\ \text{Lubricant}}\right)}\right]$$

13.6   CONCLUSIONS

In this study, the influence of the non-Newtonian behavior of lubricants governed by the power-law model on the performance indicator parameters of the four-pocketed spherical hybrid journal bearings has been examined. Based on the numerical simulation, the following salient observations are obtained:

- The value of min. fluid film thickness $\left(\bar{h}_{min}\right)$ is enhanced for a four-pocketed hybrid journal bearing system lubricated with dilatant lubricant corresponding to the similar bearing using Newtonian and pseudoplastic lubricants.
- The lubricant flow rate $\left(\bar{Q}\right)$ requirement of the multi-recess hybrid journal bearing using dilatant lubricant is decreased corresponding to the similar bearing lubricated with the Newtonian and pseudoplastic lubricants.
- Multi-recess journal bearing system operated by pseudoplastic lubricant offers a higher value of lubricant film stiffness and damping coefficients $(\bar{S}_{yy}$ and $\bar{C}_{yy})$ compared to the similar bearing system operated with dilatant or Newtonian lubricants.

Nomenclature

C_{ij}	Lubricant film damping coefficients, N s /mm
c	Radial clearance, mm
F_x, F_y, F_z	Fluid film reaction components, N
F_{RES}	Resultant fluid film reaction, N
h, h_{min}	Nominal film thickness, minimum film thickness, mm
L	Bearing length, mm
p_s, p	Supply pressure, fluid film pressure, MPa
Q	Lubricant flow rate, mm³/s
R	Radius of bearing, mm
S_{ij}	Stiffness coefficients, N/mm
t	Time, s
W_r	External radial load, N

O_j/O_b	Journal/ Bearing center
$r, \theta, \varnothing$	Spherical coordinates
X_j, Y_j, Z_j	Journal center coordinates
Non-dimensional parameters	
$\bar{C}_{ij}$	$C_{ij}(c^3/\mu R^4)$
$\bar{h}$	h/c
$\left(\bar{F}_x, \bar{F}_y, \bar{F}_z, \bar{F}_{RES}\right)$	$(F_x, F_y, F_z, F_{RES})/p_sR^2$
$\bar{Q}$	$Q(\mu/c^3ps)$
$\bar{M}, \bar{M}_c$	$M\left(\dfrac{\omega_j c^3}{\mu R^4}\right), M_c\left(\dfrac{\omega_j c^3}{\mu R^4}\right)$
$\bar{S}_{ij}$	$S_{ij}\left(\dfrac{c}{p_s R^2}\right)$
$\bar{W}_r$	W_r/p_sR^2
$\left(\bar{X}_j, \bar{Y}_j, \bar{Z}_j\right)$	$(X_j, Y_j, Z_j)/c$
$\bar{t}$	$t(c^2p_s/\mu R^2)$
$\bar{T}_f$	$T_f(1/p_scR^2)$
$\bar{p}, \bar{p}_{max}, \bar{p}_c$	$p/p_s, p_{max}/p_s, p_c/p_s$
$\bar{\omega}_{th}$	$\omega_j\left(\bar{M}_c\Big/\bar{F}\right)^{1/2}$
ε	e/c
Ω	$\omega_j(\mu R^2/c^2p_s)$
Greek symbols	
ω_j	Journal speed (rad/s)
γ	Span angle
Matrices	
$[F]$	Global fluidity matrix
$\{\bar{p}\}$	Nodal pressure vector
$\{\bar{Q}\}$	Flow vector
$\{\bar{R}_H\}$	Hydrodynamic vector term
$\{\bar{R}_{X_j}\}, \{\bar{R}_{Y_j}\}, \{\bar{R}_{Z_j}\}$	Journal velocity vector
Superscripts and subscripts	
$\cdot/\cdot\cdot$	First/second-order derivatives with time
$-$	Dimensionless terms
b/j	Bearing/Journal
Abbreviations	
min.	Minimum

REFERENCES

[1] Cusano C, Conry TF. Design of multi-recess hydrostatic journal bearings for minimum total power loss. *Journal of Engineering for Industry*. 1974;96:226–32.

[2] Ghai RC, Singh DV, Sinhasan R. Load capacity and flow characteristics of a hydrostatically lubricated four-pocket journal bearing by finite element method. *International Journal of Machine Tool Design and Research*. 1976;16:233–40.

[3] Ghosh B Load and flow characteristics of a capillary-compensated hydrostatic journal bearing. *Wear.* 1973;23:377–86.

[4] Metman KJ, Muijderman EA, van Heijningen GJJ, Halemane DM. Load capacity of multi-recess hydrostatic journal bearings at high eccentricities. *Tribology International.* 1986;19:29–34.

[5] Rowe WB. Dynamic and static properties of recessed hydrostatic journal bearings by small displacement analysis. *Journal of Lubrication Technology.* 1980;102:71–9.

[6] Nicodemus ER, Sharma SC. Influence of wear on the performance of multirecess hydrostatic journal bearing operating with micropolar lubricant. *Journal of Tribology.* 2010;132. https://doi.org/10.1115/1.4000940

[7] Singh N, Sharma SC, Jain SC, Sanjeeva Reddy S. Performance of membrane compensated multirecess hydrostatic/hybrid flexible journal bearing system considering various recess shapes. *Tribology International.* 2004;37:11–24.

[8] Tomar AK, Sharma SC. Non-Newtonian lubrication of hybrid multi-recess spherical journal bearings with different geometric shapes of recess. *Tribology International.* 2022;171:107579.

[9] Zhang Y, Yu S, Lu C, Zhao H, Liang P. An improved lumped parameter method for calculating static characteristics of multi-recess hydrostatic journal bearings. *Proceedings of the Institution of Mechanical Engineers, Part J: Journal of Engineering Tribology.* 2020;234:301–10.

[10] Tomar AK, Sharma SC. Finite element analysis of multirecess hybrid spherical journal bearing system. *Proceedings of the Institution of Mechanical Engineers, Part J: Journal of Engineering Tribology.* 2020;234:1798–821.

[11] Chiang T, Malanoski SB, Pan CHT. Spherical squeeze-film hybrid bearing with small steady-state radial displacement. *Journal of Lubrication Technology.* 1967;89:254–62.

[12] Goenka PK, Booker JF. Spherical bearings: Static and dynamic analysis via the finite element method. *Journal of Lubrication Technology.* 1980;102:308–18.

[13] Dowson D, Taylor C. Fluid-inertia effects in spherical hydrostatic thrust bearings. *Asle Transactions.* 1967;10:316–24.

[14] San Andres L Dynamic force response of spherical hydrostatic journal bearings for cryogenic applications. *Tribology Transactions.* 1994;37:463–70.

[15] Xu C, Jiang S. Analysis of the static characteristics of a self-compensation hydrostatic spherical hinge. *Journal of Tribology.* 2015;137. https://doi.org/10.1115/1.4030712

[16] Kumar A, Kakoty S. Effect of couple stress parameter on steady-state and dynamic characteristics of three-lobe journal bearing operating on TiO_2 nanolubricant. *Proceedings of the Institution of Mechanical Engineers, Part J: Journal of Engineering Tribology.* 2020;234:528–40.

[17] Kumar P, Khonsari MM. On the role of lubricant rheology and piezo-viscous properties in line and point contact EHL. *Tribology International.* 2009;42:1522–30.

[18] Naduvinamani NB, Apparao S, Gundayya HA, Biradar SN. Effect of pressure dependent viscosity on couple stress squeeze film lubrication between rough parallel plates. *Tribology Online.* 2015;10:76–83.

[19] Swamy STN, Prabhu BS, Rao BVA. Calculated load capacity of non-newtonian lubricants in finite width journal bearings. *Wear.* 1975;31:277–85.

[20] Tomar AK, Sharma SC. A study of hole-entry grooved surface hybrid spherical journal bearing operating with electrorheological lubricant. *Journal of Tribology.* 2020;142. https://doi.org/10.1115/1.4047298

[21] Tomar AK, Sharma SC. Study on surface roughness and piezo-viscous shear thinning lubricant effects on the performance of hole-entry hybrid spherical journal bearing. *Tribology International.* 2022;168:107349.

[22] Gertzos KP, Nikolakopoulos PG, Papadopoulos CA. CFD analysis of journal bearing hydrodynamic lubrication by Bingham lubricant. *Tribology International.* 2008;41:1190–204.

[23] Safar ZS. Journal bearings operating with non-newtonian lubricant films. *Wear.* 1979;53:95–100.

[24] Tayal SP, Sinhasan R, Singh DV. Finite element analysis of elliptical bearings lubricated by a non-newtonian fluid. *Wear.* 1982;80:71–81.

[25] Xie Z, Jiao J, Yang K, He T, Chen R, Zhu W. Experimental and numerical exploration on the non-linear dynamic behaviors of a novel bearing lubricated by low viscosity lubricant. *Mechanical Systems and Signal Processing.* 2023;182:109349.

[26] Tayal SP, Sinhasan R, Singh DV. Analysis of hydrodynamic journal bearings with non-newtonian power law lubricants by the finite element method. *Wear.* 1981;71:15–27.

[27] Sinhasan R, Sah PL. Static and dynamic performance characteristics of an orifice compensated hydrostatic journal bearing with non-Newtonian lubricants. *Tribology International.* 1996;29:515–26.

[28] Prashad H. The effects of viscosity and clearance on the performance of hydrodynamic journal bearings. *Tribology Transactions.* 1988;31:303–9.

[29] Wu Z, Dareing DW. Non-Newtonian effects of powder-lubricant slurries in hydrostatic and squeeze-film bearings. *Tribology Transactions.* 1994;37:836–42.

[30] Rahmatabadi AD, Nekoeimehr M, Rashidi R. Micropolar lubricant effects on the performance of non-circular lobed bearings. *Tribology International.* 2010;43:404–13.

[31] Rahmatabadi AD, Zare Mehrjardi M, Fazel MR. Performance analysis of micropolar lubricated journal bearings using GDQ method. *Tribology International.* 2010;43:2000–9.

[32] Dowson D. A generalized Reynolds equation for fluid-film lubrication. *International Journal of Mechanical Sciences.* 1962;4:159–70.

[33] Tomar AK, Sharma SC. An investigation into surface texture effect on hole-entry hybrid spherical journal bearing performance. *Tribology International.* 2020;151:106417.

14 Limiting Performance Characteristics of Partially Textured Functionally Graded Foil Journal Bearing

S. Arokya Agustin, C. Shravan Kumar, and
A. Arul Jeyakumar
SRM Institute of Science and Technology, Kattankulathur,
India

T. V. V. L. N. Rao
Assam down town University, Guwahati, India

14.1 INTRODUCTION

The sustainability of gas foil bearings lubricated with air/gas are those which are compliant with top foil, sub-foil, and bearing housing. The hydrodynamic pressure generated between the journal and the top foil deforms the foil structure. The deformation will lead to a change in the film thickness and hence hydrodynamic pressure. The characteristics of the desired gas foil bearing performance are obtained by regulating the coupling between fluid film pressure and foil deformation. The purpose of this study is to investigate limiting load capacity and stiffness coefficients of textured functionally graded foil journal bearing.

14.1.1 FOIL JOURNAL BEARINGS

Heshmat et al. (1983) discussed desirable design features (structural, geometric, and operational variables) of a gas foil journal bearing. The performance of a gas journal bearing using a spring-supported compliant foil is evaluated. Peng and Carpino (1993) calculated the linearized dynamic coefficients of a gas foil bearing coupling the flow and structure equations based on Lund's (1987) perturbation solution method. The effect of the bearing compliance on the dynamic coefficients is discussed. Peng and Khonsari (2004a) developed an analytical model

DOI: 10.1201/9781003363576-14

to predict the hydrodynamic performance of a foil journal bearing under limiting conditions. The analytical model for high bearing numbers accounts for a simplified compressible Reynolds equation for bearings operating at high speed. Peng and Khonsari (2004b) presented the load capacity over a wide range of operating speeds based on the numerical solutions. Sawicki and Rao (2005) presented the limiting dynamic performance characteristics of foil bearings following a limiting load capacity analysis by Peng and Khonsari (2004a).

Radil et al. (2002) determined the effect of radial clearance on the load capacity coefficient of foil air bearings. The radial clearance has a direct impact on the load capacity coefficient of the foil bearing. Bearings operating with radial clearances less than the optimum are prone to bearing seizure. Bearings operating with a radial clearance twice the optimum did not experience any thermal management problems but suffered decline in maximum load capacity coefficient (20%). Yu et al. (2011) developed a numerical model of foil journal bearing with protuberant foil structure and analyzed the bearing performance characteristics. Mahner et al. (2018) presented the influence of a preloaded three-pad air foil journal bearing. The assembly preload increases the elastic foil structural stiffness and the bearing damping. The overall bearing stiffness depends on the assembly preload and for lightly preloaded bearings, the fluid film affects the overall bearing stiffness considerably. Samantha et al. (2019) presented a review on the development, functionality, and safe operation of foil bearing technology and highlighted challenges in the design, analysis, and performance characteristics that must be taken into consideration.

14.1.2 Textured Foil Bearings

Yan et al. (2018) proposed the surface micro groove structure on top foil to improve the load capacity and stability performance of foil bearing. The load capacity and direct stiffness are improved with an increment of micro groove depth compared with conventional foil journal bearing. Agustin et al. (2020) derived the limiting pressure gradient solution and determined the linearized nondimensional stiffness coefficients of a foil journal bearing with a top foil texture bump profile. The extent and height of top foil texture bump on the limiting stiffness coefficients of a foil journal bearing were evaluated.

14.1.3 Functionally Graded Materials Applications

Birman and Byrd (2007) presented a review of the principal developments in theory and applications of functionally graded materials (FGMs). Nijssen and van Ostayen (2020) showed an improved potential of compliant-hydrostatic bearings with functionally graded material supports compared with conventional elastic supports. Agustin et al. (2020) derived a limiting pressure gradient solution and determined the linearized nondimensional stiffness coefficients of a foil journal bearing with a top foil texture bump profile and functionally graded functionally graded bottom foil.

In this study, the influence of partial top foil texture, partial functionally graded sub-foil, partial textured top foil with fully functionally graded sub-foil on the limiting load capacity, and limiting stiffness coefficients of a foil journal bearing are investigated.

14.2 METHODOLOGY

The configuration types of the foil journal bearing considered in this analysis are (i) textured foil journal bearing, (ii) functionally graded foil journal bearing and (iii) textured functionally graded foil journal bearing. The textured foil journal bearing consists of partial texture (bump) profile along the circumferential extent of the top foil. The functionally graded foil journal bearing consists of functionally graded sub-foil compliance configuration along the partial extent of circumferential direction. The textured functionally graded foil journal bearing consists of partial texture (bump) profile along the circumferential extent of the top foil along with full functionally graded (FG) sub-foil compliance in the bump texture region. A schematic of the texture (or FG foil) journal bearing is shown in Figure 14.1. Figure 14.1a consists of the top foil with bump texture, the functionally graded sub-foil structure, and the bearing housing. Figure 14.1b shows the extension of the bump texture region along a circumferential direction from the load line (position fixed to negative X-axis).

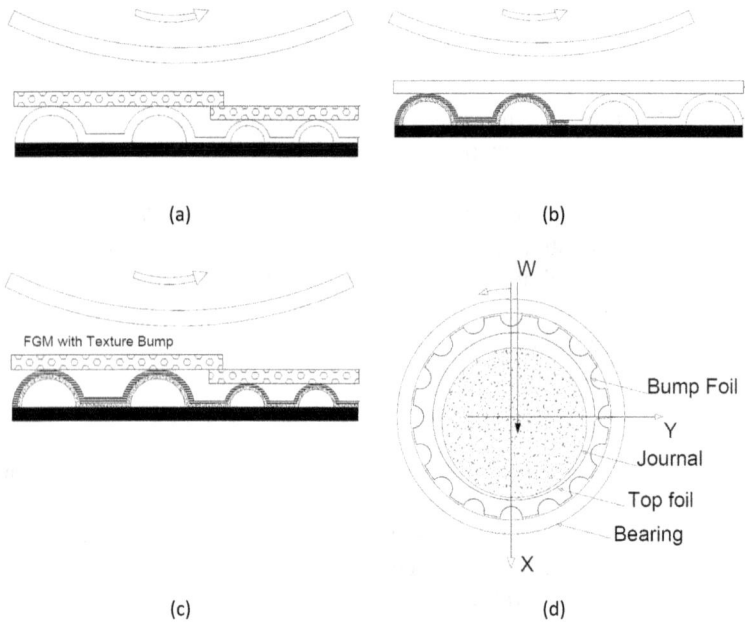

FGM with Texture Bump

W

Bump Foil

Y

Journal

Top foil

Bearing

X

(a) (b) (c) (d)

FIGURE 14.1 Schematic of texture (or FG foil) journal bearing. (a) Texture (bump) foil, (b) FG foil, (c) Textured FG foil, and (d) Foil journal bearing.

14.2.1 LIMITING DYNAMIC REYNOLDS EQUATION

The limiting load capacity and stiffness coefficients are based on the compressible Reynolds equation for foil bearings operating at high speed (large bearing numbers). The necessary conditions for very high speed operation of foil journal bearing ($\Lambda \rightarrow \infty$) obtained by simplifying the compressible Reynolds equation (Peng and Khonsari 2004a) is

$$\frac{\partial}{\partial \theta}(\mathrm{PH}) = 0 \qquad (14.1)$$

The steady state pressure and pressure gradients are obtained by infinitesimal perturbation (Lund 1987) of the simplified compressible Reynolds equation. The pressure distribution and film thickness for infinitesimally small perturbations about the journal steady state is:

$$P = P_o + P_x \Delta x + P_y \Delta y, \ H = H_o + H_x \Delta x + H_y \Delta y \qquad (14.2)$$

where

$$\begin{aligned}
H_o &= 1 + \varepsilon_o \cos(\theta - \phi_o) - H_b + f_\alpha \alpha (P_o - 1), \\
H_x &= \cos\theta + f_\alpha \alpha P_x, \\
H_y &= \sin\theta + f_\alpha \alpha P_y
\end{aligned} \qquad (14.3)$$

Substituting Eq. (14.3) in Eq. (14.1), the steady state and dynamic simplified compressible Reynolds equations are

$$\frac{\partial}{\partial \theta}(P_o H_o) = 0 \qquad (14.4)$$

$$\frac{\partial}{\partial \theta}(P_o H_x + P_x H_o) = 0, \quad \frac{\partial}{\partial \theta}(P_o H_y + P_y H_o) = 0 \qquad (14.5)$$

14.2.2 LIMITING LOAD CAPACITY

Using the analysis based on Reddy et al. (1999), the modulus of elasticity of FG sub-foil is

$$E_{fg} = E_1 (0.5 - \zeta)^\lambda + E_2 \left(1 - (0.5 - \zeta)^\lambda\right) \qquad (14.6)$$

where ζ refers to the axis at the mid-surface of the FG bottom foil. The gradient index (λ) = 0 yields the modulus of elasticity of the base material. The gradient index (λ) increases with an increasing reference material portion of functionally graded bottom foil.

The effective compliance factor of the FG sub-foil included in the nondimensional film thickness in Eq. (14.3) is considered as:

$$f_\alpha = \frac{1}{\int_{-0.5}^{0.5}\left[\left(0.5-\varsigma\right)^\lambda + E_r\left(1-\left(0.5-\varsigma\right)^\lambda\right)\right]d\varsigma} \tag{14.7}$$

In the case of functionally graded foil journal bearing with FG sub-foil compliance configuration, the effective compliance factor in Eq. (14.7) is considered for the partial extent of circumferential direction.

The texture bump foil bearing limiting pressure distribution is obtained by solving Eq. (14.4) using the pressure ($P_o = 1$) boundary conditions at inlet ($\theta = 0$) as

$$P_o = \frac{1+\varepsilon_o\cos\phi_o - H_b}{1+\varepsilon_o\cos\left(\theta-\phi_o\right)-H_b+f_\alpha\alpha\left(P_o-1\right)} \tag{14.8}$$

The limiting pressure is simplified from Eq. (14.8) as

$$P_o = \frac{\left[\begin{array}{c}-\left(1+\varepsilon_o\cos\left(\theta-\phi_o\right)-H_b-f_\alpha\alpha\right)+\\ \sqrt{\left(1+\varepsilon_o\cos\left(\theta-\phi_o\right)-H_b-f_\alpha\alpha\right)^2+4f_\alpha\alpha\left(1+\varepsilon_o\cos\phi_o-H_b\right)}\end{array}\right]}{2f_\alpha\alpha} \tag{14.9}$$

14.2.3 LIMITING DYNAMIC COEFFICIENTS

The substituting Eq. (14.3) in Eq. (14.5), $W = \sqrt{W_\varepsilon^2 + W_\phi^2}$ reduces to

$$\frac{\partial}{\partial\theta}\left(P_o\cos\theta + P_x\left(f_\alpha\alpha P_o + H_o\right)\right) = 0,$$

$$\frac{\partial}{\partial\theta}\left(P_o\sin\theta + P_y\left(f_\alpha\alpha P_o + H_o\right)\right) = 0, \tag{14.10}$$

The textured FG foil journal bearing pressure gradient distribution under limiting conditions is obtained by solving Eq. (14.10) using the pressure ($P_o = 1$) and pressure gradient ($P_x = P_y = 0$) boundary conditions at inlet ($\theta = 0$) as per Agustin et al. (2020)

$$P_x = \frac{1-P_o\cos\theta}{1-f_\alpha\alpha+2f_\alpha\alpha P_o+\varepsilon_o\cos\left(\theta-\phi_o\right)-H_b} \tag{14.11}$$

$$P_y = \frac{-P_o \sin\theta}{1 - f_\alpha \alpha + 2 f_\alpha \alpha P_o + \varepsilon_o \cos(\theta - \phi_o) - H_b} \tag{14.12}$$

The limiting nondimensional load and nondimensional bearing stiffness coefficients K_{xx}, K_{yx}, K_{xy}, K_{yy} are evaluated by integration of pressure and pressure gradients as

$$\begin{Bmatrix} F_x \\ F_y \end{Bmatrix} = \int_{\theta=0}^{\theta=2\pi} P_o \begin{Bmatrix} \cos\theta \\ \sin\theta \end{Bmatrix} d\theta, \quad \begin{Bmatrix} K_{xj} \\ K_{yj} \end{Bmatrix} = \int_{\theta=0}^{\theta=2\pi} P_j \begin{Bmatrix} \cos\theta \\ \sin\theta \end{Bmatrix} d\theta, \quad \text{for } j = x, y \tag{14.13}$$

14.3 RESULTS AND DISCUSSION

The parameters used in the analysis of textured FG foil journal bearing are: ratio of modulus of elasticity of base to reference material (E_2/E_1) of FG bottom foil (E_r) = 1, 3; nondimensional bump texture height (H_b) = 0–1.5; circumferential coordinate (from position fixed to negative X-axis) for texture (or FG foil) region (θ_t) = 0–240°; gradient index (λ) = 0–6; reference foil compliance coefficient (α) = 1, 10; eccentricity ratio (ε) = 0.1, 0.7.

14.3.1 TEXTURED FOIL JOURNAL BEARING

Figures 14.2 and 14.3 show the limiting nondimensional load capacity (W) and nondimensional stiffness coefficients (K_{ij}) of textured foil journal bearing. The influence of partial texture (bump)configuration along the circumferential direction is presented.

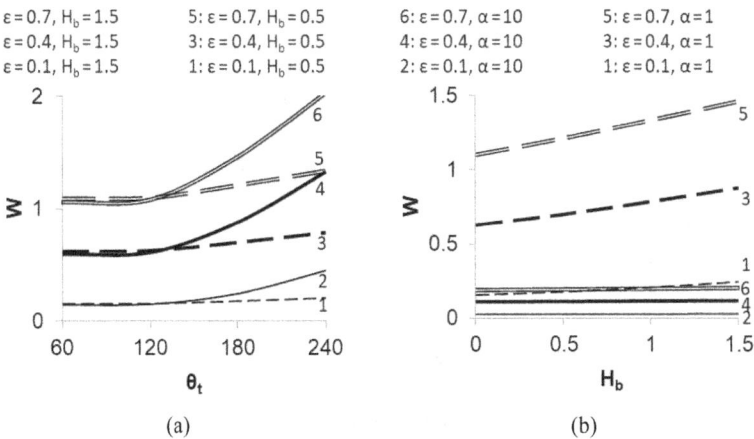

FIGURE 14.2 Nondimensional load capacity of texture foil journal bearing. (a) $\alpha = 1$ and (b) $\theta_t = 180°$.

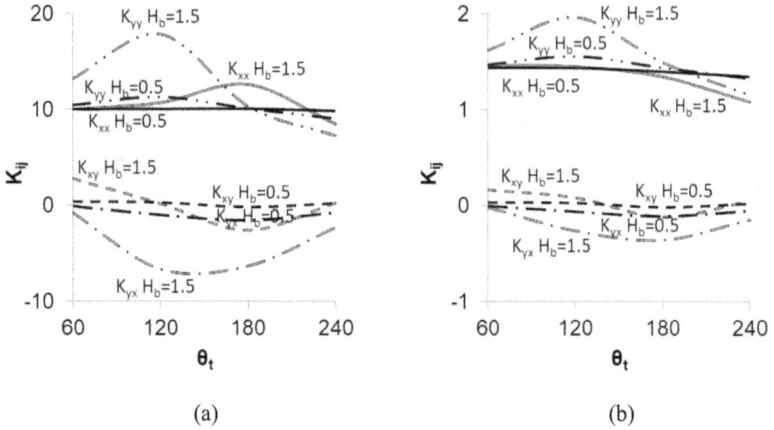

FIGURE 14.3 Nondimensional stiffness coefficients of texture foil journal bearing. (a) K_{ij} ($\varepsilon = 0.1$, $\alpha = 1$) and (b) K_{ij} ($\varepsilon = 0.7$, $\alpha = 1$).

TABLE 14.1
Limiting Performance Characteristics of Texture Foil Journal Bearing

Parameter	Limiting Performance Characteristics of Texture Foil Journal Bearing
W	W increases with increasing H_b from 0.5–1.5 for lower $\alpha = 1$ for a threshold $\theta_t = 120°$–$180°$
K_{ij}	Higher variation of K_{yj} derived from P_j, for $j = x, y$ is obtained when increasing H_b from 0.5–1.5

Figure 14.2 shows the nondimensional load capacity (W) of textured foil journal bearing. As shown in Figure 14.2(a), the nondimensional load capacity (W) increases with increasing nondimensional bump texture height (H_b) after a threshold circumferential coordinate (from position fixed to negative X-axis) for the texture bump region (θ_t).

As depicted in Figure 14.2(b), for a higher reference foil compliance coefficient ($\alpha = 10$), the nondimensional load capacity (W) increases with increasing nondimensional bump texture height (H_b) for lower reference foil compliance coefficient ($\alpha = 1$) while the same is invariant with nondimensional bump texture height (H_b).

Figure 14.3 shows the limiting nondimensional stiffness coefficients of textured foil journal bearing. As shown in Figure 14.3, higher variation of the nondimensional stiffness coefficients (K_{ij}) is observed with increasing nondimensional bump texture height (H_b). The nondimensional pressure gradients (P_j) for $j = x, y$ showed significant variation with increasing circumferential coordinate (from position fixed to negative X-axis) for texture bump region (θ_t) influencing a significant variation of nondimensional stiffness coefficients (K_{yj}) perpendicular to the load line. The nondimensional stiffness coefficients (K_{ij}) at eccentricity ratios $(\varepsilon = 0.1$ and $0.7)$ as depicted in Figures 14.3(a, b) show similar trends.

Table 14.1 shows limiting performance characteristics of texture foil journal bearing.

14.3.2 FUNCTIONALLY GRADED FOIL JOURNAL BEARING

Figures 14.4–14.5 show the limiting nondimensional load capacity (W) and nondimensional stiffness coefficients (K_{ij}) of FG foil journal bearing. The influence of partial FG foil configuration along the circumferential direction is presented.

Figure 14.4 shows the nondimensional load capacity of functionally graded (FG) foil journal bearing. The nondimensional load capacity (W) increases substantially from a threshold circumferential coordinate (from position fixed to negative X-axis) to a partial extent (θ_t) of FG foil configuration region for lower reference foil compliance coefficient $(\alpha = 1)$ as shown in Figure 14.4(a). The nondimensional load capacity (W) is invariant with gradient index (λ) for higher reference foil compliance coefficient $(\alpha = 10)$ as depicted in Figure 14.4(b). The nondimensional load capacity (W) increases with increasing gradient index (λ) until the critical value $(\lambda = 2)$ and remains nearly constant with increasing gradient index (λ).

Figure 14.5 shows the limiting nondimensional stiffness coefficients of functionally graded (FG) foil journal bearing. The nondimensional stiffness coefficients (K_{yj}) perpendicular to the load line obtained from nondimensional pressure gradients (P_j) for $j = x, y$ showed significant variation with increasing circumferential coordinate (from position fixed to negative X-axis) to a partial extent (θ_t) of FG foil configuration region as shown in Figure 14.5(a).

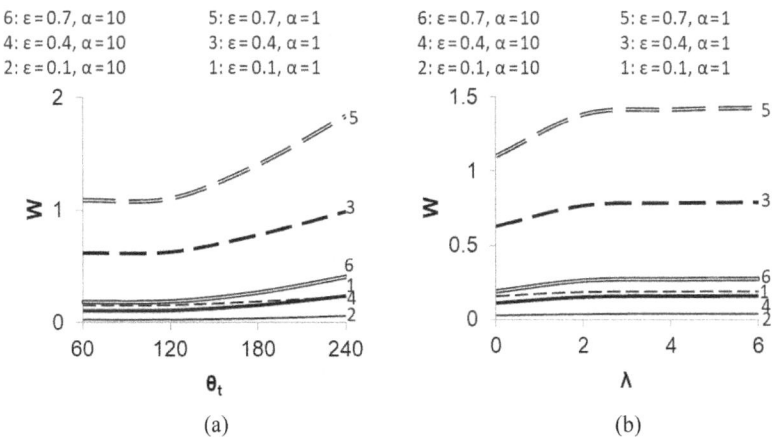

FIGURE 14.4 Nondimensional load capacity of FG foil journal bearing. (a) $E_r = 3$, $\lambda = 3$ and (b) $\theta_t = 180°$, $E_r = 3$.

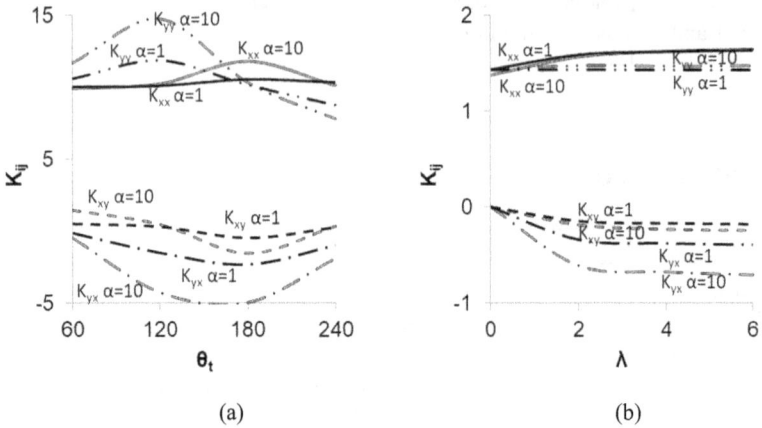

(a)   (b)

FIGURE 14.5 Nondimensional stiffness coefficients of FG foil journal bearing. (a) K_{ij} ($\varepsilon = 0.1$, $E_r = 3$, $\lambda = 3$) and (b) K_{ij} ($\varepsilon = 0.7$, $\theta_t = 180°$, $E_r = 3$).

The nondimensional stiffness coefficients (K_{ij}) varies with increasing gradient index (λ) until a threshold value ($\lambda = 2$) and remains nearly constant with further increasing gradient index (λ) as depicted in Figure 14.5(b). The nondimensional cross coupling stiffness coefficients (K_{xx} and K_{yx}) parallel and perpendicular to the load line obtained from nondimensional pressure gradients (P_x) showed significant variation with increasing gradient index (λ) until a threshold value ($\lambda = 2$) and remains nearly constant with further increasing gradient index (λ) as shown in Figure 14.5(b).

Table 14.2 shows the limiting performance characteristics of texture foil journal bearing.

14.3.3 Textured Functionally Graded Foil Journal Bearing

Figures 14.6–14.7 show the limiting nondimensional load capacity (W) and nondimensional stiffness coefficients (K_{ij}) of textured functionally graded foil journal

---

TABLE 14.2
Limiting Performance Characteristics of FG Foil Journal Bearing

Parameter	Limiting Performance Characteristics of FG Foil Journal Bearing
W	W increases for lower $\alpha = 1$ beyond a threshold $\theta_t = 120°$ for FG foil journal bearing. W increases until a threshold $\lambda = 2$ and remains invariant for $\lambda > 2$
K_{ij}	Higher variation of K_{ij} derived from P_j, for $j = x, y$ is obtained with FG foil journal bearing. Higher variation of K_{ij} is obtained until a threshold $\lambda = 2$ and remains invariant for $\lambda > 2$

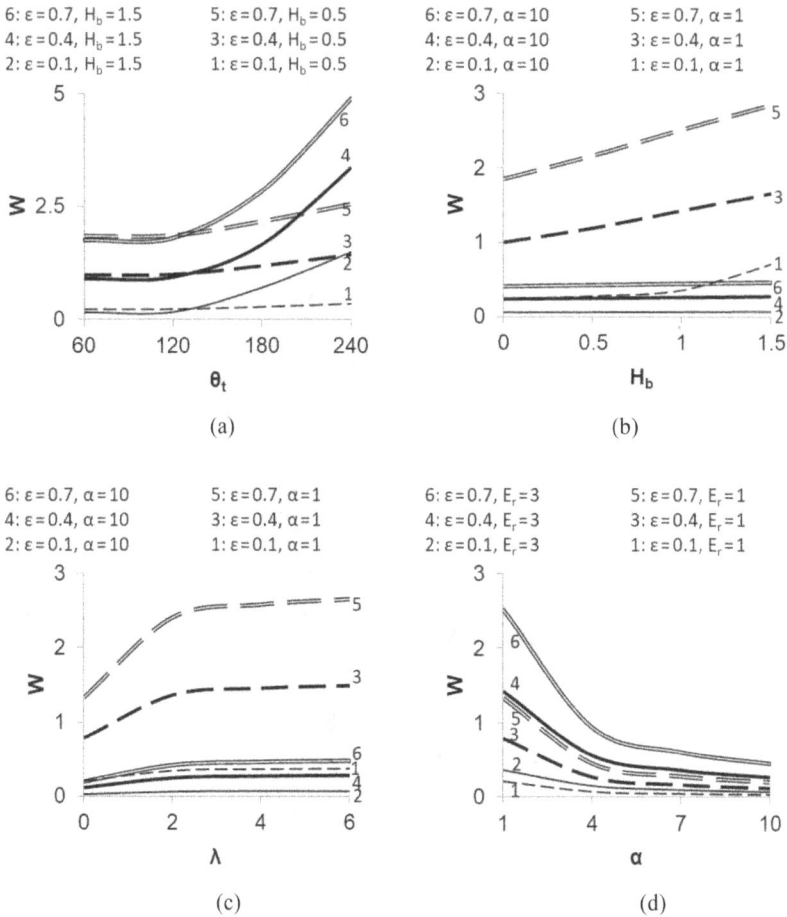

FIGURE 14.6 Nondimensional load capacity of FG texture foil journal bearing. (a) $\alpha = 1$, $E_r = 3$, $\lambda = 3$, (b) $\theta_t = 180°$, $E_r = 3$, $\lambda = 3$, (c) $\theta_t = 180°$, $E_r = 3$, $H_b = 1$, and (d) $\theta_t = 180°$, $\lambda = 3$, $H_b = 1$.

bearing. The influence of partial texture (bump)configuration along the circumferential direction is presented.

Figure 14.6 shows the nondimensional load capacity of functionally graded (FG) bump texture foil journal bearing. As shown in Figure 14.6(a), the nondimensional load capacity (W) remains constant until a threshold circumferential coordinate (from position fixed to negative X-axis) for texture region (θ_t) and then increases with increasing nondimensional bump texture height (H_b). As depicted in Figure 14.6(b) for the parameters considered in the analysis, the nondimensional load capacity (W) is invariant with nondimensional bump texture height (H_b) for higher reference foil compliance coefficient ($\alpha = 10$).

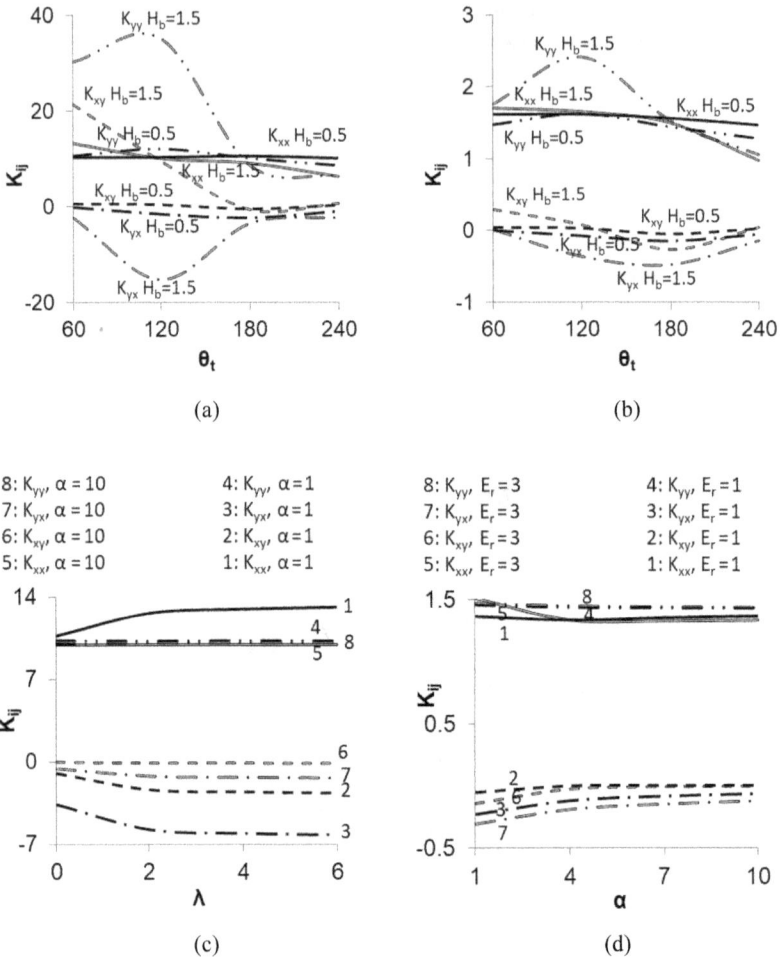

FIGURE 14.7 Nondimensional stiffness coefficients of FG texture foil journal bearing. (a) K_{ij} ($\varepsilon = 0.1$, $\alpha = 1$, $E_r = 3$, $\lambda = 3$), (b) K_{ij} ($\varepsilon = 0.7$, $\alpha = 1$, $E_r = 3$, $\lambda = 3$), (c) K_{ij} ($\varepsilon = 0.1$, $\theta_t = 180°$, $E_r = 3$, $H_b = 1$), and (d) K_{ij} ($\varepsilon = 0.7$, $\theta_t = 180°$, $\lambda = 3$, $H_b = 1$).

The nondimensional load capacity (W) increases with increasing gradient index (λ) until a critical value and remains nearly constant with increasing gradient index (λ) as plotted in Figure 14.6©. Increasing ratio of modulus of elasticity of base to reference material (E_2/E_1) of functionally graded bottom foil (E_r) increases (Figure 14.6(d)) the nondimensional load capacity (W). The increase in nondimensional load capacity (W) with increasing ratio of modulus of elasticity of base to reference material (E_2/E_1) of functionally graded bottom foil (E_r) is higher for lower reference foil compliance coefficient ($\alpha = 1$).

Figure 14.7 shows the limiting nondimensional stiffness coefficients of textured functionally graded (FG) foil journal bearing. As shown in Figure 14.7(a),

TABLE 14.3
Limiting Performance Characteristics of Textured FG Foil Journal Bearing

Parameter	Limiting Performance Characteristics of Textured FG Foil Journal Bearing
W	W increases for lower $\alpha = 1$ beyond a threshold $\theta_t = 120\,°$ for textured FG foil journal bearing. W increases until a threshold $\lambda = 2$ and remains invariant for $\lambda > 2$. W increases with increasing E_r for lower $\alpha = 1$.
K_{ij}	Higher variation of K_{yj} derived from P_j, for $j = x, y$ is obtained with textured FG foil journal bearing. Higher variation of K_{xx} and K_{yx} is obtained until a threshold $\lambda = 2$ and is invariant for $\lambda > 2$ for lower α.

higher variation of the nondimensional stiffness coefficients (K_{ij}) is observed with increasing nondimensional bump texture height (H_b). The circumferential coordinate (from position fixed to negative X-axis) for texture bump region (θ_t) significantly influences the nondimensional pressure gradients (P_j) for $j = x, y$. The nondimensional stiffness coefficients (K_{yj}) perpendicular to the load line obtained from nondimensional pressure gradients (P_j) for $j = x, y$ showed significant variation with increasing circumferential coordinate (from position fixed to negative X-axis) for texture bump region (θ_t). As depicted in Figure 14.7(b), the trends of the nondimensional stiffness coefficients (K_{ij}) at higher eccentricity ratio ($\varepsilon = 0.7$) are similar to those obtained for lower eccentricity ratio ($\varepsilon = 0.1$) in Figure 14.7(b).

As shown in Figure 14.7(c), the nondimensional cross coupling stiffness coefficients (K_{xx} and K_{yx}) parallel and perpendicular to the load line obtained from nondimensional pressure gradients (P_x) showed significant variation with increasing gradient index (λ) until a threshold value ($\lambda = 2$) and remains nearly constant with further increasing gradient index (λ) for lower reference foil compliance coefficient ($\alpha = 1$). The nondimensional stiffness coefficients (K_{ij}) showed variation with increasing reference foil compliance coefficient until a threshold value ($\alpha = 4$) and remains nearly constant with further increasing reference foil compliance coefficient (α) (Figure 14.7(d)).

Table 14.3 shows the limiting performance characteristics of FG texture foil journal bearing.

14.4   CONCLUSIONS

The present study examined the influence of FG foil journal bearing with (i) textured foil, (ii) FG foil, and (iii) textured FG foil on the limiting load capacity and stiffness coefficients. A textured FG foil journal bearing provides enhanced limiting nondimensional load capacity. The extent of (i) partial extent of texture (bump) top foil, (ii) partial extent of FG sub-foil, and (iii) partial extent of texture (bump) top foil with a full FG sub-foil influences the limiting nondimensional load capacity and stiffness coefficients. The bump texture top foil with augmented sub-foil compliance has potential to enrich the foil bearing characteristics.

REFERENCES

Agustin, S.A., Jeyakumar, A.A., Shravankumar, C. & Rao, T.V.V.L. N. 2020. Limiting Stiffness Coefficient Analysis of FGM Texture Bump Foil Journal Bearing. *Virtual International Tribology Research Symposium*, Online.

Birman, V. & Byrd, L.W. 2007. Modeling and Analysis of Functionally Graded Materials and Structures. *Applied Mechanics Reviews*, 60(5), pp. 195–216.

Heshmat, H., Walowit, J.A. & Pinkus, O. 1983. Analysis of Gas-Lubricated Foil Journal Bearings. *Journal of Lubrication Technology*, 105, pp. 647–655.

Lund, J.W. 1987. Review of the Concept of Dynamic Coefficients for Fluid Film Journal Bearings. *Journal of Lubrication Technology*, 109(1), pp. 37–41.

Mahner, M., Li, P., Lehn, A. & Schweizer, B. 2018. Numerical and Experimental Investigations on Preload Effects in Air Foil Journal Bearings. *Journal of Engineering for Gas Turbines and Power*, 140(3), p. 032505.

Nijssen, J.P.A. & van Ostayen, R.A.J. 2020. Compliant Hydrostatic Bearings Utilizing Functionally Graded Materials. *Journal of Tribology*, Nov, 142(11), p. 111801.

Peng, J.-P. & Carpino, M. 1993. Calculation of Stiffness and Damping Coefficients for Elastically Supported Gas Foil Bearings. *Journal of Tribology*, 115(1), pp. 20–27.

Peng, Z.-C. & Khonsari, M.M. 2004a. On the Limiting Load-Carrying Capacity of Foil Bearings. *Journal of Tribology*, 126(4), pp. 817–818.

Peng, Z.-C. & Khonsari, M.M. 2004b. Hydrodynamic Analysis of Compliant Foil Bearings with Compressible Air Flow. *Journal of Tribology*, 126(3), pp. 542–546.

Radil, K., Howard, S. & Dykas, B. 2002. The Role of Radial Clearance on the Performance of Foil Air Bearings. *Tribology Transactions*, 45(4), pp. 485–490.

Reddy, J.N., Wang, C.M. & Kitipornchai, S. 1999. Axisymmetric Bending of Functionally Graded Circular and Annular Plates. *European Journal of Mechanics - A/Solids*, 18, pp. 185–199.

Samanta, P., Murmu, N.C. & Khonsari, M.M. 2019. The Evolution of Foil Bearing Technology. *Tribology International*, 135, pp. 305–323.

Sawicki, J.T. & Rao, T.V.V.L.N. 2005. Limiting Stiffness and Damping Coefficients of Foil Bearing. In *Proceedings of IDETC/CIE*, ASME, pp. 1069–1073.

Yan, J., Zhang, G., Liu, Z., Zhao, J. & Xu, L. 2018. Performance of a Novel Foil Journal Bearing with Surface Micro-Grooved Top Foil. *Proceedings of the Institution of Mechanical Engineers, Part J: Journal of Engineering Tribology*, 232(9), pp. 1126–1139.

Yu, H., Shuangtao, C.,Rugang, C.,Qiaoyu, Z. & Hongli, Z. 2011. Numerical Study on Foil Journal Bearings with Protuberant Foil Structure. *Tribology International*, 44(9), pp. 1061–1070.

Index

Pages in *italics* refer to figures and pages in **bold** refer to tables.

For Product Safety Concerns and Information please contact our EU
representative GPSR@taylorandfrancis.com
Taylor & Francis Verlag GmbH, Kaufingerstraße 24, 80331 München, Germany

9 7 8 1 0 3 2 4 2 6 3 2 7